Gasification for Practical Applications

Gasification for Practical Applications

Editor

Jens Omdahl

scitus academics

Gasification for Practical Applications

Edited by **Jens Omdahl**

ISBN: 978-1-68117-299-6
Library of Congress Control Number: 2016937484

SCITUS Academics LLC,
www.scitusacademics.com
Box No. 4766, 616 Corporate Way,
Suite 2, Valley Cottage,
NY 10989

Printed in United States of America on Acid Free Paper

Preface

Some of the most promising alternative energy sources are not revolutionary ideas. We all know centuries ago –ancient windmills and water wheels. Today, a number of improvements, including innovative turbine designs, turn these old machines into state-of-the-art technologies that can help countries meet their energy needs. There is another old process, about which you probably don't know much, which is becoming increasingly popular and can combine wind and hydropower with clean, renewable energy. The process is called gasification. Gasification is a partial process of oxidation that divides a carbon source such as coal, natural gas or biomass into carbon monoxide (CO) and hydrogen (H2), plus carbon dioxide (CO2) and possibly hydrocarbon molecules such as methane (CH4). This is achieved by reacting without combustion with a controlled quantity of oxygen and/or steam at high temperatures. The resulting gas mixture is called syngas or manufacturing gas, which is a fuel. When gasified compounds are obtained from biomass, the gasification and combustion power of the resulting gas is considered to be a source of renewables. Gasification is not a new technology; it was originally developed in the 1800s and is used for lighting and cooking gas production in the city. Moreover, during the Second World War, small-scale gasifies were used to power internal combustion engine vehicles during fuel shortages. If gasification takes place at a relatively low temperature, such as 700oC to 1000oC, the hydrocarbon level of the product gas is relatively high compared to the high temperature gasification level. As a result, it can be used directly to burn for heat or electricity generation through a steam turbine or to run an internal combustion engine with appropriate gas cleaning for electricity generation.

This book provides practical information for researchers and graduate students, who want to review, upgrade and introduce a new concept to conventional gasification technology.

Table of Contents

1

CHAPTER

GASIFICATION OF WOOD BIO-OIL

Younes Chhiti[1] and Sylvain Salvador[2]

[1] *Université de Pau / LaTEP-ENSGTI, France*

[2] *Ecole des Mines d'Albi – Carmaux/RAPSODEE, France*

1. INTRODUCTION

Energy and environmental issues are two common concerns of modern society. Energy is a central part of every human being's daily life. In all its forms, such as chemical energy (food), thermal energy (heat), or electricity, energy has the ability to transform the daily lives of humans across the world by easing workloads, boosting economies and generally increasing the comfort of our lives. Worldwide energy consumption has been increasing rapidly. This has been accelerated by the improvement of the quality of life that almost directly relates to the amount of energy consumed. At present, fossil fuels based energy resources, such as coal, gas, and oil supply the majority of the total world energy requirement.

The global warming owing to the emissions of greenhouse gas is the most drastic consequence of the use of fossil fuels. According to experts in the field, global warming can disturb the natural equilibrium of the Earth's ecosystem. If CO_2 emissions are not regulated, global warming can have severe consequences for environment. These consequences, although some of them are not fully corroborated, are increasing sea and ocean levels, ocean acidification, change in rainfall patterns, hurricanes, volcanic eruptions, earthquakes and plant or animal extinctions, among others.

The development of non-conventional sources like wind, sunlight, water, biomass, etc., is inevitable. Among the renewable sources of energy, substantial focus of research is currently on the use of biomass. Besides being a renewable source of energy, there are many other advantages associated to the use of biomass. It is available abundantly in the world. Its use does not increase the net amount of CO_2in the atmosphere and can reduce the emissions of SO_2 and NO_x remarkably.

Biomass gasification is a promising technology, which can contribute to develop future energy systems which are efficient and environmentally friendly in order to increase the share of renewable energy for heating, electricity, transport fuels and higher applications.

The gasification of carbon-containing materials to produce combustible gas is an established technology. Biomass gasification is a thermochemical process that produces relatively clean and combustible gas through pyrolytic and reforming reactions. The syngas generated can be an important resource suitable for direct combustion, application in prime movers such as engines and turbines, or for the production of synthetic natural gas (SNG) and transportation fuels e.g. Fischer-Tropsch diesel.

For energy production, the major concerns about syngas are its heating value, composition, and possible contamination. The proportion of the combustible gas hydrogen (H_2), methane (CH_4), carbon monoxide (CO), and moisture determines the heating value of the gas. The composition of syngas depends on the biomass properties and gasifier operating conditions. For a specific gasification system, operating conditions play a vital role in all aspects of biomass gasification. These include carbon conversion, syngas composition, tars and soot formation and oxidation (Devi et al., 2003).

The main hurdles for large-scale implementation of energy production from solid biomass are the nature of biomass - non uniform, low-energy density, sometimes large ash content - together with the usual inconsistency between the local availability of biomass and the demand for biomass related products: heat, electricity, fuels and chemicals. Usually, import/transport of fossil fuels is cheaper. Pyrolysis may be a process to overcome these hurdles: biomass is transformed into a versatile liquid called bio-oil, easy to handle and to transport. This bio-oil would then be transported to centralized air/steam gasification units. Bio-oil is an intermediate product which is produced from relatively dry biomass via fast pyrolysis process. It is a liquid with similar elemental composition to its original feedstock and with high bulk and energy density.The high bulk and energy density of bio-oil can reduce transportation costs to large scale centralized gasification plants; these costs have been a detrimental factor in large scale use of solid biomass resource. Bio-oil can be produced where the biomass is available and then be transported over long distances to central processing units of similar scales as the current petrochemical industry. Besides technical and logistic advantages, this conversion chain may also give incentives for economic development and job creation especially in rural areas.

The essential features to obtain high yields of bio-oil (up to 75 wt% on dry basis) are a moderate pyrolysis temperature (500°C), high heating rates (10^3-10^5°C/s), short vapour residence times (<2 s) and rapid quenching of the pyrolysis vapours.

The combination of fast pyrolysis of biomass followed by transportation in large units for steam reforming has attracted considerable attention of the research community, as one of the most promising viable methods for hydrogen production.For the high temperature applications such as gasification, steam reforming or even combustion, it is of particular interest to understand the behavior of bio-oils during the very first step of its decomposition under pyrolysis conditions. However, only few works can be found on the understanding of processes occurring during thermal conversion of bio-oils.

The earliest combustion tests of bio-oil droplets were conducted in Sandia National Laboratory (Wornat et al., 1994).Streams of monodispersed droplets were injected into a laminar flow reactor. The experimental conditions were as follows: droplet diameter of about 300 m, reactor temperature of 1600 K and O_2 concentrations of 14–33%. In-situ video imaging of burning droplets reveals that biomass oil droplets undergo several distinct stages of combustion. Initially biomass oil droplets burn quiescently in a blue flame. The broad range of component volatilities and inefficient mass transfer within the viscous biomass oils bring about an abrupt termination of the quiescent stage, however, causing rapid droplet swelling and distortion, followed by a microexplosion.

Thermogravimetric analysis (TGA) is widely used to characterize the evaporation, thermal decomposition and combustion properties of bio-oils. The weight loss process of bio-oils in inert atmospheres can be divided into two stages: the evaporation of light volatiles (<150-200°C) and the subsequent thermal decomposition of unstable heavier components (<350-400°C). In the case of TGA tests performed in the presence of air, the weight loss of bio-oils can be divided into three stages. The first two stages are similar to those in inert atmospheres and the third stage is the combustion of chars formed in the first two stages (>400°C)(Ba et al., 2004a, 2004b).

Branca et al.(Branca, 2005a) studied the devolatilization and heterogeneous combustion of wood fast bio-oil. Weight loss curves of wood fast bio-oil in air have been measured, under controlled thermal conditions, carrying out two separate sets of experiments. The first, which has a final temperature of 600 K, concerns evaporation/cracking of the oil and secondary char formation. A heating rate of 0.08 °C/swas applied. The yield of secondary char varies from about 25% to 39% (on a total oil basis). After collection and milling, in the second set of experiments, heterogeneous combustion of the secondary char is carried out to temperatures of 873 K.In another study, Branca et al. (Branca et al., 2005b) found that thermogravimetric curves of bio-oil in air show two main reaction stages. The first (temperatures ≤ 600 K) concerns evaporation, formation and release of gases and formation of secondary char (coke). Then, at higher temperatures, heterogeneous combustion of secondary char takes place. They found that the pyrolysis temperature does not affect significantly weight loss dynamics and amount of secondary char (approximately equal to 20% of the liquid on a dry basis).

Hallet et al.(Hallett et al., 2006)established a numerical model for the evaporation and pyrolysis of a single droplet of bio-oil derived from biomass. The model is compared with the results of suspended droplet experiments, and is shown to give good predictions of the times of the major events in the lifetime of a droplet: initial heating, evaporation of volatile species, and pyrolysis of pyrolytic lignin to char.

Guus van Rossum et al.(Van Rossum et al., 2010) studied the evaporation of bio-oil and product distribution at varying heating rates (~1.5×10^{-2}–1.5×10^{4}°C/s) with surrounding temperatures up to 850°C. A total product distribution (gas, vapor, and char) was measured using two atomizers with different droplet sizes. A big difference is seen in char production between the two atomizers where the ultrasonic atomizer gives much less char compared to the needle atomizer, ~8 and 22% (on carbon basis), respectively. Small droplets (88-117µm generated by ultrasonic atomizer, undergoing high heating rate) are much quicker evaporated than larger droplets (~ 1.9 mm, generated by needle atomizer, undergoing low heating rate).

Calabria et al. carried out lots of studies on the combustion behaviors of fibre-suspended single bio-oil droplets. The droplet size varied between 300 and 1100 m and the furnace temperature changed in the range of 400–1200°C. The droplets were observed to undergo initial heating, swelling and microexplosion before ignition. During this stage, the temperature–time curves showed two zones with constant temperatures (100 and 450°C), which corresponded to the evaporation of light volatiles and the thermal cracking of unstable components, respectively. The droplets were ignited at around 600°C. The combustion of the droplets started with an enveloping blue flame. Then, the flame developed a yellow tail with its size increasing, which indicated the formation of soot. After that, the flame shrank and extinguished, and the remaining solid carbonaceous residues burned leading to the formation of ash (Calabria et al., 2007).

In air/steam gasification process the essential steps are pyrolysis, partial oxidation, cracking of tar, solid carbon residue gasification, reforming (steam and/or dry), and water gas shift to yield syngas, water, carbon dioxide, and unwanted products like tars, methane and carbon (Levenspiel et al., 2005). As a summary, a schematic representation of air/steam gasification of single droplet of bio-oil is proposed in Figure 1.

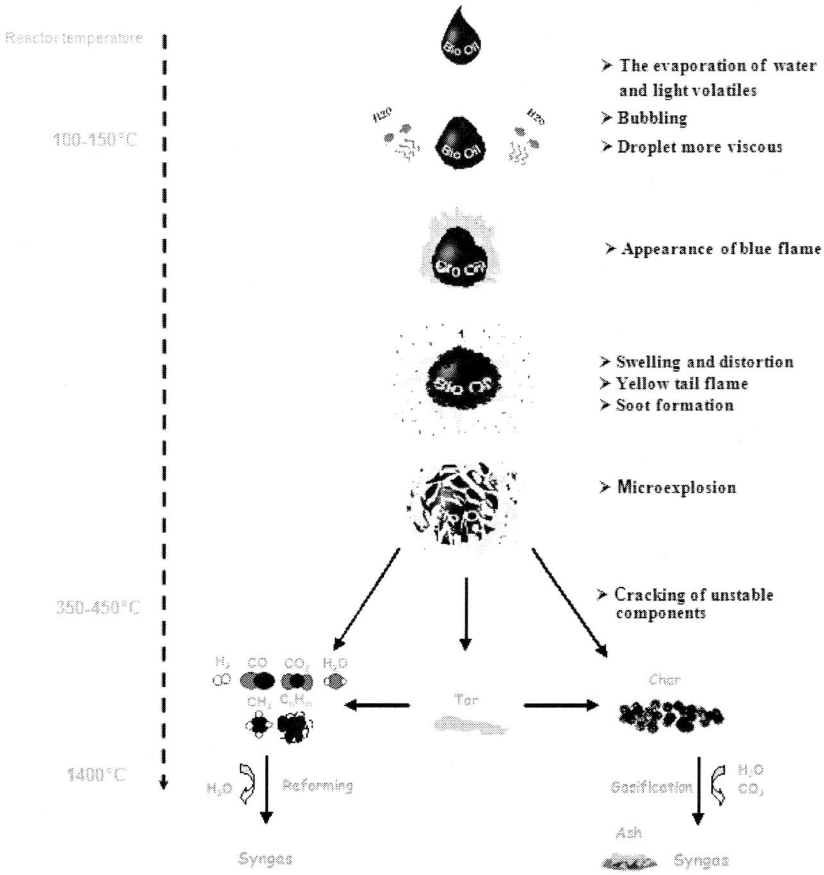

Figure 1. Schematic representation of air steam gasification of bio-oil droplet

The steam reforming of the bio-oil can be simplified as the steam reforming of an oxygenated organic compound ($C_nH_mO_k$) following:

$$C_nH_mO_k + (n - k)H_2O \leftrightarrow nCO + (n + m/2 - k)H_2 \qquad (1)$$

During the last decade, catalytic steam reforming of bio-oil components has been widely studied, focusing on acetic acid as one of the most representative compounds.

Production of hydrogen from catalytic steam reforming of bio-oil was extensively investigated by NREL (Wang et al., 1997, 1998). Czernik et al. obtained hydrogen in a fluidized bed reactor from the carbohydrate derived fraction of wood bio-oil with a yield of about 80% of theoretical maximum (Czernik et al., 2002).The catalytic steam reforming of the bio-oil or the model oxygenates (e.g., ethanol, acetic acid) has been widely explored via various catalysts, e.g., Ni-based catalysts (Sakaguchi et al., 2010), Mg-doped catalysts (Garcia et al., 2000)and noble metal-loaded catalysts (Goula et al., 2004; Rioche et al., 2005; Trimm et al., 1997).Noble metals (Pt, Ru, Rh) are more effective than the Ni-based catalysts and less carbon depositing. Such catalysts are not common in real applications because of their high cost. Catalytic steam reforming of bio-oil is a costly process and presents several disadvantages such as carbon deposit and the deactivation of catalysts due to coke or oligomer deposition even in the presence of an excess of steam (S/C > 5)(Trimm et al., 1997; Rostrup-Nielsen et al., 1997). For these reasons, there is an interest in developing non catalytic gasification of bio-oil.

Only very few works can be found on the non catalytic reforming of whole bio-oil. Bimbela et al. studied catalytic and non catalytic steam reforming of acetol (bio-oil model compound) in fixed bed at low temperature (550-750°C) in order to highlight the specific role of the catalyst in this process (Bimbela et al., 2009). The same study is carried out by Guus van Rossum et al. concerning catalytic and non catalytic gasification of bio-oil in a fluidized bed over a wide temperature range (523-914°C)(van Rossum et al., 2007). Marda et al. has developed a system for the volatilization and conversion of a bio-oil mixed with methanol to syngas via non-catalytic partial oxidation (NPOX) using an ultrasonic nozzle to feed the mixture. The effects of both temperature (from 625 to 850°C) and added oxygen (effective O/C ratio from 0.7 to 1.6) on the yields of CO and H_2 have been explored. They obtained hydrogen yield of about 75% of theoretical maximum (Marda et al., 2009). Panigrahi et al.gasified biomass-derived oil (BDO) to syngas and gaseous fuels at 800°C.They obtained syngas (H_2 + CO) yield ranging from 75 to 80 mol % (Panigrahi et al., 2003).Henrich et al. gasified lignocellulosic biomass.The first process step is a fast pyrolysis at atmospheric pressure, which produces large condensate that was mixed to slurries. The slurries are pumped into a slagging entrained flow gasifier and are atomized and converted to syngas at high operating temperatures and pressures (Henrich et al., 2004).

The objective of this work is to bridge the lack of knowledge concerning the physicochemical transformation of bio-oil into syngas using non catalytic steam gasification in entrained flow reactors. This complex process involves vaporization, thermal cracking reactions with formation of gas, tars and char that considered as undesirable product. This is followed by steam reforming of gas and tars, together with char conversion. To better understand the process, the first step of gasification (pyrolysis) and thereafter the whole process (pyrolysis + gasification) were studied. The pyrolysis study focused on the influence of the heating rate and the final pyrolysis temperature, for this aim, two complementary devices namely: a Horizontal Tubular Reactor (HTR) and a High Temperature - Entrained Flow Reactor (HT-EFR) were used to study on the one hand a wide range of heating rates, in the range from 2 to 2000°C/s and on the other hand final temperature ranging from 550 to 1000°C. Concerning gasification, the effect of temperature on syngas yield and composition wasstudiedover a wide range from 1000°C to 1400°C, for this aim HT-EFR was used.

2. MATERIALS AND METHODS

2.1. Description of the laboratory device and of the procedure

Two complementary devices, namely: a Horizontal Tubular Reactor (HTR) and a High Temperature - Entrained Flow Reactor (HT-EFR), were used to study a wide range of heating rates, in the range from 2 to 2000°C/s and final temperature from 550 to 1000°C.

2.1.1. Horizontal Tubular Reactor HTR

The experiments of fast pyrolysis were carried out in a HTR (Fig. 2). This device allowed carrying out experiments in conditions of fast pyrolysis which is not possible in a thermobalance. The reactor consisted of a double-walled quartz pipe. The length and inside diameters were 850 mm and 55 mm respectively for the inner tube, and 1290 mm and 70 mm respectively for the outer tube. The reactor temperature can reach 1100°C.

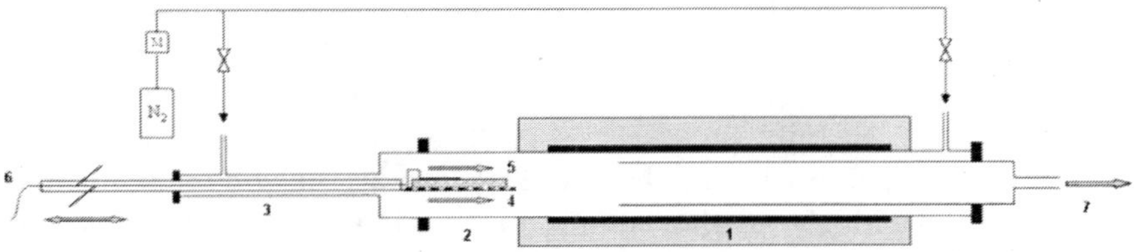

1- Furnace; 2- Quartz reactor; 3- Movable sample boat; 4- Metal grid; 5- Refractory ceramic wool soaked with 1g of bio-oil; 6-Thermocouple; 7- Gas outlet; M- Mass flow meters and controllers

Figure 2. Horizontal Tubular Reactor (HTR) ready for sample introduction

The procedure carried out for an experiment was the following. First, the furnace was heated and the gas flowrate (nitrogen) was adjusted using a mass flow meter controller. When the temperature was stabilized, the sample was placed on the metal grid at the unheated section of the reactor. This section was swept by half of the total cold nitrogen flow injected, in order to maintain it cold and under inert atmosphere, and therefore avoid its degradation. Meanwhile the second half of the nitrogen flow was preheated through the double-walled annular section of the reactor as shown in Fig. 1. The sample consisted in 1g of bio-oil was placed inside a crucible of 25 mm diameter and 40 mm height for studying the effect of temperature and of ash content. In order to achieve higher heating rates, some runs were performed with 1g of bio-oil previously soaked in a refractory ceramic wool sample of 100x20 mm length and width and 3 mm thickness. The choice of this sample holder allowed increasing the exchange surface and subsequently obtaining larger heating rates. We proved that this wool has no catalytic effect on bio-oil pyrolysis. Indeed, previous bio-oil pyrolysis experiments were carried out first with a single crucible, and secondly in the refractory ceramic wool deposited in the crucible. The wool didn't induce any change on the products yield.

The reactor outlet was first connected to an O_2 gas analyser to ensure that there is no oxygen in the reactor. Afterwards, a manual insertion enabled to move the sample in the furnace at different velocities, ranging between 0.06 and 30 cm/s. The sample temperature evolution was measured using a thermocouple placed in the middle of the sample in order to determine a heating rate for each experiment. Variation of the heating rate was obtained by varying the sample introduction through the tubular reactor. Four different durations have been used: 16, 8, 4 and finally 0.03 min resulting in four different heating rates. The sample then remained in the middle of the reactor for a definite time and is brought back out of the furnace; the solid residue was weighed after cooling. Even after several experiments, no char deposit was observed inside the reactor. Only tar deposits were observed in the cold outlet of the reactor. The reactor outlet was connected to a sampling bag at t = 0 just before sample introduction. The gases formed by pyrolysis were collected in the bag. The duration of all experiments was 10 min with a 2 NL/min N_2 flowrate which enabled to know accurately the volume of N_2 sampled in the bag. In HTR reactor, the volume of formed gas never exceeded 1% of the volume of N_2 sampled in the bag. After the experiment the bag was disconnected from HTR, and connected to the micro-chromatograph analyser (µGC). From the total volume of gas in the bag and measure of the gas concentration, the quantity of each gas formed by 1g of bio-oil can be precisely calculated.

2.1.2. Entrained Flow Reactor EFR

A laboratory scale high temperature entrained flow reactor HT-EFR was used in this work. It consisted in a vertical tubular reactor electrically heated by a total of 18 kW three-zones electrical furnace, and was able to reach 1600°C in a 1m long isothermal reaction zone, as illustrated in Fig. 3.

The atmosphere gas is generated by feeding the controlled flow of nitrogen in a 2 kW electrical steam generator. This atmosphere gas is then preheated to 900°C using a 2.5 kW electrical battery of heating elements before reaching the isothermal reaction zone. The HT-EFR was initially set up to achieve high heating-rate gasification of solid biomass, and was equipped for the present work with a specially designed bio-oil pulverization feeder, in order to obtain a very constant mass flowrate spray.

The feeder consists of a 1 m long and 14 mm o.d. probe cooled with water at 30°C. At its end a commercial stainless steel nozzle is integrated. This allows uniform distribution with fine atomization. Nozzle type (DELAVAN WDB) is a solid cone, with orifice diameter of 0.46mm and a spray angle of 60°.

The oil is fed with a syringe which is automatically pushed. The expected mass flowrate of 0.3 g/min was too low for direct pulverization. Therefore, a 3.5 NL/min N_2 flowrate was used to entrain oil in the feeding probe and to ensure a thin spray of the oil. The spray of droplets is dispersed on the section of a 75 mm i.d. alumina reactor swept by 15 $NL.min^{-1}$ of atmosphere gas. The steam gasification experiments were carried out in HT-EFR with steam to fuel mass ratio (fuel includes inherent water in bio-oil) of S/F=4.5.

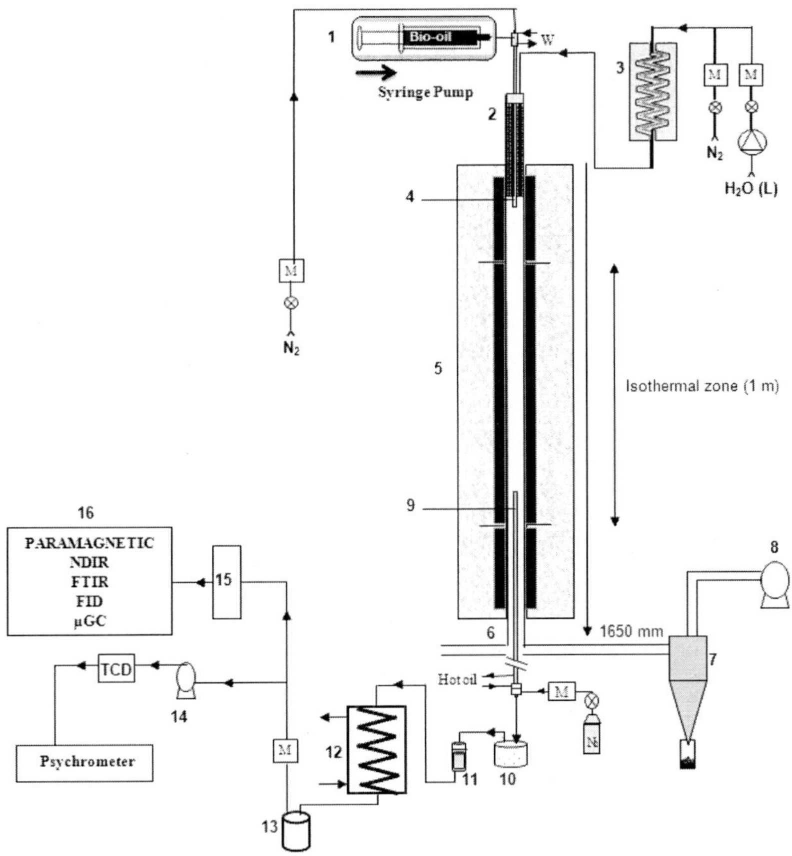

1- Injection system; 2- Electrical preheater; 3- Steam generator; 4- Water cooled feeding probe; 5- Three zones electrical furnace; 6- 75 mm i.d. alumina reactor; 7- Cyclone collector; 8- Exhaust fan; 9- Oil cooled sampling probe; 10- Hot settling box; 11- Hot particle collector (filter); 12- Water cooler; 13- Condensate collector; 14- Sampling pump; 15- Gas dryer; 16- Gas analyser; M- Mass flow meters and controllers; N_2- Nitrogen; W- Water (probes cooling)

Figure 3. Entrained flow reactor

Reactions take place along the reactor during a controlled gas residence time, which was of about 3-4s. The residence time of droplets or solid residue after reaction is assumed to be similar to that of the gas because of the very small particle size. The gas residence time was calculated as the ratio of the reaction zone length to the average gas velocity in the reactor. At 1650 mm downstream of the injection point, gases and solid residue were sampled by a hot-oil cooled probe at 150°C. Gas and solid residue were separated using a settling box and a filter, both heated to avoid water condensation. The water and potential remaining tars were first condensed in a heat exchanger, and non-condensable gases were forwarded to a micro-chromatograph analyser (μGC) to quantify H_2, CO, CO_2, CH_4, C_2H_2, C_2H_4, C_2H_6, C_3H_8 and C_6H_6. The μGC offers excellent resolutions of all analyze species at higher concentrations with repeatability of ± 2 percent relative standard deviation, the system offers also a minimum detectable quantity of about 10 ppm for most gases species.

Gases were also analyzed by other analyzers that allowed checking the absence of O_2, to confirm the analysis and to control continuously gas production: a Fourier Transform InfraRed (FTIR) analyser, a Non-Dispersive InfraRed (NDIR) analyser coupled with a paramagnetic analyser for O_2 and a Thermal Conductivity Detector (TCD) to quantify H_2.

2.2. Feedstock

The feedstock used for all experiments was a bio-oil produced by fast pyrolysis of softwood on an industrial-scale fluidized bed unit (Dynamotive, West Lorne, Ontario) and provided by CIRAD, France. Its physico-chemical properties have been measured (see Table 1). The water content of the bio-oil measured by Karl Fischer method (ASTM E203) is around 26 wt % which is in agreement with the average values reported in the literature. It can be noticed that the solid particles content is rather high (2.3 wt.%) while the ash content remains low (around 0.06 wt.%). This confirms that the solid particles mainly consist of high-carbon content char particles. These particles were entrained during bio-oil production by the gas stream to the bio-oils condensers. Ultimate analysis and LHV of the bio-oil are very similar to those of wood. From the ultimate analysis, the chemical formula of the bio-oil can be established as $CH_{1.18}O_{0.48}.0.4H_2O$.

3. RESULTS AND DISCUSSIONS

3.1. Thermal decomposition of bio-oil: focus on the products yields under different pyrolysis conditions

3.1.1. Preliminary runs of bio-oil pyrolysis at two final reactor temperatures (The experiments were carried out in a HTR)

Two reactor temperatures were tested in order to evaluate the effect of the final pyrolysis temperature on devolatilization process affecting the yield of gas, condensate and residual solid:

- Moderate temperatures at 550°C;

- High temperature 1000°C to approach the severe conditions of gasification.

The yields of final products are listed in Figure 4. With temperature increasing from 550 to 1000°C, the total gas yield sharply increases from 12.2 to 43.0 wt.%, while condensate (tar + water) decreases from 73.2 to 47.5 wt.%. Varying temperature shows a great influence on the gas composition as well.

Table 1. Ultimate analysis and several characteristics of bio-oil derived from hardwood fast pyrolysis

Ultimate analysis (wt.%)					H_2O	Ash	Solids	LHV	Kinematic viscosity
C	H	O	S	N	(wt.%)	(wt.%)	(wt.%)	($MJ.kg^{-1}$)	at 20 °C ($mm^2.s^{-1}$)
42.9	7.1	50.58	<0.10	<0.10	26.0	0.057	2.34	14.5	103

Figure 5 shows that the main gas products are H_2, CO, CO_2, CH_4 and some C_2 hydrocarbons (C_2H_2, C_2H_4 and C_2H_6). Among them, the H_2 and CO content increased significantly from 0.056 wt.% to 1.65 wt.% and from 5.9 to 23.9 wt.% respectively as temperature increased from 550 to 1000°C. Yields of CH_4 also increased from 1.2 to 5.0 wt.% whilst that of CO_2 increased from 4.2 to 10.8 wt.%. The yields of C_2H_2, C_2H_4 and C_2H_6 are relatively small. The specie C_2H_6 only appears at 550°C while C_2H_2 only appears at 1000°C. The thermal cracking of gas-phase hydrocarbons at high temperature might explain the variation of gas product composition observed.

Finally, with increasing temperature from 550°C to 1000°C, the char yield decreased significantly from 14.5 to 9.4 wt.%. However changing the reactor temperature implies a change of both the heat flux density imposed to bio-oil (and hence its heating rate) but also the final temperature reached by the char produced. Therefore the later trend observed might be due to two reasons:

- The char formed at 550°C contains residual volatile matters which are released when theemperature increases to 1000°C;
- Increasing the heating rate results in the decrease of the char yield. This is actually in good agreement with what is usually observed in the literature from pyrolysis of biomass (Ayllón et al., 2006; Haykiri-Acma et al., 2006; Mani et al.,2010).

To check the first assumption, a char first prepared at 550°C was submitted to a second heating step at 1000°C. During this second step, the mass of char did not change, which excluded the first assumption, and highlighted actually the effect of heating rate. In order to confirm this trend, additional experiments were carried out to separate the effect of these two parameters. This is studied in details in the following section.

3.1.2. Effect of heating rate and final temperature on the product yields

The temperature profiles obtained in the HTR are illustrated in Figure 6. A calculation of the highest heating rate is then made taking into account only the linear part of curves. Details of the calculated heating rates and products yields obtained from experiments are given in Table 2.

The temperature profiles curves show that the heating rate ranges from 2 to 14°C/s at the final pyrolysis temperature of 550°C, and from 2 to 100°C/s at the final pyrolysis temperature of 1000°C. The response time of temperature measurement system was characterized by placing the thermocouple alone and the thermocouple placed in the refractory ceramic wool without bio-oil sample together inside the reactor in 0.03 min. The results are also plotted in Figure 6. At 1000°C we can notice that the response of the thermocouple and refractory ceramic wool does not exceed 100°C/s. But, it appears that the actual heating rate for the sample introduced in 0.03min may be still higher than 100°C/s. This is further illustrated on Figure 6.

Table 2. Product yield of bio-oil pyrolysis at different temperatures and heating rates

	t (min)[a]	Heating rate °C/s	Solid %wt	Total gas %wt	condensate %wt
Pyrolysis at 550 °C	16	2	14.4	14.1	71.4
	8	5	12.4	13.7	73.8
	4	10	11.4	13.3	75.2
	0.03	14	10.5	13.3	76.0
	flash	>2000	1.2	13.6	85.1
Pyrolysis at 1000 °C	16	2	11.5	41.6	46.8
	8	5	10.4	41.7	47.8
	4	14	8.6	40.9	52.2
	0.03	100	3.8	43.8	53.4
	flash	>2000	0.9	40.3	58.7

In order to highlight the effect of heating rate and final temperature on the yields of char, they were plotted in figure 7, with the heating rate as the x scale, using a log scale. The low heating rate experiments gave higher yields of char. Char yield then decreased significantly: from 14.4 wt.% down to 10.5 wt.% when heating rate increased from 2 to 14°C/s at the final temperature of 550°C, and from 11.5 to 3.8 wt.% when heating rate was increased from 2 to 100°C/s at the final temperature of 1000°C.

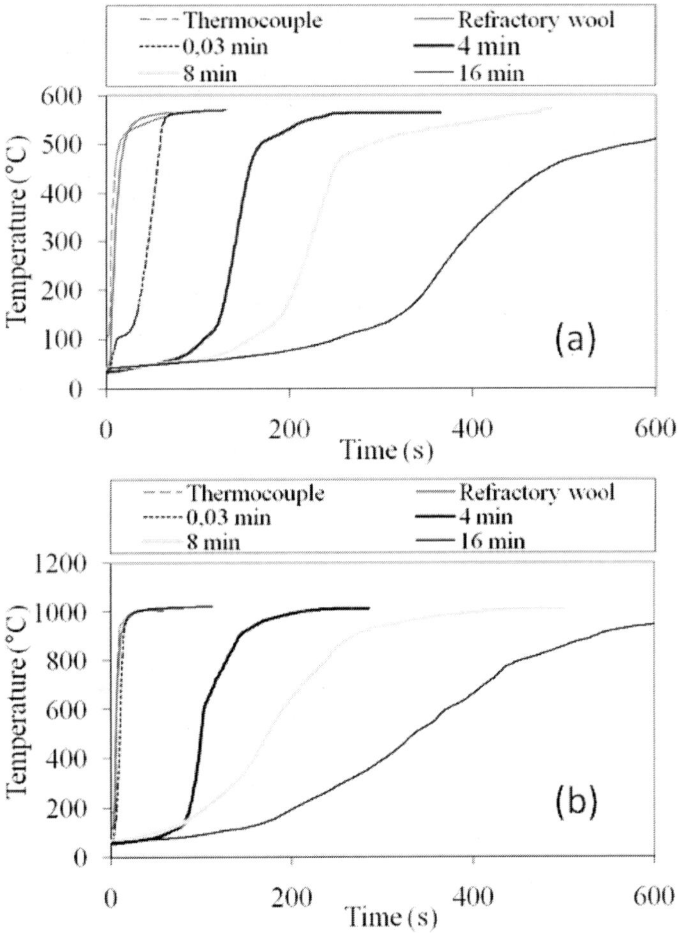

Figure 4. Temperature evolution of the sample during bio-oil pyrolysis in HTR at different heating rates and two final temperatures. a: 550°C, b: 1000°C

In order to increase still the heating rate and reach the flash pyrolysis conditions, we have performed additional experiments in the HT-EFR. This process allows achieving very high heating rate.

Indeed it is shown that when a particle or droplet is transported by a cold spraying gas, its heating rate is controlled by mixing of the cold gas with the hot gas in the reactor. CFD modeling was used and derived this order of magnitude. Heating rate was estimated at 2000°C/s(Van de Steene et al., 2000). Under these conditions, the char yield measured is very low: < 1 wt.%. As can be seen in Figure 7, the char yield obtained with HT-EFR is in rather good agreement with the values obtained in HTR and extrapolated to high heating rates. This result is in agreement with the work carried out by Guus van Rossum et al.(Van Rossum et al., 2010). They found that small droplets (undergoing high heating rate) are much quicker evaporated and give fewer char compared to larger droplets (undergoing low heating rate pyrolysis).

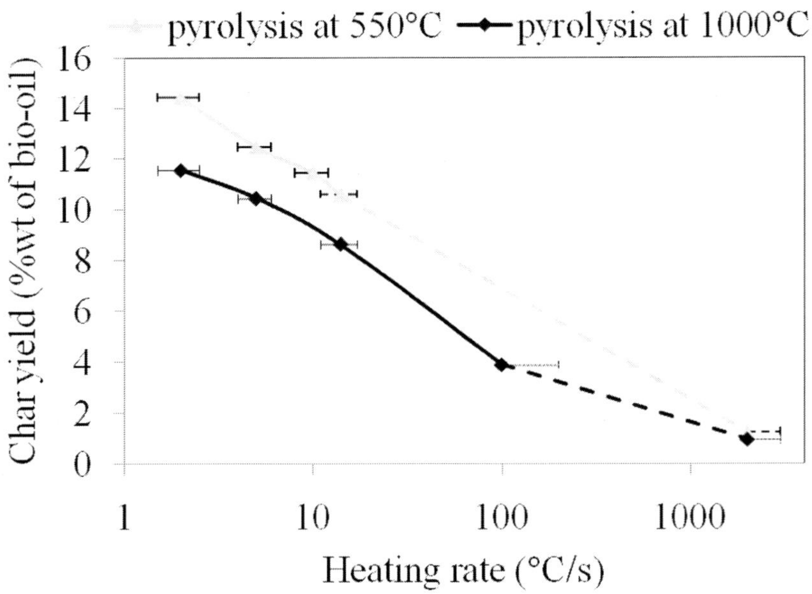

Figure 5. Char yield obtained from pyrolysis of bio-oil at two final temperatures: 550°C and 1000°C - effect of heating rate

Globally from all the data collected, the char yield depends very much on the heating rate, and less on the final temperature, confirming the observation from section 3-1. These results give important information for understanding the pathways occurring during gasification of bio-oil in reactors such as EFR: the amount of char formed by pyrolysis and submitted to subsequent steam-gasification reactions will be very low whereas the main reactions will occur in the gas phase (reforming, partial oxidation...). Considering that solid gasification is rate-limiting, this might be an advantage of using bio-oil instead of biomass as feedstock for EFR gasification.

Figure 8 shows the effect of heating rate on the product yields at two final pyrolysis temperatures. There is no apparent impact of the heating rate but a drastic influence of the temperature on the total gas yield which remains of about 13–14 wt.% and 40–43 wt.% at 550°C and 1000°C, respectively.

On the other hand, we can notice that the total condensate yield increased when the heating rate increased and when the final temperature decreased. A value of 76 wt.% is obtained at 14°C/s and a final temperature of 550°C, which is about 5 wt.% higher than that obtained at 2°C/s. In the same manner, at 1000°C the total condensate yield increased with the heating rate, up to 53.4 wt.% at 100°C/s. This value was about 6 wt.% higher than that of 2°C/s.

All these trends can be summarized and explained as follows.

i.Pyrolysis inside the sample

The volatile matters yield increases with the heating rate of bio-oil, to the detriment of the char yield as reported earlier. The primary volatiles may undergo secondary reactions through two competitive pathways (Zaror et al., 1985; Seebauer et al., 1997):

- re-polymerizing to form char;
- cracking to form lighter volatiles which implies less tar repolymerisation.

The re-polymerization pathway is probably favored by lower heating rates. Indeed, low heating rates lead to longer volatiles residence times inside the sample, and favor secondary reactions of re-polymerization to form solid residue. These conditions are known to favor the formation of secondary char from biomass pyrolysis experiments (Zaror et al., 1985)and apparently, this could be extended to the case of bio-oil pyrolysis.

ii.Gas phase reactions outside the sample

Once the volatiles have escaped from the sample, they can undergo additional secondary gas-phase cracking reactions as previously presented.

The conversion rate of this reaction highly increases with the gas temperature, leading to higher gas yields to the detriment of condensates. This result is in agreement with number of pyrolysis works carried out on biomass (Seebauer et al., 1997).

Let's notice that due to the procedure described, higher heating rate leads to lower residence time of tars in the hot zone because the bio-oil sample is introduced more rapidly to the centre of the heated zone. The estimate of the gas residence time in the HTR was calculated, from their release at the sample position (which varies with time according to the duration of sample introduction) to the exit of the reactor. It varies from 8 to 16s at 550°C and from 5 to 10s at 1000°C.

3.2. Gasification of wood bio-oil (The experiments were carried out in a HT-EFR)

3.2.1. Effect of temperature

The first objective was to study the influence of temperature - over a wide range - on the syngas yield and composition.

Generally the gas mixture formed from catalytic reforming of bio-oil is composed of hydrogen, carbon monoxide and dioxide, methane, acetylene, unconverted steam, coke (carbon) and soot. Figure 9, presents the mole fraction of the gaseous products from this work (in dry basis and without N_2) as a function of temperature in the range 1000 to 1400°C. Error bars were established by repeating each test 2 or 3 times. The species C_2H_4, C_2H_6, C_3H_8 and C_6H_6 are not detected by chromatography. Whatever the operating temperature between 1000°C and 1300°C, bio-oil is mainly decomposed to H_2, CO, CO_2, CH_4 and C_2H_2. Above 1300°C C_2H_2 disappears, while CH_4 disappears above 1400°C. As the temperature rises, the fraction of H_2 increases monotonically at the expense of carbon monoxide, methane and acetylene. Above 1300°C the hydrogen content remains almost stable. At 1400°C hydrogen mole fraction reaches the maximum value of 64 mol% of the syngas.

The reactions that may explain the increase of hydrogen with temperature are:

The steam reforming of CH_4 and C_2H_2 into H_2 and CO (2)

The water gas shift reaction $CO + H_2O \leftrightarrow CO_2 + H_2$ (3)

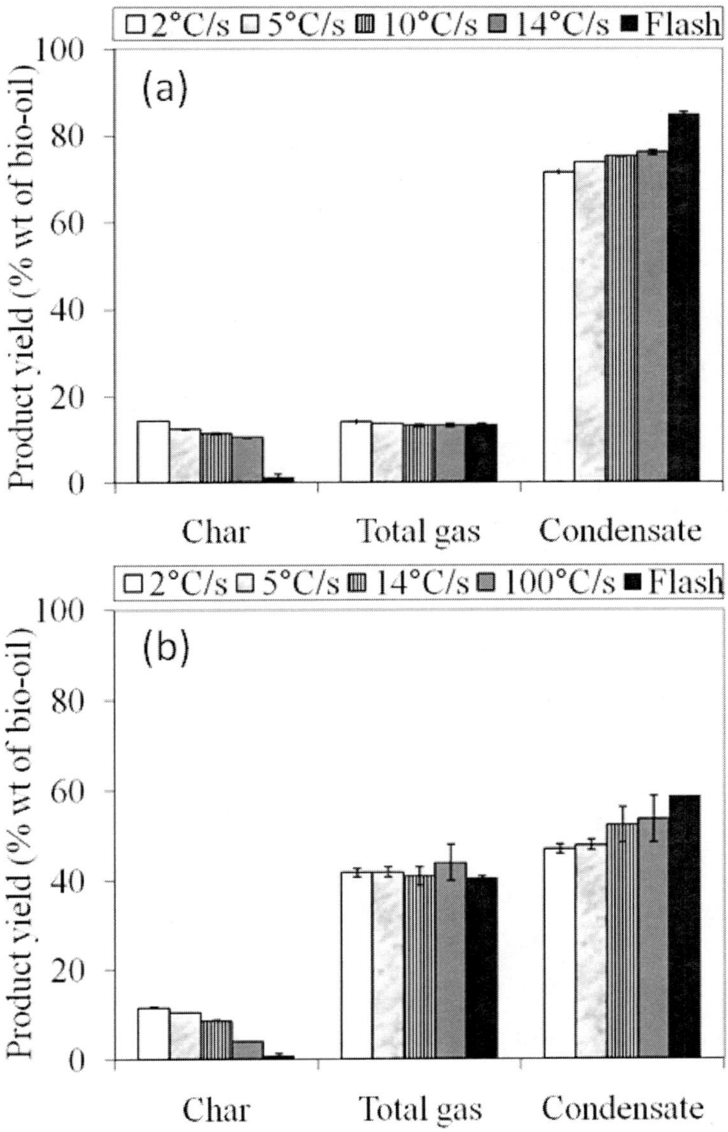

Figure 6. Product yield obtained from bio-oil pyrolysis at two final temperatures. a: 550°C, b: 1000°C- effect of heating rate

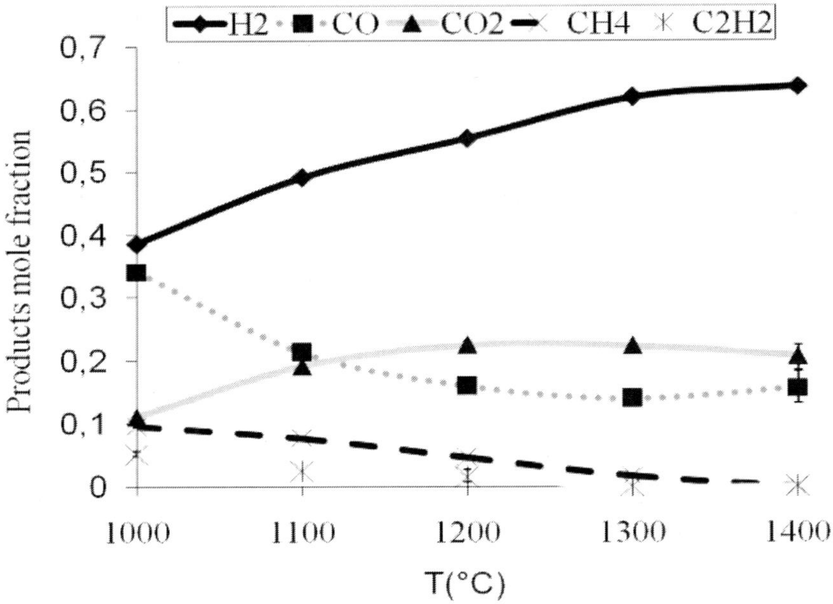

Figure 7. Composition of the produced syngas (dry basis and without N_2) - effect of temperature, at S/F=4.5

The water gas shift reaction can also explain the increase of carbon dioxide and the decrease of carbon monoxide between 1000 and 1200°C. Above 1200°C, carbon monoxide slightly increases. This may be explained by steam gasification of the solid carbon residue (char and soot) resulting from the pyrolysis of oil dropletsto yield carbon monoxide and hydrogen following the reaction:

$$C + H_2O \leftrightarrow CO + H_2 \tag{4}$$

and potentially following the Boudouard reaction which would explain the slight decrease of CO_2:

$$C + CO_2 \rightarrow 2CO \tag{5}$$

It was observed that as the temperature increases the amount of collected solid decreases significantly above 1000°C. The process allows achieving very high heating rate estimated at 2000°C/s(Van de Steene et al., 2000). Under these conditions, the char yield measured is very low: < 1 wt.%. At 1400°C more than the 99.9% the bio-oil is converted to gas.

3.2.2. Equilibrium calculation

The thermodynamic equilibrium calculation is independent of reactor and predicts the yield of final products, based on the minimization of the Gibbs free energy of the system. It was conducted here using FactSage software 5.4 to establish whether the syngas was close or not to equilibrium at the different temperatures. Operating temperature varied from 1000°C to 1400°C; pressure was fixed at 1 atm. The software is not presented in detail in this paper; details of the thermodynamic calculation could be found on FactSage web site.

The results of prediction are presented in Figure 10, expressed in g of gas produced per g of crude bio-oil injected. As the temperature increases from 1000 to 1400°C the calculated equilibrium yield of H_2 remains approximately constant at 0,11g/g, while the yield of CO increases from 0.3 at 1000°C to 0.45g/g at 1400°C. The CO_2 yield decrease from 1.1 g/g at 1000°C down to 0.9 at 1400°C.

At 1000°C the calculation yields are far away from the experimental results. The deviation from equilibrium at lower temperatures is also reported by Sakaguchi et al (Sakaguchi et al., 2010). At 1200°C the thermodynamic equilibrium begins to establish. The calculation nevertheless does not retrieve the presence of CH_4 and C_2H_2. At 1400°C the experimental yields are very close to the equilibrium calculation yields: 0.11 and 0.12 respectively for H_2, 0.45 and 0.45 respectively for CO, and 0.86 and 0.88 respectively for CO_2. It can be concluded that at this temperature the equilibrium is reached.

It is also interesting to compare the obtained experimental yields at 1400°C to the theoretical yields corresponding with complete gasification of oil that would follow:

$$CH_{1.18}O_{0.48}.0.4H_2O + 1.12\ H_2O \rightarrow CO_2 + 2.11\ H_2 \qquad (6)$$

The maximum stoichiometric H_2 yield for this oil would be 0.150g per 1g crude bio-oil while a value of 0.126 g was obtained experimentally. This shows that under our experimental conditions and at 1400°C steam reforming of bio-oil lead to a production of H_2 with a yield of about 84% of theoretical yields corresponding with complete gasification of oil (reaction 6).

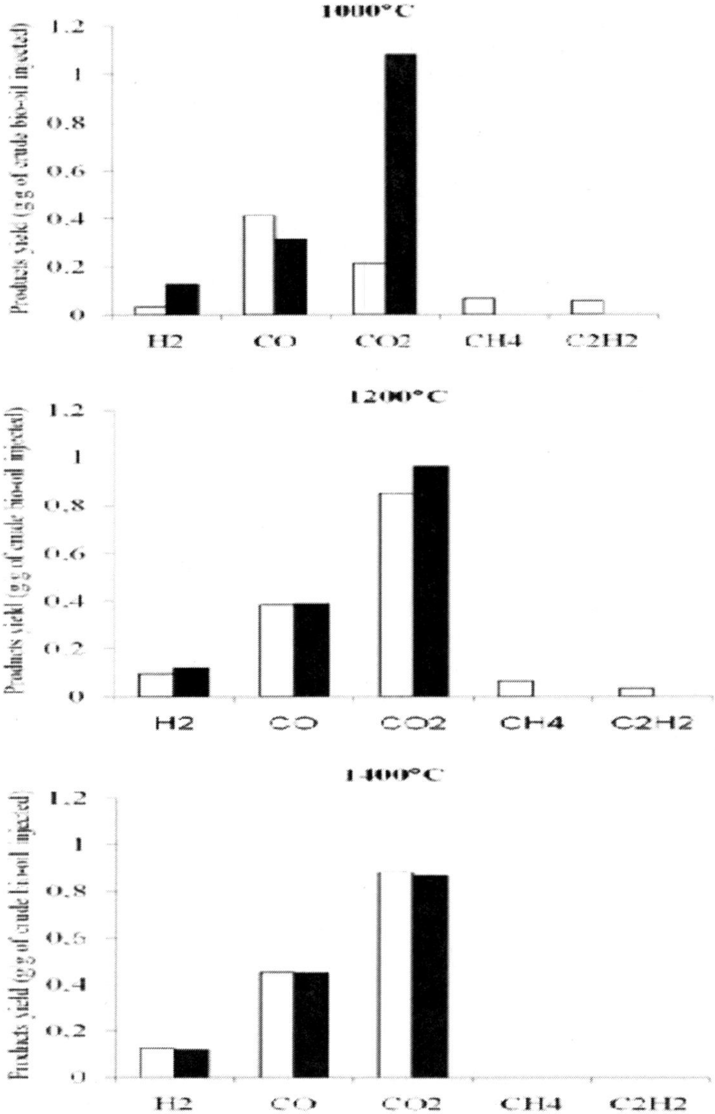

Figure 8. Gas yield from bio-oil reforming at 1000, 1200 and 1400°C, S/F=4.5. □ Experiments; ■ Equilibrium calculation

4. CONCLUSION

Gasification of biomass is one of the leading near-term options for renewable energy production. When large scale units are considered, bio-oil shows lots of advantages compared to solid biomass. The combination of decentralized fast pyrolysis of biomass followed by transportation and gasification of bio-oil in bio-refinery has attracted great attention.

The overall purpose of this research was to investigate the feasibility of a whole bio-oil non catalytic steam gasification process for the production of high quality syngas in entrained flow reactor.

From a chemical point of view, bio-oil gasification process is quite complex and consists of the following main stages: vaporization, thermal cracking reactions with formation of gas, tars and char that considered as undesirable products. This is followed by steam reforming of gas and tars, together with char oxidation. To better understand the process, the first step of gasification (pyrolysis) and thereafter, the whole process (pyrolysis+gasification) were separately studied.

In the pyrolysis step, a temperature increase from 550°C to 1000°C greatly enhanced the gas yield, whilst solid and liquid yields decreased significantly in agreement with the literature. The heating rate of bio-oil has little impact on the gas yield, but plays a major role on the char yield. Hence the char yield decreases from 11 wt.% with a heating rate of 2°C/s down to 1 wt.% for flash heating rate of 2000°C/s at a final temperature of 1000°C. At very high heating rate, the final temperature has little influence on the char yield. These results show that for gasification under industrial EFR conditions, the quantity of char is very small. Thus the gasification process mainly consists in gas/tar reforming. Nevertheless, the production of clean syngas will require either complete gasification of char or its removal from the gas produced by the gasifier.

In steam gasification process, whole bio-oil was successfully steam gasified in HT-EFR. An increase in the reaction temperature over a wide range from 1000°C to 1400°C implies higher hydrogen yield and higher solid carbon conversion. A thermodynamic equilibrium calculation showed that equilibrium was reached at 1400°C. At this temperature steam reforming of bio-oil leads to yield of equal 84% of theoretical yields corresponding with complete gasification of oil.

ACKNOWLEDGEMENT

The authors gratefully acknowledge the financial support from EnerBio Program of Fondation Tuck France, and express their gratitude to Mr. Bernard AUDUC technician in Ecole des Mines d'Albi-Carmaux for his assistance and contribution to experimental device design and operation.

REFERENCES

1. M. Ayllón, M. Aznar, J. L. Sȧnchez, G. Gea, J. Arauzo, 2006Influence of temperature and heating rate on the fixed bed pyrolysis of meat and bone meal. Chemical Engineering Journal, 1212-3June 2006), 85961385-8947
2. T. , A. Chaala, M. G. Pérez, D. Rodrigue, C. Roy, (2004a, 2004aColloidal properties of bio-oils obtained by vacuum pyrolysis of softwood bark. Characterization of water soluble and water-insoluble fractions. Energy Fuel, 185August 2004), 7047120887-0624
3. T. , A. Chaala, M. G. Pérez, C. Roy, (2004b, Colloidal properties of bio-oils obtained by vacuum pyrolysis of softwood bark. Storage stability. Energy Fuel,181December 2003), 1882010887-0624
4. C. Branca, C. Di Blasi, R. Elefante, (2005a, Devolatilization and Heterogeneous Combustion of Wood Fast Pyrolysis Oils. Ind. Eng. Chem. ResVol. 44, 4January 2005), 7998100888-5885
5. C. Branca, C. Di Blasi, C. Russo, (2005b, Devolatilization in the temperature range 300-600 K of liquids derived from wood pyrolysis and gasification. Fuel, 841August 2004), 37450016-2361
6. F. Bimbela, M. Oliva, J. Ruiz, L. Garcia, J. Arauzo, (2009, Catalytic steam reforming of model compounds of biomass pyrolysis liquids in fixed bed: Acetol and n-butanol. J. Anal. Appl. Pyrolysis,851-2December 2008), 2042130165-2370
7. R. Calabria, F. Chiariello, P. Massoli, (2007, Combustion fundamentals of pyrolysis oil based fuels. Exp Therm Fluid Sci, 315July 2006), 4134200894-1777
8. S. Czernik, R. French, C. Feik, (2002, Hydrogen by catalytic steam reforming of liquid byproducts from biomass thermoconversion

processes. Ind Eng Chem Res, 4117July 2002), 420942150888-5885

9. L. Garcia, R. French, S. Czernik, E. Chornet, (2000, Catalytic steam reforming of bio-oils for the production of hydrogen: effects of catalyst composition. Appl. Catal., 2012January 2000), 2252390092-6860X

10. L. Devi, K. J. Ptasinski, F. J. J. G. Janssen, (2003, A. Review, the. of, Measures. Primary, Tar. for, in. Elimination, Gasification. Biomass, Processes, n Biomass Gasification Processes. Biomass and Bioenergy,242September 2002), 1251400961-9534

11. Goula, S. K. Kontou, P. E. Tsiakaras, (2004, Hydrogen production by ethanol steam reforming over a commercial Pd/γ-Al2O3 catalyst. Appl. Catal., 492February 2004), 1351440926-3373

12. W. L. H. Hallett, N. A. A. Clark, (2006, Model for the evaporation of biomass pyrolysis oil droplets. Fuel, 854September 2005), 5325440016-2361

13. H. Haykiri-Acma, S. Yaman, S. Kucukbayrak, (2006, Effect of heating rate on the pyrolysis yields of rapeseed. Renewable Energy, 316May 2005), 8038100960-1481

14. E. Henrich, F. Weirich, (2004, Pressurized Entrained Flow Gasifiers for Biomass. Environmental Engineering Science, 211July 2004), 53641092-8758

15. O. Levenspiel, (2005, What will come after petroleum. Ind. Eng. Chem. Res., 4414February 2005), 507350730888-5885

16. T. Mani, P. Murugan, J. Abedi, N. Mahinpey, (2010, Pyrolysis of wheat straw in a thermogravimetric analyzer: Effect of particle size and heating rate on devolatilization and estimation of global kinetics. Chemical Engineering Research and Design, 888August 2010), 9529580263-8762

17. J. R. Marda, J. Di Benedetto, S. Mc Kibben, R. J. Evans, S. Czernik, R. J. French, A. M. Dean, (2009, Non-catalytic partial oxidation of bio-oil to synthesis gas for distributed hydrogen production. International journal o f hydrogen energy, 3420October 2009), 851985340360-3199

18. S. Panigrahi, A. K. Dalai, S. T. Chaudhari, N. N. Bakhshi, (2003, Synthesis Gas Production from Steam Gasification of Biomass-Derived Oil. Energy Fuels, 173April 26, 2003), 6376420887-0624

19. C. Rioche, S. Kulkarni, F. C. Meunier, J. P. Breen, R. Burch, (2005, Steam reforming of model compounds and fast pyrolysis bio-oil on supported noble metal catalysts. Appl. Catal., 611-2October 2009), 1301390926-3373

20. J. R. Rostrup-Nielsen, Industrial relevance of coking. (1997Catal. Today, 373August 1997), 2252230920-5861

21. M. Sakaguchi, Watkinson. A. Paul, E. Naoko, (2010, Steam Gasification of Bio-Oil and Bio-Oil/Char Slurry in a Fluidized Bed Reactor. Energy Fuels, 249August 2010), 518151890887-0624

22. V. Seebauer, J. Petek, G. Staudinger, (1997, Effects of particle size, heating rate and pressure on measurement of pyrolysis kinetics by thermogravimetric analysis. Fuel, 7613October 1997), 127712820016-2361

23. D. L. Trimm, (1997, Coke formation and minimization during steam reforming reactions. Catal. Today, 373August 1997), 2332380920-5861

24. L. Van de Steene, S. Salvador, G. Charnay, (2000, Controlling powdered fuel combustion at low temperature in a new entrained flow reactor. Combustion Science and Technology, 1591January 2000), 2552790010-2202

25. G. Van Rossum, S. R. A. Kersten, W. P. M. van Swaaij, (2007, Catalytic and Noncatalytic Gasification of Pyrolysis Oil. Ind. Eng. Chem. Res., 4612March 2007), 395939670888-5885

26. G. Van Rossum, Güell, Ramachandran. R. P. Balegedde, K. Seshan, L. Lefferts, W. P. M. Van Swaaij, S. R. A. Kersten, (2010, Evaporation of pyrolysis oil: Product distribution and residue char analysis. AIChE Journal, 568August 2010), 220022100001-1541

27. D. Wang, S. Czernik, D. Montane, (1997, Biomass to hydrogen via pyrolysis and catalytic steam reforming of the pyrolysis oil and its fractions. Ind Eng Chem Res, 365May 1997), 150715180888-5885

28. D. Wang, S. Czermik, E. Chornet, (1998, Production of hydrogen from biomass by catalytic steam reforming of fast pyrolytic oils. Energy Fuels, 121January 1998), 19240887-0624

29. Wornat, B. G. Porter, N. Y. C. Yang, (1994, Single droplet combustion of biomass pyrolysis oils. Energy Fuel, 85September 1994), 11311421520-5029

30. Zaror, I. S. Hutchings, D. L. Pyle, H. N. Stiles, R. Kandiyoti, (1985, Secondary char formation in the catalytic pyrolysis of biomass. Fuel, 647july 1985), 9909940016-2361

2

CHAPTER

GASIFICATION OF PLASTIC WASTE AS WASTE-TO-ENERGY OR WASTE-TO-SYNGAS RECOVERY ROUTE

Anke Brems[1*], Raf Dewil[1], Jan Baeyens[2], Rui Zhang[2]

[1] *Department of Chemical Engineering, Chemical and Biochemical Process Technology and Control Section, Katholieke Universiteit Leuven, Heverlee, Belgium; *Corresponding Author: anke.brems@cit.kuleuven.be*

[2] *College of Life Science and Technology, The Key Laboratory of Bioprocess of Beijing, Beijing University of Chemical Technology, Beijing, China*

ABSTRACT

The disposal of plastic solid waste (PSW) has become a major worldwide environmental problem. New sustainable processes have emerged, i.e. either advanced mechanical recycling of PSW as virgin or second grade plastic feedstock, or thermal treatments to recycle the waste as virgin monomer, as synthetic fuel gas, or as heat source (incineration with energy recovery). These processes avoid land filling, where the non-biodegradable plastics remain a lasting environmental burden. Within the thermal treatments, gasification and pyrolysis gain increased interest. Gasification has been widely studied and applied for biomass and coal, with results reported and published in literature. The application to the treatment of PSW is less documented. Gasification is commonly operated at high temperatures (>600°C to 800°C) in an airlean environment (or oxygen-deficient in some applications): the air factor is generally between 20% and 40% of the amount of air needed for the combustion of the PSW. Gasification produces mostly a gas phase and a solid residue (char and ashes). The use of air introduces N2 in the product gases, thus considerably reducing the calorific value of the syngas, because of the dilution. The paper will review the existing literature data on PSW gasification, both as the result of laboratory and pilot-scale research. Processes developed in the past will be illustrated. Recently, the use of a sequential gasification and combustion system (at very high temperatures) has been applied to various plastic-containing wastes, with atmospheric emissions shown to be invariably below the legal limits. Operating results and conditions will be reviewed in the paper, and completed with recent own lab-scale experimental results. These results demonstrate that gasification of PSW can be considered as a first order reaction, with values of the activation energy in the order of 187 to 289 kJ/mol as a function of the PSW nature.

Keywords: Plastic Waste; Re-Use; Gasification; Syngas; Char; Sustainability; Reaction Rate

1. INTRODUCTION: PSW AND ITS TREATMENT OPTIONS

Plastics are light-weight, durable, and versatile, allowing their incorporation into a diverse range of applications. In recent years, the environmental, social and economic impact of plastics has been the topic of the political agenda, with a focus on sustainable production and the decoupling of adverse environmental effects from waste generation. The disposal of waste plastics has become a major worldwide environmental problem. The USA, Europe and Japan generate about 55 million tons of post consumer plastic waste [1]. These waste products were previously dumped in landfill sites, a non-sustainable and environmentally questionable option. The number of landfill sites and their capacity are moreover decreasing rapidly and in most countries the legislation on landfills is becoming increasingly stringent.

New sustainable processes have emerged, i.e. 1) the advanced mechanical recycling of post-consumer plastic waste as virgin or second grade plastic feedstock; and 2) thermal treatments to recycle the waste as virgin monomer, as synthetic fuel gas, as hydrocarbon feedstock, or as a heat source (incineration with energy recovery). These processes avoid land filling, where the non-biodegradable plastics remain a lasting environmental burden. Plastic solid waste (PSW) treatment can be divided in four methods, as illustrated in Figure 1, each individual method providing a unique set of advantages making it particularly suited and beneficial to a specific location, application or product requirement [2,3]. The purpose of recycling is to minimize the consumption of finite natural resources, and this is especially relevant in the case of plastics which account for the use of 4% - 8% of the global oil production [3]: re-using plastics is the required course of action, with the additional benefit of reducing emissions associated with plastic production [3]. The most appropriate recovery method is chosen considering the environmental, economic and social impact of a particular technique. Figure 1 illustrates the position of each recycling method within the production chain.

2. THERMO-CHEMICAL RECYCLING OF PSW

Thermo-chemical recycling refers to advanced technology processes which convert plastic materials into smaller molecules, usually liquids or gases, which are suitable for use as a feedstock for the production of new petrochemicals and plastics [1]. Products have moreover been proven to be useful as fuel. The technology behind its success is the de-polymerisation process that can result in a very profitable and sustainable industrial scheme, providing a high product yield and a minimum residual waste. Processes of pyrolysis, gasification, hydrogen-nation, and steam/catalytic cracking have been previously reported upon [1].

Recently, much attention has been paid to chemical recycling (mainly pyrolysis, gasification and catalytic degradation) as a method of producing various hydrocarbon fractions from PSW. By their nature, a number of polymers are advantageous for such treatment. Thermolysis is the treatment of PSW in the presence of heat at controlled temperatures and under a controlled environment. Thermolysis processes can be divided into pyrolysis (thermal cracking in an inert atmosphere), gasification (in the sub-stoichiometric presence of air usually leading to CO and CO_2 production) and hydrogenation (hydrocracking) [4].

Pyrolysis can be successfully applied to Polyethylene teraphthalate (PET), polystyrene (PS), polymethylmetacrylate (PMMA) and certain polyamides such as nylon, efficiently depolymerising them into constitutive monomers [5-7]. Polyolefins, and in particular polyethylene (PE), has been targeted as a potential pyrolysis feedstock for fuel (gasoline) production, or to produce waxes as feedstock for synthetic lubricants, albeit with a limited success.

The development of value added recycling technologies is highly desirable as it would increase the economic incentive to recycle polymers. Several methods for thermochemical recycling are presently in use, such as direct gasification, and degradation by liquefaction [8]. Various degradation methods for obtaining petrochemicals are presently under investigation, and conditions suitable for pyrolysis and gasification are being researched extensively [9].

Catalytic cracking and reforming facilitate the selective degradation of waste plastics. The use of solid catalysts such as silica-alumina, ZSM-5 or zeolites, effect tively converts polyolefins into liquid fuel, giving lighter fractions as compared to thermal cracking.

Gasification has recently been attracting increased attention as thermo-chemical recycling technique. Its main advantage is the possibility of treating heterogeneous and contaminated polymers with limited use of pre-treatment, whilst the production of syngas creates different applications in synthesis reactions or energy utilisation. Gasification has been widely studied and applied for biomass and coal, with results reported and published in literature. The application for the treatment of plastic solid waste is less documented, although the number of publications increases exponentially.

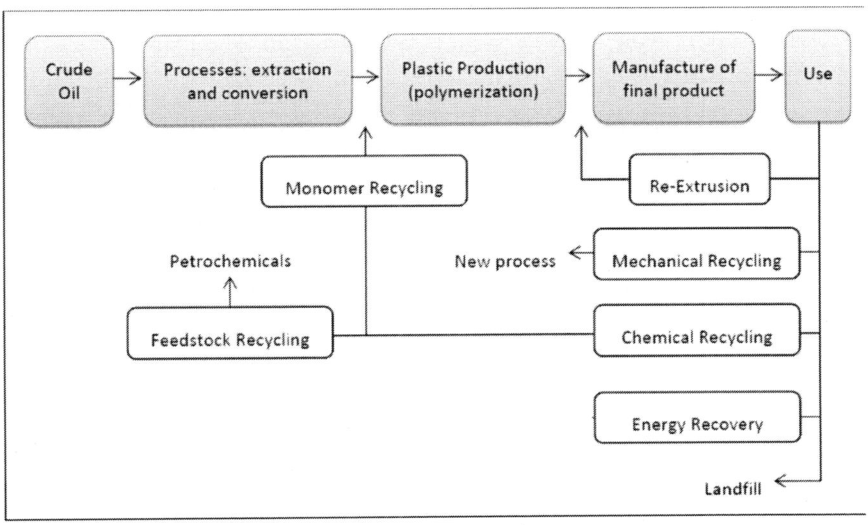

Figure 1. Schematic of recycling methods and their position within the processing line (Adapted from [2]).

Figure 2 shows different thermolysis schemes, main technologies and their main products obtained, as initially presented by Mastellone [10].

3. GASIFICATION

3.1. Gasification of PSW

Gasification or partial oxidation of plastic waste is commonly operated at high temperatures (>600°C - 800°C). Air (or oxygen in some applications) is used as a gasification agent, and the air factor is generally 20% - 40% of the amount of air needed for the combustion of the PSW. The process essentially oxidizes the hydrocarbon feedstock in a controlled fashion to generate the endothermic depolymerisation heat. The primary product is a gaseous mixture of carbon monoxide and hydrogen, with minor percentages of gaseous hydrocarbons also formed. This gas mixture is known as syngas and can be used as a substitute for natural gas or in the chemical industry as feedstock for the production of numerous chemicals. For most of the PSW components, the ash and char production is limited [3,11-13].

A hydrogen production efficiency of 60% - 70% from polymer waste has been reported [14]. Gasification is an attractive alternative to direct incineration of PSW, since it reduces the formation of dioxins and aromatic compounds.

Gasification efficiently utilizes the chemical energy and recoverable raw materials inherent in unsorted domestic waste, industrial and special waste (e.g. medical waste), and is capable of transforming almost all of the total waste input into technically usable raw materials and energy [12].

Co-gasification of biomass with polymers has also been shown to increase the amount of hydrogen produced while the CO content is reduced [15].

A 40 MW fluidized bed gasifier has been installed by Corenso in Varkaus (Finland) for processing polyethylene plastics with metallic aluminium recovery from recycling of Tetrapak cartons.

A typical circulating fluidized bed gasifier is illustrated in Figure 3" target="_self"> Figure 3, with the syngas being thermally used in a boiler arrangement [16].

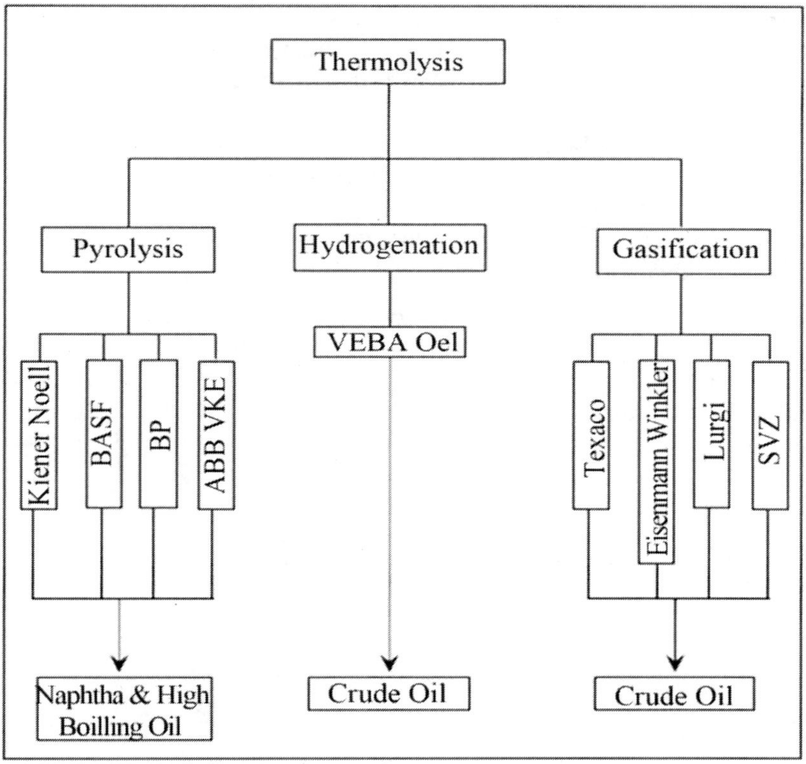

Figure 2. Different thermolysis schemes with reference to the main technologies [10].

Air is mostly used as a gasification agent: the main advantage of using air instead of O_2 alone is to simplify the process and reduce the cost. But a disadvantage is N_2 being present causing the reduction in the calorific value of resulting syngas due to the dilution.

Several types of gasification processes have previously been developed and reported. Their practical performance data, however, have not necessarily been satisfactory for universal application. If char is produced in the gasification, it needs to be further processed and/or burnt. Other gasification schemes (mainly in pilot scale) use a great deal of expensive pure oxygen, whilst others necessitate considerable amounts of expensive materials such as coke and limestone. An ideal gasification process for PSW should produce a high calorific value gas, completely combusted char, produce an easily recoverable ash and should not require any additional installations for air/water pollution abatement.

Early gasification attempts of plastics present in municipal solid waste have been reported since the 1970s [17,18]. The gasification with high calorific value fuel gas obtained from PSW was demonstrated in research stages and results were reported and published in literature for PVC [19], PP [20] and PET [21]. The need for alternative fuels has lead for the co-gasification of PSW with other types of waste, mainly biomass. Pinto et al. [15,22] studied the fluidized bed co-gasification of PE, pine and coal and biomass mixed with PE. Slapak et al. [23] designed a process for steam gasification of PVC in a bubbling fluidized bed.

Xiao et al. [24] co-gasified five typical kinds of organic components (wood, paper, kitchen waste, PE-plastic, and textile) and three representative types of simulated municipal solid waste (MSW) in a fluidized-bed (400°C - 800°C). It was determined that plastic should be gasified at temperatures in excess of 500°C to reach a lower heating value (LHV) of 10,000 kJ/Nm^3. Figure 4" target="_self"> Figure 4 represents the process of co-gasification used in their study.

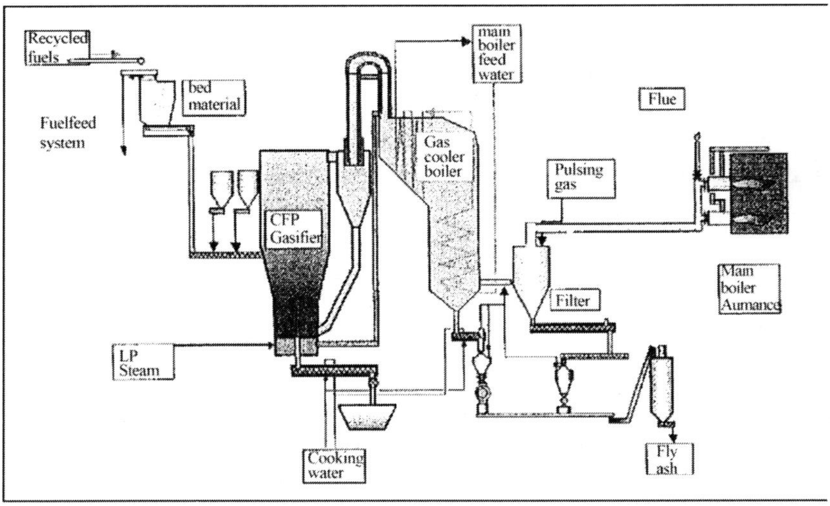

Figure 3. Fluidized bed gasification plant [16].

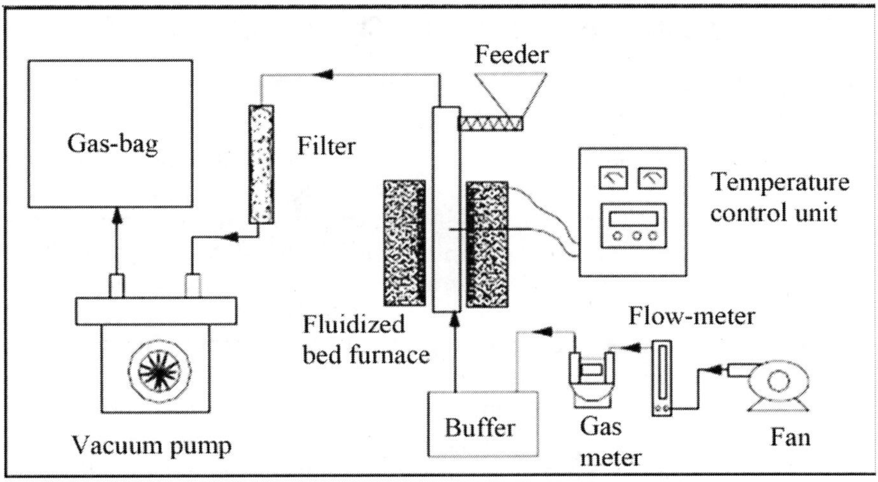

Figure 4. Illustration of the Xiao et al. [24] fluidized bed gasifier.

Gasification produces three different phases: a solid phase (char), a liquid phase (tars) and a gas phase [25]. First products yielded are usually in the range of C_{20} to C_{50}. These products are cracked in the gas phase to obtain lighter hydrocarbons, as ethene and propene, which are unstable at high temperatures and react to form aromatic compounds as benzene or toluene. In thermo-chemical treatment of polyolefins (mainly PE and PP), products obtained mainly depend on cracking reactions in the gas phase. Long residence time of volatiles in reactors and high temperatures decreases tar production but increases char formation [26].

It is believed that increasing temperatures above 500°C and prolonging gas residence time, reduces the tar content in the gas product from gasification of PSW, ASR, MSW and even mixtures of coal, biomass and PSW [22, 27-30].

At temperatures above 800°C, larger paraffins and olefins produced from decomposition of plastics are cracked into H_2, CO, CO_2, CH_4 and lighter hydrocarbons [31].

In PSW gasification, the endothermic gasification reactions involving steam and CO_2 [32,33] (and high heating rates create a char which is more reactive and easier to deal with [34,35]. As a result of these reactions, high gasification temperatures have been reported to increase the H_2 concentration, gas yield [15] and sometimes LHV [36] for a wide range of gasification configurations and oxidizing media.

3.2. Gasification of Automotive Shredder Residue (ASR)

A specific and recent development involves the gasification of ASR, where plastics are significantly present.

The study of Harder and Forton [37] describes the process developed by Schwarze Pumpe, producing methanol as a fuel. Sequential gasification and incineration tests were reported by Mancini et al. [38] and make use of a rotary kiln operated between 850°C and 1120°C with an air factor <1. Combustion of the gases is completed in a secondary afterburner chamber. The system includes a boiler (steam at 43 bar, 430°C) and turbine. The capacity was on average 2400 kg/h during the tests. The paper fully describes the characteristics of the ASR used, the combustion properties, the ash analyses, the composition of the exhaust gases, the process operational problems, the residue management, and the energy efficiency. It was concluded that the process requires minor modifications. Atmospheric emissions were invariably considerably below the legal limits.

A similar full report of a sequential gasification and combustion unit to treat ASR, using a fluidized bed gasifier (590°C) followed by a cyclonic afterburner (1400°C) is described by Viganò et al. [39] and by Cho et al. [40]. All operational details and properties of the different process streams are included in the publications. Operation of the afterburner at 1400°C moreover produces a vitrified (and hence inert) slag. It is concluded that the sequential gasification and combustion system reaches appealing energy and environmental performances, despite the unfavourable characteristics of ASR.

A catalytic gasification of ASR with hydrogen generation is presented by Lin et al. [41], using a lab-scale fixed-bed downdraft gasification process. A 15 wt% NiO/Al_2O_3 catalyst is used at 760°C - 900°C. It is predicted that such a process, conducted at 46.2 atm would yield 87% of syngas, with a 0.27 m^3 reactor allowing to ultimately produce 100 kW of electricity starting from 220 kg/h of ASR. Further tests are however needed.

Sequential microwave pyrolysis and high temperature agent gasification (HTAG) experiments were performed by Donaj et al. [42,43]. The research suggests using the liquid and solid residue from the microwave pyrolysis as fuel for the HTAG process. In this process a gasifying agent (steam, air or an air/steam mixture) is heated to temperatures above 900°C, providing all the heat needed for gasification. The use of HTAG can lead to higher conversion of fuel to gas, higher hydrogen yields and lower tar content in comparison to conventional gasification [42,43]. Although the results are fairly supportive for the gasification of the liquid pyrolysis residue, additional process optimization is required towards the pyrolysis of the solid residue.

The use of a sequential gasification and combustion system (at very high temperatures) as reported by Mancini et al. [38], by Viganò et al. [39] and by Cho et al. [40] demonstrates that atmospheric emissions were considerably below the legal limits. In the specific Japanese case, the operation of the afterburner at 1400°C moreover produces a vitrified (and hence inert) slag. It is concluded that the sequential gasification and combustion system reaches appealing energy and environmental performances, despite the unfavourable characteristics of ASR.

3.3. Large-Scale Gasification Technologies

One of the most common technologies is the Waste Gas Technology UK Limited (WGT) process (Figure 5). Different kinds of wastes (PSW, MSW, sludges, rubbers, wood and straw) are dried and mechanically pre-treated, sorting out incombustibles and granulated to optimum sized particles and fed into a cylindrical reactor for gasification at 700°C - 900°C to yield a HCV gas [44]. Upon discharge and subsequent separation of gas and char, the latter may be utilized via combustion in a boiler to raise steam while the gas is quenched and cleaned of contaminants prior to its use in a gas engine or turbine and possibly CCGT applications.

The Texaco gasification process is by far the most common and well known technology. First pilot scale experiments (10 tons/day) were carried out in the US [45]. Figure 6 reviews the process which consists of two parts, a liquefaction step and an entrained bed gasifier. In the liquefaction step the plastic waste is mildly thermally cracked (depolymerisation) into synthetic heavy oil and some condensable and non-condensable gas fractions. The non-condensable gases are reused in the liquefaction as fuel (together with natural gas). Oil and condensed gas produced are injected to the entrained gasifier [46]. The gasification is carried out with oxygen and steam at a temperature of 1200°C - 1500°C. After a number of cleaning processes (amongst others, HCl and HF removal), a clean and dry synthesis gas is produced, consisting predominantly of CO and H_2, with smaller amounts of CH_4, CO_2, H_2O and some inert gases [47]. Table 1 summarizes the products from the input criteria and process.

In the case of PSW severely contaminated with other waste products (including contaminated wood, waste water purification sludge, waste derived fuel, paper fractions, etc.), the SVZ process constitutes the optimum solution. The input is fed into a reactor (kiln), together with lignite (in the form of briquettes) and waste oil. Oxygen and steam are used as gasification media, and are supplied in counter flow with the input materials [47]. This processes synthesis gas (a mixture of hydrogen and CO), liquid hydrocarbons, and effluent. The gas is used mainly for methanol production and around 20% is used for electricity production. One of the main advantages of this process is its tolerance for various input parameters. Tukker et al. [47] stated a number of acceptance criteria for the SVZ process, summarized below:

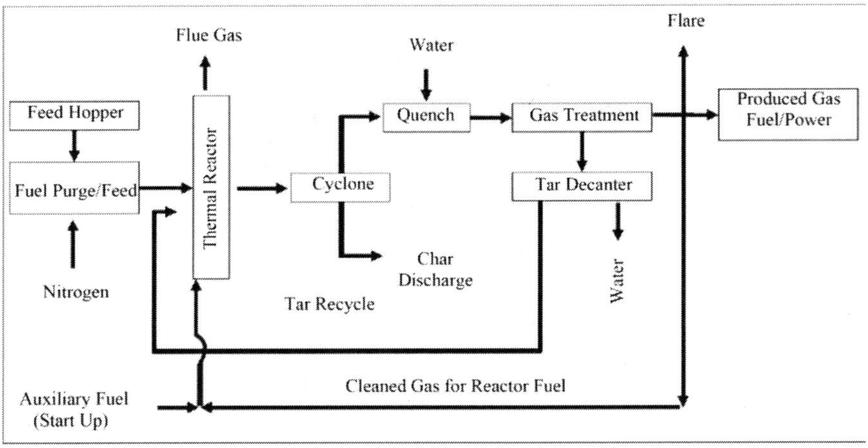

Figure 5. WGT process schematic [44].

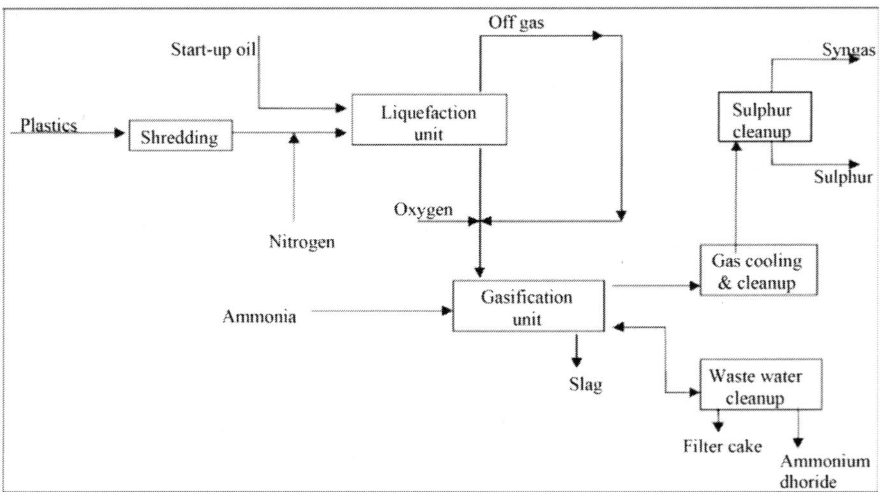

Figure 6. Texaco gasification process schematic diagram, showing both stages involved (liquefaction and gasification) [45].

Table 1. Input criteria and expected output for the Texaco gasification process [47].

Input criteria	
Feedstock	Dry to the touch, not sticky, shredded or chipped to <10 cm, with <1% under 250 μm
PSW content	>90 wt%
Free metals	<1 wt%
PVC content	<10 wt%
Ash content	<6 wt%
Residual moisture	<5 wt%
Paper content	<10 wt%
Expected products	
Synthesis gas	350.000 Nm^3/day (predominantly H_2/CO) out of 150 tonnes of PSW/day
Vitrified slag	-
Fines	High quality, equivalent to fly ash

- Particle size: >20 to 80 mm;
- Chlorine content: 2% as default, though higher concentrations are tolerable;
- Ash content: up to 10% or more.

As a producer of chlorine and vinylchloride, Akzo Nobel started a process for mixed PSW gasification. The process consists of two separate circulating fluid bed (CFB) reactors at atmospheric pressure (Figure 7). The first is a gasification reactor in which waste (usually rich with PVC) is converted at 700°C - 900°C into product gas (fuel and HCl gas) and tars. The second unit is a combustion reactor that burns the residual tar to provide heat for the gasification process. Circulating sand between the gasifier and combustor transfers heat between the two reactors. Both reactors are of the riser type with a very short residence time. This type of reactor allows a high PVC waste throughput. If the input contains a lot of PE and PP, the output will contain a significant amount of propylene and ethylene [47].

4. EXPERIMENTAL WORK

4.1. Feedstock Material

As PSW samples, PET, PE, PP and PS were selected. They were either chopped into 1 - 2 cm long chips with maximum thickness below 0.5 mm, or used as pellets of about 3 to 5 mm.

4.2. Fluidized Bed Gasification Reactor

The gasification experiments were performed in a bubbling fluidized reactor. The experimental set-up is illustrated in Figure 8.

The bubbling fluidized bed is manufactured from Incoloy steel, and externally insulated. The air flow rate was controlled by flow meter. The heat-input from the electrical heater (50 kW) was set by the bed temperature. Vapours, gases and fluidization gas were exhausted by an induced draft fan, passing a water-cooled condenser and a liquid collection tank. Life steam (110°C, 5 bar) could be added in the windbox of the fluidized bed, i.e. below the distributor plate, built as a sintered metal plate with pressure drop of about 3000 Pa at the operating flow rate and temperature.

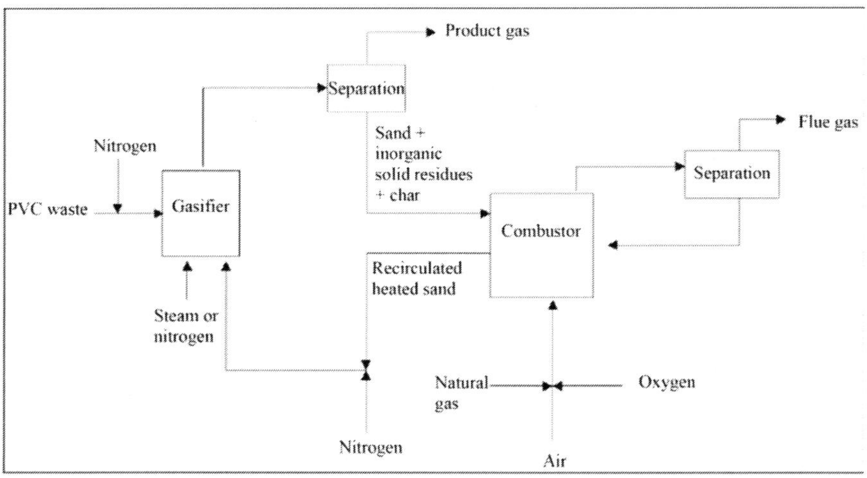

Figure 7. Akzo Nobel process schematic diagram [47].

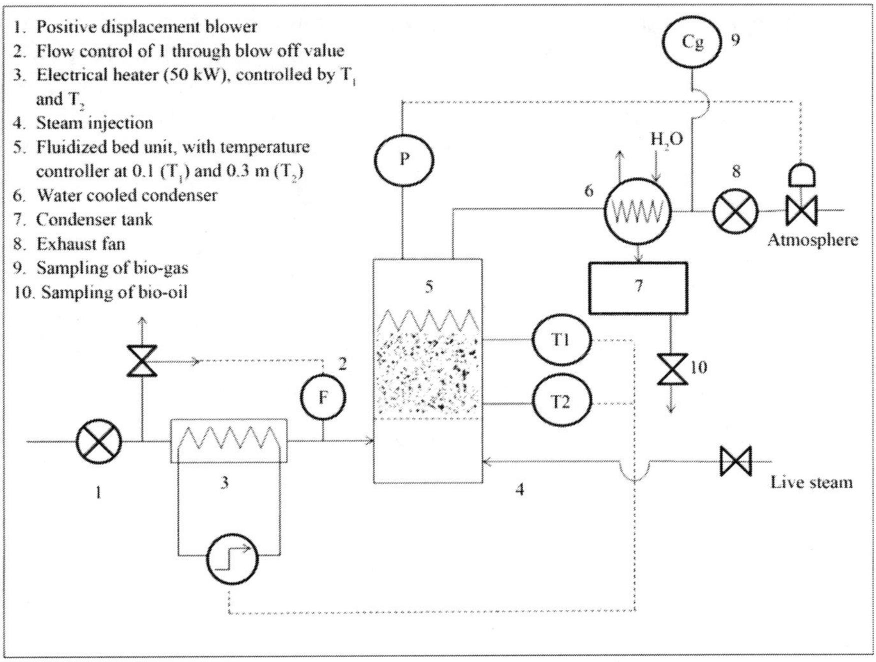

Figure 8. Schematic diagram of the gasification set-up.

The air flow rate was maintained at about 20% to 25% of the stoichiometric air flow needed for the combustion of each specific PSW material (Equivalence Factor, EF, 20% - 25%), and the velocity in the bed varied from 2 to 4 times the minimum fluidisation velocity, Umf. For some of the experiments, a steam dilution was also applied, reducing EF to below 15%.

A known weight (between 0.3 and 0.5 kg) of the PSW sample was placed into a perforated basket. The perforations enabled the fluidized bed sand to fluidize within the voidage of the PSW chips.

The fluidized bed consisted of about 15 kg of quartz sand (150 - 300 μm, 2600 kg/m^3), for a total static bed height of about 35 cm.

Gases were exhausted to the atmosphere, after appropriate analysis of CO, CO_2, H_2 and CH_4. Temperatures between 550°C and 800°C were tested.

In additional experiments, live-steam (110°C, 5 bar) was directly injected directly into the fluidization air flow, thus further reducing the partial pressure of O_2 being present. All experiments were twice repeated under identical operating conditions, and average results were used to determine the kinetics of the PSW gasification.

The progress of the gasification reaction was monitored through the on-line measurement of the CO production. A first order gasification reaction was assumed to be valid, allowing the calculation of the reaction rate constants from PSW conversion as measured by CO production. The order of the reaction was determined by Brems et al. [7], and indeed found to be close to 1 for all polymers.

4.3. Experimental Results

The optimum temperatures of gasification (max. gas yield) are defined in Table 2" target="_self"> Table 2, together with the activation energy, E_a, and the pre-exponential factor, A, of the Arrhenius kinetic expression.

The reaction rate constant, k, can be determined using the Arrhenius equation, reproduced below:

$$k = Ae^{-E_a/RT} \tag{1}$$

Using the experimental data, the reaction rate constant for the different PSW materials can be predicted at any temperature.

At the respective optimum T of PE, k equals 5.35 s^{-1}, representing a fast reaction rate.

The required residence time of PE in the reactor can be determined from the overall reaction kinetics, as given for a first order reaction:

$$\frac{-dW}{dt} = kW \tag{2}$$

Integration of Eq.2 and introducing X_p, the degree of conversion (the weight percentage of initial solid that is successfully converted into the product), yields:

$$-\ln\left(1 - X_p\right) = kt \tag{3}$$

From experimentation it was seen that the gasification of PSW is very efficient and a conversion factor, X_p, of 0.99 is easily achieved.

The required residence time, t, can then be evaluated as:

$$-\ln\left(1 - 0.99\right) = 5.35t \tag{4}$$

For the case of PE, this leads to a required residence time, t, of 0.86 s.

Gasification is hence a very fast reaction at the optimum temperatures, and ideally suited for a fluidised bed gasifier.

5. CONCLUSIONS

From the previous review and experimental work, the gasification of PSW can certainly be developed into a valid recycling route for PSW, producing a syngas, rich in H_2 and CO. Although references of industrial scale application are given in the literature, the future breakthrough of the process will require further experimental work to improve the equipment design and product optimisation. Advances in that area will aid in the improvement and more widespread use of gasification reactors.

Table 2. Experimental values of T, E_a and A.

PSW	T (K)	A (s^{-1})	E_a (kJ/mol)
PS	953	2.94×10^{14}	212
PET	968	2.94×10^{16}	238
PE	1008	5.12×10^{15}	289
PP	1018	1.99×10^{13}	187

6. ACKNOWLEDGEMENTS

The present research was supported by the Fundamental Research Funds for the Central Universities (Program RC1101).

REFERENCES

1. Baeyens, J., Brems, A. and Dewil, R., (2010) Recovery and recycling of post-consumer waste materials—Part 2. Target wastes (glass beverage bottles, plastics, scrap metal and steel cans, end-of-life tyres, batteries and household hazardous waste). International Journal of Sustainable Engineering, 3, 232-245.

2. Al-Salem, S.M., Lettieri, P. and Baeyens, J. (2009) Recycling and recovery routes of plastic solid waste (PSW): A review. Waste Management, 29, 2625-2643.

3. Al-Salem, S.M., Lettieri, P. and Baeyens, J., (2010) The valorization of plastic solid waste (PSW) by primary to quaternary routes: From re-use to energy and chemicals. Progress in Energy and Combustion Science, 36, 103-129.

4. Ahrenfeldt, J. (2007) Characterisation of biomass producer gas as fuel for stationary gas engines in combined heat and power production. Ph.D. Thesis, Technical University of Denmark, Lyngby.

5. Yoshioka, T., Gause, G., Eger, C., Kaminsky, W. and Okuwaki, A. (2004) Pyrolysis of polyethylene terephthalate in a fluidised bed plant. Polymer Degradation and Stability, 86, 499-504.

6. Smolders, K. and Baeyens, J. (2004) Thermal degradation of PMMA in fluidised beds. Waste Management, 24, 849-857.

7. Brems, A., Baeyens, J., Beerlandt, J. and Dewil, R. (2011) Thermogravimetric pyrolysis of waste polyethylene-terephthalate and polystyrene: A critical assessment of kinetics modeling. Resources, Conservation and Recycling, 55, 772-781.

8. Steiner, C., Kameda, O., Oshita, T. and Sato, T. (2002) EBARA's fluidized bed gasification: Atmospheric 2 × 225 t/d for shredding residues recycling and two-stage pressurized 30 t/d for ammonia synthesis from waste plastics. Proceedings of Second International Symposium on Feedstock Recycle of Plastics and Other Innovative Plastics Recycling Techniques, Ostend, 8-11 September 2002.

9. Aguado, J., Serrano, D.P., Miguel, G.S., Escola, J.M. and Rodriguez, J.M. (2007) Catalytic activity of zeolitic and mesostructured catalysts in the cracking of pure and waste polyolefins. Journal of Analytical and Applied Pyrolysis, 78, 153-161.

10. Mastellone, M.L. (1999) Thermal treatments of plastic wastes by means of fluidised bed reactors. Ph.D. Thesis, Second University of Naples, Naples.

11. Arena, U. and Mastellone, M.L., (2006) Fluidized bed pyrolysis of plastic wastes. In: Scheirs, J. and Kaminsky, W., Eds., Feedstock Recycling and Pyrolysis of Plastic Wastes: Converting Waste Plastics into Diesel and Other Fuels, John Wiley & Sons, Chichester.

12. Scheirs, J. (1998) Polymer recycling. Wiley, New York.

13. Vermeulen, I., Van Caneghem, J., Block, C., Baeyens, J. and Vandecasteele, C. (2011) Automotive shredder residue (ASR): Reviewing its productions from end-of-life vehicles (ELVs) and its recycling, energy and chemicals valorization. Journal of Hazardous Materials, 190, 8-27.

14. Wallmann, P.H., Thorsness, C.B. and Winter, J.D. (1998) Hydrogen production from wastes. Energy, 23, 271-278.

15. Pinto, F., Franco, C., Andre, R.N., Miranda, M., Gulyurtlu, I. and Cabrita, I. (2002) Co-gasification study of biomass mixed with plastic wastes, Fuel, 81, 291-297.

16. VTT (2004) Power production from waste and biomass IV. Proceedings of the VTT Symposium, Finland, 8-10 April 2002.

17. Buekens, A.G. (1978) Resource recovery and waste treatment in Japan. Resource Recovery and Conservation, 3, 275-306.

18. Hasegawa, M., Fukuda, X. and Kunii, D. (1974) Gasification of solid waste in a fluidized bed with circulating sand. Conservation and Recycling, 3, 143-153.

19. Borgianni, C., Filippis, P.D., Pochetti, F. and Paolucci, M. (2002) Gasification process of wastes containing PVC. Fuel, 14, 1872-1833.

20. Xiao, G., Jin, B., Zhou, H., Zhong, Z. and Zhang, M. (2007) Air gasification of polypropylene plastic waste in fluidized bed gasifier. Energy Conversion and Management, 48, 778-786.

21. Matsunami, J., Yoshida, S., Yokota, O., Neuzka, M., Tsuji, M. and Tamaura, Y. (1999) Gasification of waste tyre and plastic (PET) by solar thermochemical process for solar energy utilization. Solar Energy, 65, 21-23.

22. Pinto, F., Franco, C., Andre, R.N., Tavares, C., Dias, M. and Gulyurtlu, I. (2003) Effect of experimental conditions on co-gasification of

coal, biomass and plastic wastes with air/steam mixture in a fluidized bed system. Fuel, 82, 1967-1976.

23. Slapak, M.J.P., Kasteren, J.M.N.V. and Drinkenburg, A.A.H. (2000) Design of a process fors team gasification of PVC waste. Resources, Conservation and Recycling, 30, 81-93.

24. Xiao, G., Ni, M., Chi, Y., Jin, B., Xiao, R., Zhong, Z. and Huang, Y. (2009) Gasification characteristics of MSW and ANN prediction model. Waste Management, 29, 240- 244.

25. Aznar, M.P., Caballero, M.A., Sancho, J.A. and Francs, E. (2006) Plastic waste elimination by co-gasification with coal and biomass in fluidized bed with air in pilot plant. Fuel Processing Technology, 87, 409-420.

26. Cozzani, V., Nicolella, C., Rovatti, M. and Tognotti, L., (1997) Influence of gas phase reactions on the product yields obtained in the pyrolysis of polyethylene. Industrial and Engineering Chemistry Research, 36, 342-348.

27. Stiles, H.N. and Kandiyoti, R. (1989) Secondary reactions of flash pyrolysis tars measured in a fluidized bed pyrolysis reactor with some novel design features. Fuel, 86, 275-282.

28. Zolezzi, M., Nicolella, C., Ferrara, S., Iacobucci, C. and Rovatti, M. (2004) Conventional and fast pyrolysis of automotive shredder residues (ASR). Waste Management, 24, 691-699.

29. Miscolczi, N., Bartha, L., Deák, G. and Jóver, B. (2004) Thermal degradation of municipal solid waste for production of fuel-like hydrocarbons. Polymer Degradation and Stability, 86, 357-366.

30. Ciliz, N.K., Ekinci, E. and Snape, C.E. (2004) Pyrolysis of virgin and waste polyethylene and its mixtures with waste polyethylene and polystyrene. Waste Management, 2, 173-181.

31. Ponzio, A., Kalisz, S. and Blasiak, W. (2006) Effect of operating conditions on tar and gas composition in high temperature air/steam gasification (HTAG) of plastic containing waste. Fuel Processing Technology, 3, 223-233.

32. Franco, C., Pinto, F., Gulyurtlu, I. and Cabrita, I. (2003) The study of reactions influencing the biomass gasification process. Fuel, 82, 835-842.

33. Marquez-Montesinos, F., Cordero, T., Rodriguez-Mirasol, J. and Rodriguez, J.J. (2002) CO_2 and steam gasification of grapefruit skin char. Fuel, 81, 423-429.

34. Zanzi, R., Sjöström, K. and Björnbom, E. (1996) Rapid high-temperature pyrolysis of biomass in a free fall reactor. Fuel, 75, 545-550.

35. Zanzi, R., Sjöström, K. and Björnbom, E. (2002) Rapid pyrolysis of agricultural residues at high temperature. Biomass and Bioenergy, 23, 357-366.

36. Narvaez, A., Orio A., Aznar, M.P. and Corella, J. (1996) Biomass gasification with air in an atmospheric bubbling fluidized bed: effect of six operational parameters. Industrial and Engineering Chemistry Research, 35, 2110- 2120.

37. Harder, M.K. and Forton, O.T. (2007) A critical review of developments in the pyrolysis of automotive shredder residue. Journal of Analytical and Applied Pyrolysis, 79, 387-394.

38. Mancini, G., Tamma, R. and Viotti, P. (2010) Thermal process of fluff: Preliminary test on a full scale treatment plant. Waste Management, 30, 1670-1682.

39. Vigano, F., Consonni, S., Grosso, M. and Rigamonti, L., (2010) Material and energy recovery from automotive shredder residue (ASR) via sequential gasification and combustion. Waste Management, 30, 145-153.

40. Cho, S.J., Jung, H.Y., Seo, Y.C. and Kim, W.H. (2010) Studies on gasification and melting characteristics of automotive shredder residue. Environmental Engineering Science, 27, 577-586.

41. Lin, K.S., Chowdhury, S. and Wang, Z.P. (2010) Catalytic gasification of automotive shredder residues with hydrogen generation. Journal of Power Sources, 195, 6016- 6023.

42. Donaj, P., Yang, W., Blasiak, W. and Forsgren, C. (2010) Recycling of automotive shredder residue with a microwave pyrolysis combined with high temperature steam gasification, Journal of Hazardous Materials, 182, 80-89.

43. Donaj, P., Blasiak, W., Yang, W. and Forsgren, C. (2011) Conversion of microwave pyrolysed ASR's char using high temperature agents. Journal of Hazardous Materials, 185, 472-481.

44. WGT (2002) Waste gas technology energy from waste. http://www.wgtuk.com/ukindex.html

45. Weissman, R. (1997) Recycling of mixed plastic waste by the Texaco gasification process. In: Hoyle, W. and Karsa, D.R., Eds., Chemical Aspects of Plastics Recycling, The Royal Society of Chemistry Information Services, Cambridge.

46. Croezen, H. and Sas, H. (1997) Evaluation of the Texaco gasification process for treatment of mixed household waste. Final Report of Phase 1 and 2, CE, Delft, The Netherlands.

47. Tukker, A., de Groot, H., Simons, L. and Wiegersma, S. (1999) Chemical recycling of plastic waste: PVC and other resins. European Commission, DG III, Final Report, STB-99-55 Final, Delft, The Netherlands.

3

CHAPTER

STEAM GASIFICATION OF DIFFERENT BROWN COALS CATALYSED BY THE NATURALLY OCCURRING CALCIUM SPECIES

Peter N. Kuznetsov, Svetlana M. Kolesnikova, Ludmila I. Kuznetsova

Institute of Chemistry and Chemical Technology, Siberian Branch of Russian Academy of Sciences, Krasnoyarsk, Russia

ABSTRACT

The effects of the constituents of mineral matter in brown coals from different deposits of Kansk-Achinsk, Lena and from Yallourn Basins on the structural parameters and steam gasification reactivities of respective coal chars at moderate temperature and at low and high pressure were studied in this paper. The data on how the preliminary decationization with diluted hydrochloric, acetic and sulphuric acids affect char gasification reactivities are presented. The importance of surface area and crystallinity of chars and the presence of naturally occurring metals on gasification reactivity is considered. Quantitative correlations between the calcium contents and the extents of gasification are revealed. The gasification results obtained in a flow reactor with steam stream and in an autoclave reactor at high pressure of gaseous products are compared. The catalytic effect of dispersed calcium oxide-carbonate particles produced from the naturally occurring calcium containing carboxylates was shown to be a key factor for char gasification reactivity, the effect in the flow reactor being much larger as compared to that in the autoclave reactor. This was mainly related to different forms of catalytically active calcium species and to the composition of the gaseous reaction mixture.

Keywords: Coal; Mineral Matter; Steam Gasification; Calcium; Catalysis

1. INTRODUCTION

Gasification is one of the most important multipurpose and efficient technologies to convert coal to energy carriers and to a large variety of valuable products, such as synthetic natural gas, synthesis gas, hydrogen, ammonia, methanol, hydrocarbons and also various carbon materials, catalyst support and sorbents [1]. Coal gasification will continue to play an important role in the foreseeable future. However, this process results in the emissions of greenhouse gases and a variety of air pollutants including SO_x, NO_x, and large amount of ash. Ultrafine particulates and trace metals emitted from coal are potentially carcinogenic. Therefore efforts must be intensified in the short term to better understand the key factors affecting coal gasification reactions and to develop advanced technologies for clean coal utilization with higher efficiency and near-zero emissions.

It is known that different ranked coals can be gasified successfully by using an appropriate technology. However, the reactivity of coals has been shown [2-6] to tend to decrease with an increase in the rank. The application of catalysts allows coal reactivity to be increased significantly [2-11]. The following sequence of activity of alkali and alkaline-earth metal catalysts for steam gasification of subbituminous coal char was found: $K_2CO_3 > Na_2CO_3 > KCl > NaCl > CaCl_2 \sim CaO$ [7]. Also, other metal species, such as Mg, Ba, Fe, Ni, Mn can catalyse the gasification reaction [8-10].

Many researchers [3,4,6,8,11-17] studied the effects of the mineral matter naturally occurring in the coals on the gasification reaction. Miura and Hashimoto et al. [3] reviewed in details a large body of the literature data for the reactivities of coal chars for gasification with H_2O, CO_2 and O_2 to understand the controlling factors of reaction. The authors concluded that the gasification rates of chars from low rank coals can be affected by mineral components such as these containing Ca, K, Na, Fe and Mg acting as the gasification catalysts. Partial demineralization of coals resulted in decrease in the gasification rates by factors of 5 to 25 as compared to that of respective parent coals [11,13,14]. However, an anomalous increase in H_2O,

CO_2 and air gasification reactivity was obtained when the high volatile bituminous coal has been demineralized with acids [18].

It should be noted that the correlations between the content of specific metals and char gasification reactivity were generally revealed by using the coals (or chars) doped with metals. The authors [4] attempted to correlate gasification conversions of a large number of coals with the specific naturally occurring inorganic constituents. It was reported that though inorganic matter affects conversions during gasification, the systematic patterns regarding the effect of specific metals in coals were difficult to correlate quantitatively. The effects of impregnated salts on CO_2 gasification reactivity changed with reaction pressure and coal type in a specific manner. Hence catalysis by mineral matters has not always been found to provide adequate explanation of the gasification kinetics. One should not assume the metals to only catalyse reactions if for no other reason than that they can affect the physical structure and consequently the reactivity of char [19,20].

Because of the complexity of the reactions involved, detailed characterization of coal and char composition, crystallinity and morphology of organic matter, and quantifying their effects on reaction rates is crucial for accurate modeling of gasification systems and for determining key factors. There are currently insufficient data available on the relationships between the coals and the development of char structure.

The comparative experimental investigation of the composition, structural characteristics and gasification behaviour of a suite of the chars produced from different brown coals to find the quantitative correlations was performed in this paper. The effects of the composition of mineral matter on the specific surface area and crystallinity of chars and on their reactivities for steam gasification under atmosphere and high pressure in two experimental reactors were studied. The brown coals from the large Kansk-Achinsk and Lena basins were used. KanskAchinsk brown coals are of particular interest for gasification because of rather low contents of mineral matter in which calcium is abundant [21,22]. It has been reported [23-25] that these are Ca cations which contribute a major effect to cross-linking of the organic matter of brown coals resulting in suppressed reactivity for hydrogenation. On the other hand, calcium species affected significantly [26] the properties of steam activated carbons. A sample of brown coal from Yallourn deposit in Australia was also used for comparison, since it contains little inorganic matter which is mainly occurring in the form of metal ions [27].

2. EXPERIMENTAL

2.1. Coals

The brown coal samples from the Borodino and Berezovo deposits of Kansk-Achinsk Basin, Kangalas deposit of Lena Basin and a sample of brown coal from Yallourn deposit in Australia were studied. The coals were crushed, and a required fraction of 0.5 - 1.0 mm dried in a vacuum drying oven at 85°C was used.

2.2. Acid Treatment of Coals

Selected coal samples were treated with deluted acid aqueous solutions to remove the specific metal cations and to identify their effects on the properties of respective chars. Hydrochloric, sulfuric, and acetic acid sulutions of 0.1, 0.2 and 1.0 N concentrations were used. An approximately 200 g dried coal sample was treated in 2 l of aqueous acid solutions in a glass flask (filled with an inert gas) with continuous stirring using a magnetic stirrer at room temperature. In 3 h, the solution was decanted, and the treatment with the fresh portion of acid solution was repeated. Then the coal was filtered off, thoroughly washed with distilled water and dried in a vacuum oven at 85°C.

2.3. Pyrolysis and Gasification Procedures

The coals were pyrolyzed in a quartz reactor of 20 mm in diameter. Weighed dried coal (15 g) was placed into the reactor which was then closed (not hermetically for the gases to be allowed to escape) and heated at a controlled rate of 7°C - 8°C/min. The pyrolysis was carried out at the temperature of 700°C for 1 h.

The steam gasification of char produced was carried out at relatively low temperature 700°C by using two experimental units equipped with a quartz flow tubular reactor and with an autoclave reactor. The tubular reactor of 20 mm in diameter was charged with 5 g of char and heated at the rate of 5°C/min under nitrogen flow (2.5 l/h). On attaining the reaction temperature of 700°C, the water feeding (20 g/h) was added to nitrogen flow and gasification was carried out for 45 min at atmospheric pressure.

The autoclave unit was equipped with five 90 cm^3 stainless steel autoclaves inserted into a rotated heated sand-bath. Each autoclave was charged with 3 g of char and 4 g of distilled water. The autoclaves were purged with nitrogen from the air and hermetically sealed. The heating rate was 7°C - 8°C/min and the speed of rotation 90 rev/min. The reaction was carried out at 700°C for 2 h under autogenous pressure of water vapour and gaseous products generated. Then the autoclaves were allowed to cool and the gaseous products were analysed by a gas chromatograph. This autoclave unit allowed thus five char samples to be tested simultaneously under strictly the same conditions. Carbon burn-off was evaluated based on the quantity of organic matter in the char before and after gasification.

2.4. Coal and Char Characterization

The composition of coals was characterized by proximate and ultimate analysis. The separate coal samples (both the parent and acid treated) were subjected to oxygen-containing functional group analysis. The content of hydroxyl groups was determined by acetylation in pyridine and the carboxylic oxygen by reaction with calcium acetate and titration of the liberated acetic acid. The contents of metals were determined by atomic absorption spectrometry and X-ray fluorescence analysis of coal ashes and the data obtained were recalculated based on dry coal.

The phase composition of mineral substances and the structural characterization of the organic matter were monitored by X-ray diffraction. The parameters of the stacking structure of char carbon matter were estimated based on the (002) and (10) bands, the profiles were corrected with adsorption, polarization, and atomic scattering factors according to recommendations described in the references [28-32]. The distance between the aromatic layers d_{002} in the stacking structure was calculated from the (002) peak position by means of Bragg's equation. The average crystallite thickness L_c was evaluated from the width at the half maximum of the (002) peak using Scherrer's equation, and the number of aromatic layers n in the stacks by means of the Equation $(L_c/d_{002}) + 1$. The average diameter of layers L_a was evaluated from the width at the half maximum of the (10) peak. The XRD patterns of powdered char samples packed into aluminium holder were recorded under strictly the same conditions by using a DRON-3 diffractometer with CuKa radiation and a step-scanning method ($D2\theta = 0.2°$, 25 s/step) between 2θ from 8 and 55°. The calculated relative crystallite dimensions were sufficiently reliable to monitor the difference in gross-molecular structure of carbon in the chars produced from different coals.

The BET surface areas of chars were measured by low temperature nitrogen adsorption using ASAP 2020 and GCH-1 analizators.

3. RESULTS AND DISCUSSION

3.1. Coal Characterization

Shown in Table 1 are the list of the brown coal samples from different deposits and the characterization of their mineral matters. The content of ash ranged from 1.4 to 14.8 wt%. The parent coals showed large variations in the concentration of Ca (from 0.15 to 1.80 wt%), which prevailed in most of the coals alongside the silicon and aluminium. The concentrations of Mg were less than that of Ca by a factor of 2 - 8, with the exception of Yallourn coal which featured almost equal weight amount of Mg and Ca. The contents of Fe varied from 0.06 to 0.55 wt%. The alkali metals (Na and K) occurred in very low concentrations: 0.15 wt% in Kangalas coal, 0.06 wt% in Yallourn coal and near 0.01 in Kansk-Achinsk coals [24].

The amounts of most minor elements in Borodino and Berezovo coals were much less than that for coal Clarke numbers (i.e. the average content in the world's coals), except for the alkaline earth Sr and Ba, and also for Mn [22,24], showing a typical feature of Kansk-Achinsk brown coal.

The treatment with the acid solutions decreased the ash content by a factor of 1.2 - 6.0 depending on the coal samples (Table 1). The alkali cations and alkaline earth metals Mg and Ca were almost completely removed by HCl, H_2SO_4 and CH_3COOH solutions. Also the cations of minor elements such as Mn, V, Be, Y and Yb readily passed into HCl solution (95% and more) [24]. The recovery of Fe varied from 50 to 93% when coals were treated with HCl and H_2SO_4 solutions, however no Fe was removed with 1 N CH_3COOH solution.

The data in Table 2 show rather close elemental composition of organic matter of representative coal samples, with the exceptions of Yallourn coal, which features a lowered carbon content and increased oxygen content, and of Kangalas coal with enhanced hydrogen content. The nitrogen and sulphur contents were near to 0.8 and 0.3% in all the samples, respectively. HCl treatment hardly affects significantly the ultimate analysis data of coals [24].

Shown in Table 3 are the contents of hydroxyl and carboxyl groups in the representative parent and in the respective acid treated coals. It is seen that the coals differed moderately with the content of hydroxyl groups (in the range of 4.1 - 5.32 mg-equiv per 1 g of daf coal). On the other hand, the parent coals differed remarkably from each other with the content of free acids (from 0.78 to 2.06 mg-equiv./g). Acid treatment affected moderately and nonregularly the content of hydroxyls, however, greatly increased the content of free acids (by 1.2 - 2.3 times depending on coal samples). Both parent and acid treated Yallourn coal samples show an increased hydroxyl and carboxyl groups due to the increased total oxygen content.

It is known that alkali and alkaline earth metallic species (mainly Na, Mg and Ca) can exist in the brown coals *Difference between COOH contents in the acid treated and in the parent coals.

mainly in two forms: as ion-exchangeable cations associated mainly with the carboxyl groups of the organic coal matter or in the forms of soluble and insoluble minerals. This aspect of metal forms is both attractive and interesting to understand the role of metal species in the chemical reaction of organic matter.

Table 1. The characterization of mineral matters of brown coals from different deposits.

Coal deposit sample	Treatment	Ad wt%	Content, wt% on daf coal		
			Ca	Mg	Fe
Parent coals					
Borodino, Bor_1	No	4.1	0.80	0.10	0.10
Borodino, Bor_2	No	6.5	1.25	0.30	0.06
Borodino, Bor_3	No	7.7	1.22	0.20	0.08
Borodino, Bor_4	No	5.4	1.08	-	-
Borodino, Bor_5	No	4.4	1.30	0.10	0.10
Borodino, Bor_6	No	5.3	1.13	0.30	0.15
Borodino, Bor_7	No	4.9	0.73	0.19	0.18
Borodino, Bor_{21}	No	4.9	1.20	0.19	0.18
Borodino, Bor_{22}	No	4.2	1.80	0.45	0.46
Borodino, Bor_{23}	No	14.8	1.43	0.80	0.23
Berezovo, Ber_9	No	4.3	1.40	0.20	0.50
Kangalas, Ka_{30}	No	7.9	1.22	0.20	0.44
Yallourn, Ya_{30}	No	1.4	0.15	0.16	0.40
Acid-treated coals					
Borodino, Bor_2D	0.2 N HCl	3.3	0.003	≤10 - 5	0.03
Borodino, Bor_3D	0.2 N HCl	4.8	≤0.0004	≤10 - 5	0.02
Borodino, Bor_5D	0.2 N HCl	1.1	≤0.0001	≤10 - 5	0.01
Borodino, Bor_6D	0.2 N HCl	1.3	≤0.001	≤10 - 5	0.01
Borodino, $Bor_{22}D$	0.1 N HCl	0.9	0.009	0.0009	0.138
Borodino, $Bor_{22}D$	1.0 N HCl	0.7	0.004	0.0005	0.008
Borodino, $Bor_{22}D$	1.0 N H_2SO_4	1.0	0.07	0.001	0.015
Borodino, $Bor_{22}D$	1.0 N CH_3COOH	1.7	0.04	0.002	0.46
Borodino, $Bor_{23}D$	1.0 N HCl	12.6	0.015	не обн.	0.035
Berezovo, Ber_9D	0.2 N HCl	0.8	≤0.0001	≤10 - 5	0.004

Table 2. Ultimate analysis data for representative brown coal samples.

Coal sample	Wt% on daf coal				
	C	H	N	S	O
Borodino, Bor_1	69.7	4.9	0.8	0.3	24.3
Borodino, Bor_2	70.8	4.8	0.8	0.3	23.3
Borodino, Bor_3	71.3	4.8	0.9	0.2	22.8
Borodino, Bor_6	70.0	4.8	0.8	0.2	24.2
Borodino, Bor_{21}	71.9	4.6	0.7	0.3	22.1
Borodino, Bor_{22}	71.4	5.4	0.8	0.2	22.1
Borodino, Bor_{23}	69.3	5.2	0.7	0.3	23.9
Berezovo, Ber_9	70.1	4.9	0.9	0.3	23.8
Kangalas, Kan_{20}	71.0	5.5	0.8	0.4	21.8
Yallourn, Ya_{10}	66.6	4.7	0.6	0.3	27.8
Borodino, Bor_6D (0.2 HCl)	71.3	4.9	0.8	0.3	22.7

Table 3. The data on oxygen functional analysis of the parent and HCl treated brown coal samples (mg-equiv./g coal).

Coal sample	Parent coals			Acid treated coals		
	OH	COOH	Ca	OH	COOH	ΔCOOH*
Borodino, Bor_2	4.06	0.87	0.62	4.40	1.72	0.85
Borodino, Bor_3	4.10	1.12	0.61	4.34	1.91	0.79
Borodino, Bor_5	4,80	1.03	0.65	4.50	1.72	0.69
Borodino, Bor_6	5.09	1.22	0.56	4.71	1.76	0.54
Borodino, Bor_7	4.50	1.37	0.36	4.82	1.78	0.41
Borodino, Bor_{22}	4.70	0.91	0.90	4.38	1.87	0.96
Borodino, Bor_{23}	4.95	1.03	0.71	4.63	1.84	0.81
Berezovo, Ber_9	5.00	0.78	0.70	4.90	1.81	1.03
Yallourn, Ya_{10}	5.18	2.06	0.08	5.32	2.53	0.47

The treatment of brown coals with deluted acids (HCl, H_2SO_4 and acetic acid) resulted in the exhaustive extraction of the alkali and alkaline earth metallic species, calcium being predominant in all the coals, except for Yallourn coal. As regards Kansk-Achinsk coals, the previous systematic investigations [24,33,34] have documented calcium to occur predominantly in the organic matter in the ion-exchangeable form of the carboxylate complexes. In Table 3, the contents of calcium cations (in mg-equiv. quantity) in the parent coals are shown also in order to compare quantitatively the reaction stoichiometry. One can see that the increase in the number of COOH groups as a result of acid treatment is nearly equal or slightly more than the quantity of calcium extracted. Some excess in increase in the number of COOH may be due to the fact that it has been not considered some contribution of other cations (Mg^{2+}, in the first place). The data confirm thus again that in the coals studied, calcium occurs predominantly in the form of carboxylate complexes. Only little or no calcium minerals (calcite, in the first place) may occur in the parent coals though they were not visible by XRD. It should be noted, however, that calcite, sulfate and other calcium based minerals can be found in the parent brown coals when containing higher amount of ash, for example, in the heavy part of brown coal after its benefication.

3.2. Coal Carbonization

The char yields from the pyrolysis of the parent and acid treated Kansk-Achinsk coals varied from 53 to 62 % and those from Kangalas and Yallourn coals did 50% (Table 4). BET surface areas of chars produced from the parent and acid treated coals ranged from 35 to 262 m^2/g, and Yallourn char showed much larger surface area of 425 m^2/g.

The X-ray diffraction patterns of the coals and respective chars were obtained to identify the minerals and to characterize the structural composition of organic matter. Predominantly quartz with different quantities was detected in the XRD patterns. No distinct reflections related to other mineral species (such as calcite) were reliably observed and identified because of probably low content, low crystallinity, or high dispersion.

The XRD patterns showed typical broad asymmetrical reflection in the 2θ region from 8° to 34° due to a specific inter-molecular ordering of carbon matter and to a weak reflection with a maximum centered at 2θ of about 44° due to intra-molecular ordering (10 reflection). It was shown by using the computer treatment that broad asymmetrical

Table 4. The characterization of chars produced on pyrolysis of different brown coals at 700°C.

Char	A^d, wt%	Ca, wt%	Char yield, wt% on daf	BET surface area, m^2/g on daf char
Bor_1	6.9	1.40	57	140
Bor_2	11.2	2.27	55	224
Bor_3	12.9	2.12	57	74
Bor_4	8.7	1.80	60	118
Bor_5	7.4	2.45	53	140
Bor_6	8.6	1.91	59	77
Bor_7	6.8	1.30	56	262
Bor_{21}	7.9	2.03	59	262
Bor_{22}	6.9	2.8	64	162
Bor_{23}	23.3	2.3	62	239
Ber_9	7.5	2.55	55	180
Ka_{20}	14.7	2.44	50	49
Ya_{10}	2.8	0.30	50	425
Bor_2D (0.2 HCl)	5.3	0.005	61	120
Bor_3D (0.2 HCl)	7.7	0.002	59	-
Bor_5D (0.2 HCl)	1.4	0.002	56	114
Bor_6D (0.2 HCl)	2.1	0.002	60	156
$Bor_{22}D$ (0.1 N HCl)	1.9	0.02	53	42
$Bor_{22}D$ (1.0 N HCl)	1.4	0.008	50	35
$Bor_{22}D$ (1.0 N H_2SO_4)	2.0	0.137	51	46
$Bor_{22}D$ (1.0 N CH_3COOH)	2.7	0.06	62	71
$Bor_{23}D$ (1.0 N HCl)	19.8	0.03	58	101
Ber_9D (0.2 HCl)	1.4	0.002	55	220

reflection attributed to inter-molecular ordering can be well simulated by a superposition of two Gaussians (Figure 1). According to the references [28-31], they were referred to ordered carbon portion in which polyaromatic molecules are stacked in parallel (2q reflection centered at 23.0° - 24.5°) and to disordered other molecules of g-component located at the periphery of the polyaromatic clusters (2θ centered at 19° - 21°) and not included in the stacking structure [28,29]. Large broadening of the 002 reflection and 10 reflection is interpreted in terms of small dimensions of crystallites. The small crystallites of stacked polyaromatics and the disordered other molecules represent the crystallite and noncrystallite char components, respectively, and their combination forms char microtexture. The proportion between the content of these two structures was evaluated based on the areas of the respective Gaussian peaks.

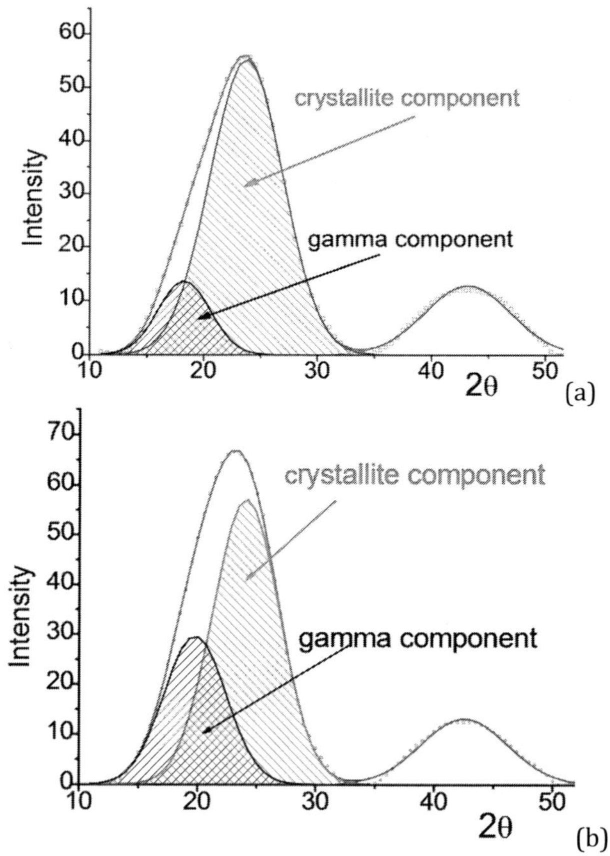

Figure 1. The fragments of the typical X-ray diffraction patterns for the chars from the parent (a) and from the 0.2 N HCl treated brown coals (b).

Table 5. The proportion between the X-ray structural carbon portions and structural parameters for the crystallites of chars produced from different brown coals.

Coal char	Proportion, %		d_{002}	Number	L_c	L_a
	Crystallite component	γ-component	nm	of layers	nm	nm
From parent coals						
Bor_1	79	21	0.391	3.9	1.15	1.70
Bor_2	88	12	0.392	3.6	1.04	1.71
Bor_3	83	17	0.379	4.0	1.14	1.57
Bor_4	74	26	0.383	3.7	1.04	1.57
Bor_5	84	16	0.379	3.9	1.09	1.69
Bor_6	80	20	0.375	4.0	1.12	1.61
Bor_7	89	11	0.375	3.9	1.10	1.64
Bor_{22}	85	15	0.390	3.8	1.08	1.95
Ber_9	93	7	0.373	3.7	1.01	1.82
Kan_{20}	78	22	0.367	4.0	1.11	1.65
Ya_{10}	80	20	0.388	4.1	1.22	1.84
From HCl treated coals						
Bor_2D	60	40	0.379	4.3	1.25	1.68
Bor_3D	61	39	0.393	4.5	1.37	1.61
Bor_5D	72	28	0.373	4.4	1.28	1.75
Bor_6D	64	36	0.372	4.2	1.21	1.63
$Bor_{22}D$ (H_2SO_4 treated)	72	28	0.370	4.6	1.33	2.24
Ber_9D	63	37	0.361	4.8	1.36	2.08

The structural parameters evaluated from the X-ray diffraction patterns are summarized in Table 5. One can see that the chars from the parent coals consist of mainly crystallites (74% - 93%) with a small portion (7% - 26%) of disordered g-component. On the other hand, acid treated chars differ with less content of crystallites (60% - 72%) and with the increased portion of γ-component (28% - 40%).

The polyaromatic layers in the stacks in all the chars have the diameters mainly from 1.57 to 1.95 nm (with the occasional exceptions) with the average statistical value of 1.69 ± 0.10 nm and inter-layer spacing of 0.379 ± 0.009 nm (Table 5). The thicknesses of the stacks L_c range from 1.01 nm to 1.37 nm depending on the char sample, the chars from the parent coals having the stacks consisted of 3.6 - 4.1 polyaromatic layers. The acid treated chars consist of stacks with L_c of 1.01 - 1.22 nm and 4.2 - 4.8 number of layers.

No regular correlations were found between the specific surface areas, X-ray structural parameters and the composition of coals, such as proximate and ultimate analytical data and the contents of specific metals.

3.3 Char gasification

3.3.1. Gasification in Flow Reactor

Shown in Table 6 are the data on char gasification in a flow tubular reactor at 700°C for 45 min reaction period. One can see that the extents of burn-off of the chars produced from the parent coals changed from 27.8 to 67.3% wt% on a dry ash-free basis, and the chars from the acid treated coals showed much less burn-offs (2.9% - 11.3%). For example, the burn-off of char produced from Bor_{22} coal treated with 1.0 N HCl solution was as low as 3.2% that contrasted with 67.3% for char from the parent Bor_{22} coal. As the gasification time of this HCl treated char was increased to 150 min (i.e., by a factor of >3), the burn-off extent increased to only 9.0%. The CH_3COOH treated sample showed slightly higher reactivity (11.3%) as compared to the samples treated with HCl and H_2SO_4 acids (3.2% - 6.2%), however much less then that of the parent one. Last column in Table 5 shows how the chars differed from each other with the extents of burn-off normalized to the unit surface areas. One can see that the surface normalized burn-off for chars from the parent coals vary from 0.23 to 0.42 wt%/m^2, and the acid treated chars show much less values (from 0.09 to 0.16 wt%/ m^2).

A rational explanation for the differences in the surface normalized reactivities of chars is the reaction to occur not on total carbon surface area but only on the surface in the immediate vicinity of mineral species which can act as a catalyst. To clarify which species were more important in determining the reactivity, the correlations between the burn-off and mineral matter com position were analysed. The comparison of the data on gasification in Table 6 and on the mineral matter contents in Table 1 showed no correlations. A certain correlation was observed (with the determination coefficient of $R^2 = 0.84$) when the values for surface normalized carbon burn-off were plotted versus the content of calcium only (Table 4). However, good correlation (with the determination coefficient of $R^2 = 0.95$) proved to take place between the carbon burn-off based on wt% daf char and the content of calcium (Figure 2):

Burn-off extent, wt% = $6.2 - 6.7\ Ca + 10.3\ Ca^2$; $R^2 = 0.95$.

Table 6. Steam gasification of chars from different coals for 45 min at 700°C in flow reactor. Water feeding of 20 g/h, nitrogen flow of 2.5 l/h.

Char	A^d of char residue, wt%	Burn-off, wt% on daf	Burn-off per unit surface area, wt%/m^2 daf
Bor_3	18.1	27.8	0.38
Bor_{21}	12.6	37.6	0.38
Bor_{22}	18.7	67.3	0.42
Bor_{23}	39.7	54.6	0.23
$Bor_{22}D$ (0.1 N HCl)	1.85	5.3	0.12
$Bor_{22}D$ (1.0 N HCl)	1.55	3.2	0.09
$Bor_{22}D$ 1.0 N H_2SO_4)	2.1	6.2	0.14
$Bor_{22}D$ (1.0 N CH_3COOH)	3.1	11.3	0.16
$Bor_{23}D$ (1.0 N HCl)	20.8	7.6	0.09

This correlation suggests the reactivity of brown coal chars for steam gasification to be affected in some way mainly by calcium species in agreement with most of the previous works [8,9,11,16,17]. Other metals, such as Mg and Fe, could affect the reaction. However their effects were not evident because of probably less contents and lower specific activities as compared to those of calcium. For example, CH_3COOH solution extracted calcium almost completely from Bor_{22} coal, but no extraction of iron was observed (Table 1). The char produced was much less reactive as compared to the char from the parent coal (11.3 and 67.2%, respectively, Table 6).

Displayed in Table 7 are the gasification data obtained in the autoclave reactor at 700°C for 2 h. The extents of carbon burn-off ranged from 18.4 to 31.8 wt% and the total yields of gaseous products from 23.3 to 61.0 wt% based on dry ash-free chars. Again, acid treated chars showed less reactivities as compared to those produced from the parent coals. The accumulation of gaseous products gradually increased the pressure in the autoclave (up to 11 - 13 MPa at the reaction conditions) depending on carbon burn-off. Good correlations were observed between the values of carbon burn-off and gas-

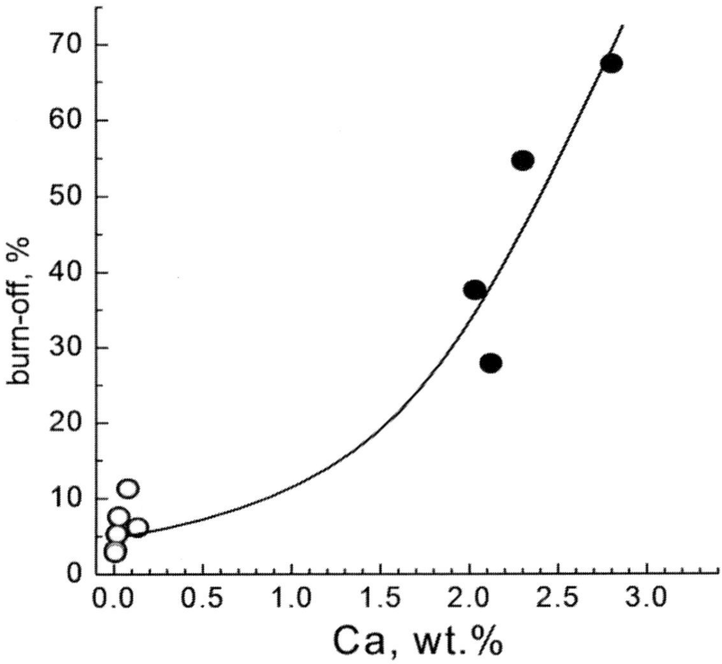

Figure 2. Carbon burn-off dependant on calcium content in the chars. Flow reactor. Dark symbols—chars from parent coals; open symbols—from HCl treated coals.

Table 7. Steam gasification of chars from different coals at 700°C in the autoclave reactor.

Char	A^d, wt%	Burn-off, wt% on daf	Gas yield, wt% on daf	Gas yield, wt%/m^2 daf
Bor_1	6.8	27.0	48.3	0.34
Bor_2	11.2	27.4	51.6	0.23
Bor_3	12.9	30.2	54.9	0.74
Bor_4	8.7	28.1	52.0	0.44
Bor_5	7.4	28.0	54.2	0.39
Bor_6	8.6	29.7	55.5	0.72
Bor_7	6.8	26.5	54.5	0.21
Bor_{21}	8.4	26.3	54.0	0.55
Ber_9	7.5	31.8	61.0	0.34
Kan_{20}	14.6	26.9	55.1	1.11
Ya_{10}	2.8	20.2	34.7	0.08
Bor_2D (0.2 HCl)	5.2	20.0	31.2	0.26
Bor_3D (0.2 HCl)	7.6	18.5	23.3	0.22
Bor_5D (0.2 HCl)	1.4	18.4	28.8	0.25
Bor_6D (0.2 HCl)	2.1	20.9	34.5	0.22
Ber_9D (0.2 HCl)	1.4	19.3	33.5	0.15

yield. The values of burn-off measured based on the quantity of organic matter in the char before and after gasification and by measuring the carbon content in the carbonaceous gases were also well corroborated.

One can see from Table 7 that again the extents of char gasification were quite insensitive to the total surface areas. For example, Yallourn char with the largest surface area showed least reactivity as compared to other parent chars with much less surface area. Up to 14 fold difference (from 0.08 to 1.11 wt%/m^2) is observed between the surface normalized reactivities which correlate too weakly to the contents of calcium. On the other hand, HCl treated chars showed rather close reactivities.

Not all surface carbon atoms are well known [3,35-38] to have the same reactivity for gasification because of structural defects and differences in the strain energy within the carbon network besides the effects of catalytic species on the carbon surface. Carbon atoms out of the crystallites show usually higher specific reactivity and are easily gasified as compared to those in the crystallites.

It follows from the comparison of the XRD structural parameters of organic matter in Table 5 and gasification reactivities in Table 7 that relatively ordered chars from the parent coals with only 11% - 26% of disordered g-fraction proved to show higher extents of burn-off than less ordered acid treated chars enriched with 28% - 40% of g-fraction. This may mean that structural properties (surface area and crystallinity) of carbon matter of chars could hardly control steam gasification reactivities under the given conditions.

From the data in Tables 1 and 7, fairly good correlations between the values for carbon burn-off and gas yield based on daf wt. and the content of calcium with the determination coefficients of R^2 = 0.89 and 0.92 were revealed (Figure 3):

Carbon burn-off, wt% = $19.2 + 7.2C_{Ca} - 1.4C_{Ca}^2$; $R^2 = 0.89$;

Gas yield, wt% = $29.5 + 18.2\ C_{Ca} - 3{,}5C_{Ca}^2$; $R^2 = 0{,}92$.

Some deviations from the correlation curves in Figure 3 can arise from difference in the forms and dispersions of calcium species and/or from nonuniform distribution within the organic matter of char. Also, other metals, such as iron and magnesium could contribute to catalysis of the reaction. However their effects were no evidently because of probably less contents and lower specific catalytic activities as compared to those of calcium.

3.3.2. The Composition of Gaseous Products

The composition of the gaseous products was dependent on coal chars. The gaseous autoclave products consisted of mainly CO_2 (32 - 37 mol%), H_2 (20 - 29 mol%), and CH_4 (31 - 41 mol%) with least amount of CO (1% - 2%). Figure 4 shows the yields of CO_2 and also of H_2 to correlate with Ca contents:

1) CO_2 yield (mmol/g) = $3.07 + 1.16\ C_{Ca} - 0.21\ C_{Ca}^2$; $R^2 = 0.90$;

2) H_2 yield (mmol/g) = $2.14 + 0.91\ C_{Ca} - 0.13\ C_{Ca}^2$; $R^2 = 0.74$.

It seems reasonable to suggest CO_2 and H_2 to be originated from the same gross-reaction catalyzed by calcium species:

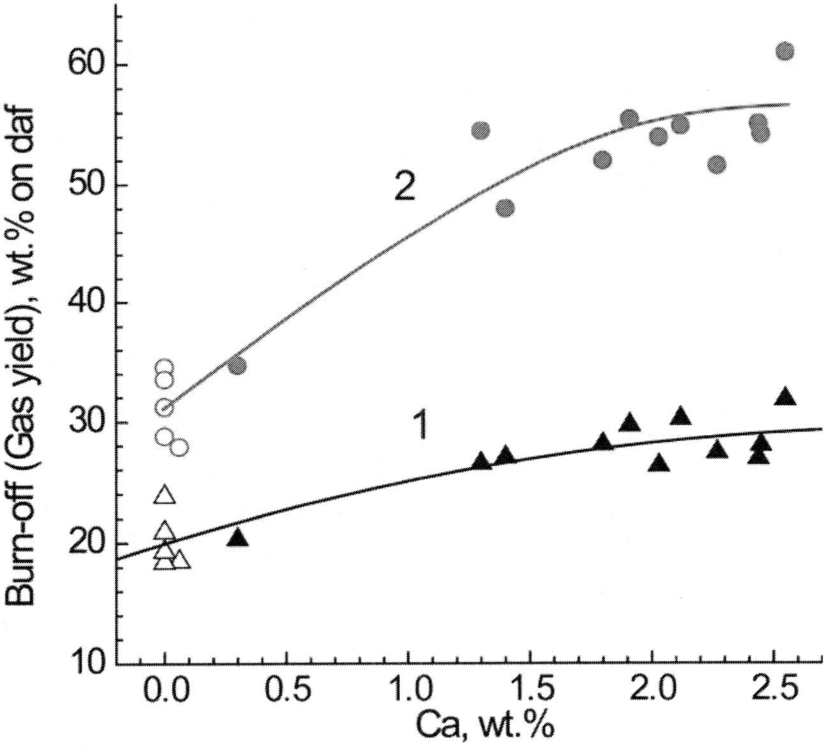

Figure 3. Carbon burn-off (1) and yield of gases (2) dependant on calcium content in the chars. Autoclave reactor. Dark symbols—chars from parent coals; open symbols— from HCl treated coals.

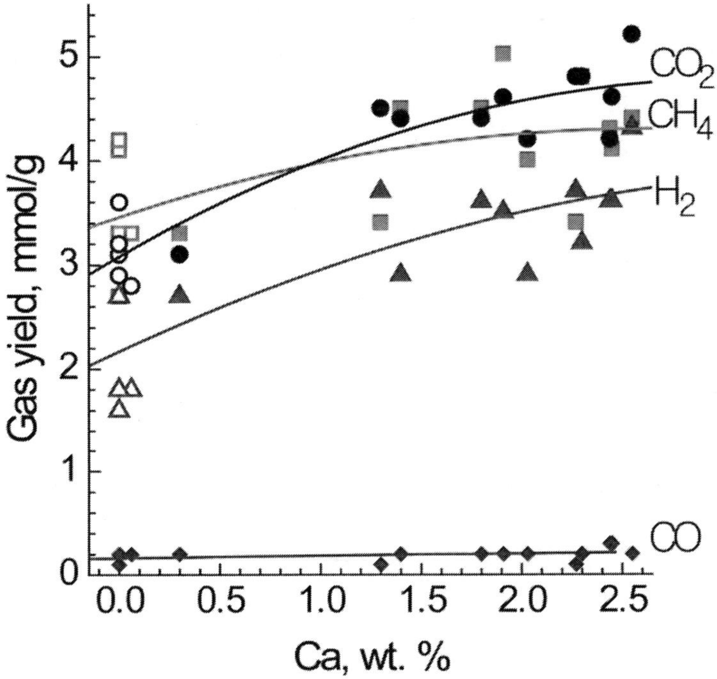

Figure 4. The yields of gaseous products dependant on the calcium content in the chars. Autoclave reactor.

$$C + H_2O = CO + H_2 \tag{1}$$

$$CO + H_2O = CO_2 + H_2 \tag{2}$$

However, the actual H_2 yields were much less than those of CO_2 in consideration of stoichiometry of reactions. This could result from partial hydrogen consumption in a methane generation reaction. However, large scatter for methane yields with no distinct trend and with too unsatisfactory correlation (R^2 = 0.34) is observed in Figure 4, HCl treated chars yielding essentially same amount of methane as the chars from the parent coals. It follows from this that alkali and alkaline metal species have no relevance to methanation under given conditions. The catalysis of CO hydrogenation into methane by Fe species [2,10] can not be excluded, however its effect is difficult to observe because of low Fe concentrations.

The yield of CO was very low in all cases (0.05 - 0.3 mmol/g) irrespectively of metal contents. This may mean that water gas shift reaction (2) and CO hydrogenation reaction occured rapidly under these conditions producing H_2, CO_2 and CH_4.

3.3.3. XRD Analysis of Char Residues

XRD patterns of the char residues after ashing showed complicated mixtures of various crystalline mineral substances. Shown in Figure 5 are the XRD patterns of the ashes from the representative parent coals and also from HCl treated one, as examples. Major minerals identified in the ashes from Kansk-Achinsk and Kangalas coals were quartz (a-SiO_2), calcite ($CaCO_3$), lime (CaO), anhydrite ($CaSO_4$), hematite (α-Fe_2O_3), $Ca_3Al_6O_{12}$-$CaSO_4$, $Ca(OH)_2$ and trace amounts of some other nonidentified minerals. Yallourn coal ash consisted of mainly $FeFe_2O_4$ (may be $MgFe_2O_4$), $CaSO_4$ and much less $CaCO_3$ and CaO species. Mainly a-SiO_2, much less a-Fe_2O_3 and only trace amount of $CaSO_4$ were detected in the ashes from the 0.1 N HCl treated coals. The excess of sulphur

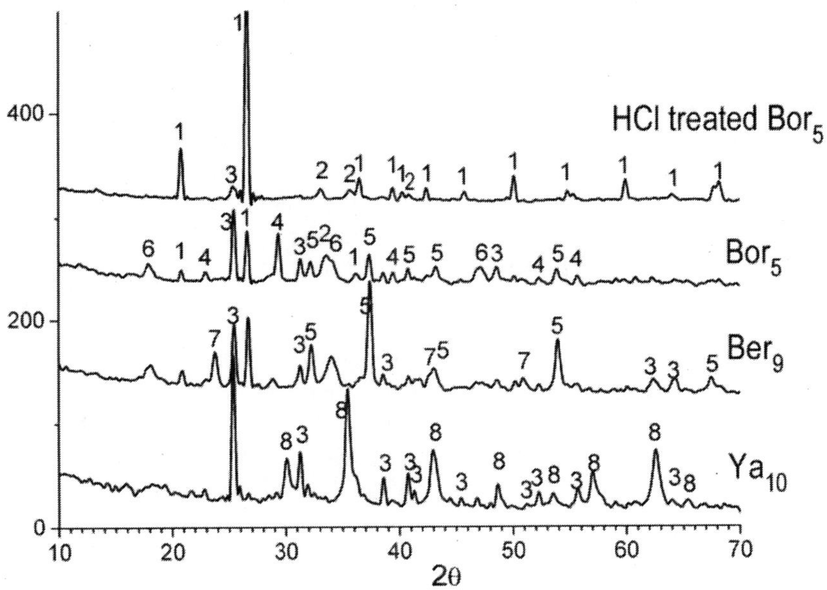

Figure 5. X-ray diffraction patterns of the ashes from the demineralized and parent coal chars. 1—α-SiO_2; 2— α-Fe_2O_3; 3—$CaSO_4$; 4—$CaCO_3$; 5—CaO; 6—$Ca(OH)_2$; 7— $Ca_3Al_6O_{12}\cdot CaSO4$; 8—$FeFe_2O_4\cdot(MgFe_2O_4)$.

over calcium in the acid treated chars was favorable for the binding of calcium in the form of $CaSO_4$. The sizes of the crystallites of the main $CaCO_3$ and CaO particles estimated from the width of XRD peaks were less than 30 nm.

4. CONCLUSIONS

The conclusion should be drawn thus from the results obtained in this study that steam gasification reactivities of chars produced from Kansk-Achinsk, Kangalas and Yallourn brown coals at the moderate temperature of 700°C under atmospheric pressure in a flow reactor and under a high pressure of reaction products in the autoclave reactor obey the common regularity controlled by catalysis on the well-dispersed CaO or/and $CaCO_3$ particles. The latter were produced during the carbonization and gasification steps from the calcium carboxylate derivatives naturally occurring in the brown coals. These nanosized calcium species catalysed gasification mainly into CO_2 and H_2 gases. The production of CH_4 seems to occur by another way irrespectively of calcium species. The structural properties of chars (surface area and crystallinity) hardly affect the gasification reactivity.

However, from the comparison of the gasification behaviors of chars under different reaction conditions, the question may be raised as to why much larger differences between the reactivities of chars from the parent coals and from the respective acid treated ones were obtained in the flow reactor with steam stream at atmospheric pressure as compared to those in the autoclave reactor under high pressure of gaseous products. One reason could be caused by different forms of calcium species produced in different reactors and this can be interpreted in terms of a calcium carbonate—calcium oxide cycle mechanism [8]. Under reaction conditions at 700°C, a mobile thermodynamic equilibrium between these calcium species is strongly controlled by CO_2 pressure. Therefore at the atmospheric pressure in a tubular reactor with the steam in nitrogen flow where the gaseous products were removed continuously, calcium oxide was probably the predominant species that effectively catalysed reaction. On the other hand, calcium carbonate (which are probably catalytically less active as compared to calcium oxide) predominanted during the gasification in the autoclave under high pressure of CO_2. Apart from this, the gaseous products CO_2, H_2 and CH_4 accumulated in the autoclave may inhibit steam gasification due to stronger adsorption and lower specific reactivity [11,39].

The results obtained allow Kansk-Achinsk brown coals to be considered as the preferable and environmentally freindly raw material naturally charged with the precursors of the catalytically active species for effective gasification. The chars derived from the acid treated coals differed with very low gasification reactivities. This fact may be in the way in developing the effective processes for ultraclean gasification of ash-free coals.

5. ACKNOWLEDGEMENTS

This work was supported in part by the Siberian Branch of the Russian Academy of Sciences (Integration Program, Project no. 13).

REFERENCES

1. T. A. Adams and P. I. Barton, "Combining coal Gasification and Natural Gas Reforming for Efficient Polygeneration," Fuel Processing Technology, Vol. 92, No. 3, 2011, pp. 639-655.
2. A. Sharma, T. Takanohashi, K. Morishita, T. Takarada and I. Saito, "Low Temperature Catalytic Steam Gasification of Hyper Coal to Produce H_2 and Synthesis Gas," Fuel, Vol. 87, No. 4-5, 2008, pp. 491-497.
3. K. Miura, K. Hashimoto and P. L. Silveston, "Factors Affecting the Reactivity of Coal Chars during Gasification, and Indices Representing Reactivity," Fuel, Vol. 68, No. 11, 1989, pp. 1461-1467.
4. L. Lemaignen, Y. Zhuo, G. P. Reed, D. R. Dugwell and R. Kandiyoty, "Factors Governing Reactivity in Low Temperature Coal Gasification. Part II. An Attempt to Correlate Conversions with Inorganic and Mineral Constituents," Fuel, Vol. 81, No. 3, 2002, pp. 315-326.
5. I. M. K. Ismail and P. L. Walker, "D.s.c. and TGA Measurements of O_2 Interaction with Coal Chars," Fuel, Vol. 68, No. 11, 1989, pp. 1456-1460.
6. T. Takarada, Y. Tamai and A. Tomita, "Reactivities of 34 Coals under Steam Gasification," Fuel, Vol. 64, No. 10, 1985, pp. 1438-1442.

7. Z.-I. Liu and H.-H. Zhu, "Steam Gasification of Coal Char Using Alkali and Alkaline-Earth Metal Catalysts," Fuel, Vol. 65, No. 10, 1986, pp. 1334-1338.

8. Y. Ohtsuka and A. Tomita, "Calcium Catalysed Steam Gasification of Yallourn Brown Coal," Fuel, Vol. 65, No. 12, 1986, pp. 1653-1657.

9. F. Kapteijn, H. Porre and J. A. Moulijn, "CO_2 Gasification of Activated Carbon Catalyzed by Earth Alkaline Elements," AIChE Journal, Vol. 32, No. 4, 1986, pp. 691-695.

10. Y. Ohtsuka, Y. Tamai and A. Tomita, "Iron-Catalyzed Gasification of Brown Coal at Low Temperatures," Fuel and Energy, Vol. 1, No. 1, 1987, pp. 32-36.

11. A. Linares-Solano, E. J. Hippo and P. L. Walker, "Catalytic Activity of Calcium for Lignite Char Gasification in Various Atmospheres," Fuel, Vol. 65, No. 6, 1986, pp. 776-779.

12. C.-Z. Li, "Some Recent Advances in the Understanding of the Pyrolysis and Gasification Behaviour of Victorian Brown Coal," Fuel, Vol. 86, No. 12-13, 2007, pp. 1664- 1683.

13. T. Kyotani, S. Karasawa and A. Tomita, "TPD Study of Coal Chars in Relation to the Catalysis of Mineral Matter," Fuel, Vol. 65, No. 10, 1986, pp. 1466-1469.

14. P. Samaras, E. Diamadopoulos and G. P. Sakellaropoulos, "The Effect of Mineral Matter and Pyrolysis Conditions on the Gasification of Greek Lignite by Carbon Dioxide," Fuel, Vol. 75, No. 9, 1996, pp. 1108-1114.

15. P. J. J. Tromp, F. Kapteijn and J. A. Moulijn, "NATO ASI Series. Series C. Mathematical and Physical Sciences," No. 370, Kluwer, Amsterdam, 1992, p. 84.

16. K. Matsuoka, T. Yamashita, K. Kuramoto and A. Tomita, "Transformation of Alkali and Alkaline Earth Metals in Low Rank Coal during Gasification," Fuel, Vol. 87, No. 6, 2008, pp. 885-893.

17. K. Mitsuoka, S. Hayashi, H. Amano, K. Kayahara, E. Sasaoka and A. Uddin, "Gasification of Woody Biomass Char with CO_2: The Catalytic Effects of K and Ca Species on Char Gasification Reactivity," Fuel Processing Technology, Vol. 92, No. 1, 2011, pp. 26-31

18. A. Linares-Solano, C. Salinas-Martínez de Lecea, F. Rodríguez-Reinoso and M. Almela-Alarcón "Anomalous' Increase in CO_2

Reactivity, Relative to Steam and Air on Acid Treatment of Coals," Fuel, Vol. 65, No. 10, 1986, pp. 1345-1348.

19. S. Zhang, J. I. Hayashi and C.-Z. Li, "Volatilisation and Catalytic Effects of Alkali and Alkaline Earth Metallic Species during the Pyrolysis and Gasification of Victorian Brown Coal. Part IX. Effects of Volatile-Char Interactions on Char-H_2O and Char-O_2 Reactivities," Fuel, Vol. 90, No. 11, 2011, pp. 1655-1661.

20. C. Sheng, "Char Structure Characterised by Raman Spectroscopy and Its Correlations with Combustion Reactivity," Fuel, Vol. 86, No. 15, 2007, pp. 2316-2324.

21. D. Yanovsky, V. Sharypov, M. Schipko and P. Kuznetsov, "Pyrolysis and Steam Gasification of Solid Residues from Borodino Coal Hydrogenation," Khimiya Tverdogo Topliva, No. 5, 1989, pp. 77-81 (Rus).

22. K. V. Gavrilin and A. Yu. Ozerskii, "Kansko-Achinskii ugol'nyi Bassein (Kansk-Achinsk Coal Basin)," Nedra, Moscow, 1996 (Rus).

23. P. Kuznetsov, "The Chemical Modification, Swelling of Coals, and Reactivity in Hydrogenation," Khimiya Tverdogo Topliva No. 3, 1998, pp. 53-68 (Rus).

24. P. Kuznetsov and L. Kuznetsova, "Effect of the Mineral Components of Brown Coals on the Properties of Organic Matter in the Interaction with Solvents," Solid Fuel Chemistry, Vol. 42, No. 6, 2008, pp. 372-380.

25. P. Kuznetsov, L. Kuznetsova, J. Bimer, et al., "Quantitative Relation between the Macromolecular Characteristics of Brown Coal and Its Reactivity in Conversion with Tetralin," Fuel, Vol. 76, No. 2, 1997, pp. 189-193.

26. P. Kuznetsov, L. Kuznetsova and E. Kutikhina, "Effect of the Decationization of Brown Coal from the KanskAchinsk Basin on the Physicochemical Properties of the Resulting Sorbents," Solid Fuel Chemistry, Vol. 42, No. 3, 2008, pp. 153-159.

27. D. J. Allardice, L. M. Clemow and W. R. Jackson, "Determination of the Acid Distribution and Total Acidity of Low-Rank Coals and Coal-Derived Materials by an Improved Barium Exchange Technique," Fuel, Vol. 82, No. 1, 2003, pp. 35-40.

28. M. N. Siddiqiu, M. F. Ali and J. Shirikoff, "Use of X-Ray Diffraction in Using the Aging Pattern of Asphalt Fractions," Fuel, Vol. 81, No. 1, 2002, pp. 51-58.

29. V. Saranchuk, A. Airuni and K. Kovalev, "Nadmolekulyarnaya Organizatsiya, Struktura i Svoistva Uglya (Supramolecular Organization, Structure, and Properties of Coal)," Naukova Dumka, Kiev, 1988 (Rus).

30. A. F. Lukovnikov, Yu. M. Korolev, G. S. Golovin, et al., "X-Ray Diffraction of Kuznetsk Coals," Khimiya Tverdogo Topliva, No. 5, 1996, pp. 3-13 (Rus).

31. H. Takagi, K. Maruyama, N. Yoshizawa, Y. Yamada and Y. Saito, "XRD Analysis If Carbon Stacking Structure in Coal During Heat Treatment," Fuel, Vol. 83, No. 17-18, 2004, pp. 2427-2433.

32. M. Siddiuqui, M. Ali and J. Shorokoff, "Use of X-Ray Diffraction in Assessing the Aging of Asphalt Fractions," Fuel, Vol. 81, No. 1, 2002, pp. 51-58.

33. E. N. Loskutova, N. M. German, K. I. Bochkareva and K. N. Schiryaeva, "Pyrolysis of Brown Coals," Nauka, Novosibirsk, 1968.

34. R. Ya. Birgaus and T. A. Kuharenko, "Chemistry and Processing of Fuels," Vol. 27, International Gemological Institute, Moskwa, 1971, pp. 3-12.

35. J. F. Byrne and H. Marsh, "Introductory Overview," In: J. W. Patrick, Ed., Porosity in Carbons: Characterization and Applications, Edward Arnold, London, 1995, pp. 7- 12.

36. H. Marsh, "The Determination of Surface Areas of Coals—Some Physicochemical Considerations," Fuel, Vol. 44, No. 2, 1965, pp. 253-260.

37. N. R. Laine, F. J. Vastola and P. L. Walker, "The Importants of Active Surface Area in the Carbon Oxygen Reaction," The Journal of Physical Chemistry, Vol. 67, No. 10, 1963, pp. 2030-2034.

38. J. B. Lewis, "Modern Aspects of Graphite Technology," Academic Press, New York, 1970, p. 129.

39. D. G. Roberts and D. J. Harris, "Char Gasification in Mixtures of CO_2 and H_2O: Competition and Inhibition," Fuel, Vol. 86, No. 17-18, 2007, pp. 2672-2678.

4

CHAPTER

GASIFICATION A DRIVER TO STRANDED RESOURCE DEVELOPMENT

Yuli Grigoryev

[1] *CEPMLP, University of Dundee, UK*

1. INTRODUCTION

When evaluating natural gas projects, stakeholders and decision-makers have been traditionally limited by the requirement for large reserves to be recoverable before any investment can be committed. Technological advancement and commercial know-how has unlocked many potential reserves that would technically be considered stranded or entirely overlooked by IOCs (International Oil Companies) and NOCs (National Oil Companies). This has important implications for policymakers as it can affect the way natural gas is utilised as an indigenous supply, an export resource or substitution fuel.

There have been some structural changes in the way the gas sector operates over the last decade. This has been an exciting time for market observers, as political, economic, financial and technical inputs have driven major changes in the industry.

One of the ways that the industry has changed is the attention to using gas in less traditional methods. To understand how this has developed, this chapter analyses the major recent trends, the traditional market characteristics and discusses the outlook of potential future changed in the field of natural gas development, production and processing.

When looking at gas field development, stakeholders evaluate a number of options to monetise gas. These options are limited by a host of factors, each with a unique position with respect to geography, government, market and political dynamics.

This chapter acts as an overview of the monetisation routes and the options that are becoming available to decision-makers.

2. CURRENT AND FUTURE TRENDS OF OIL AND GAS MARKET

Over the past decade, there has been a paradigm shift in the behaviour of the oil and natural gas markets. Compared to the traditional model, where gas production was secondary to the production and marketing of oil, and prices of gas were naturally linked to the price of oil (or a basket of oil products), we have seen gas emerge as in increasingly important fuel with a decoupling of prices. This has been particularly evident in the North American market, where competitive forces and regulation of the midstream sector allowed for an emergence of a separate gas market, marked by consistently high liquidity. There has also been a discrepancy in the regional gas price, which lends to arbitrage activity by spot traders, and some LNG (Liquefied Natural Gas) cargoes have been redirected from initial destinations to other markets, even whereby destination clauses in contracts have been broken.

There is some expectation that natural gas consumption may over take oil consumption by 2030, which will result from a number of pressures on oil consumption, ranging from economic and environmental, to issues relating to security of supply.

In Europe, this shift has seen a slower uptake. A primary reason for this is the separation of supply and demand centres, with the EU zone increasingly relying on imported gas. The biggest supplier to this market is Russia, also the world's largest producer of natural gas, which as been supplying gas to Europe via soviet built high pressure interregional trunk pipelines.

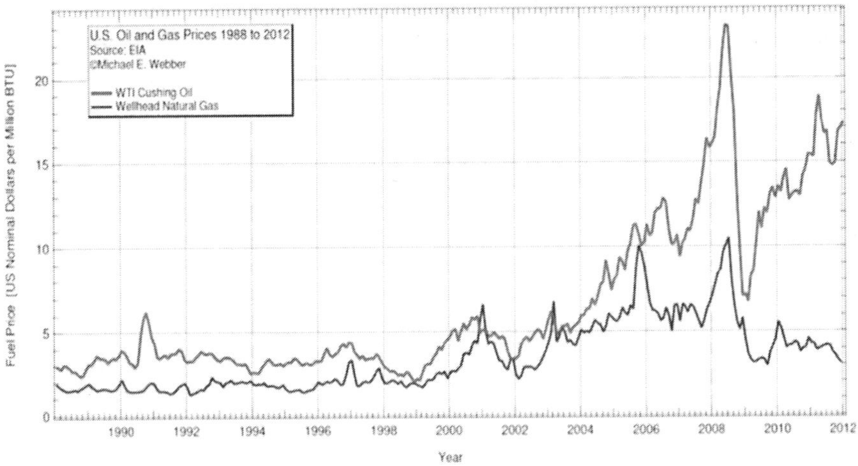

Source: IEA, Michael E Webber, 2012Figure 1

Figure 1.

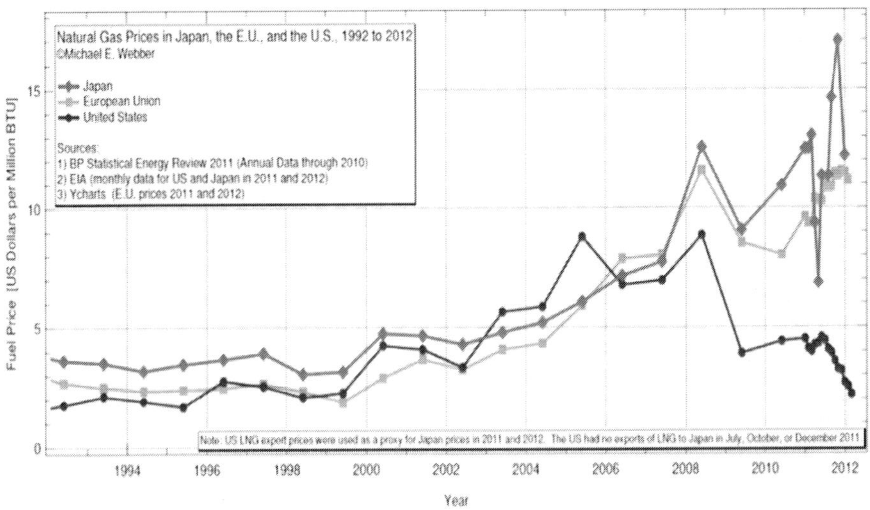

Source: IEA, BP Stats Review, Michael E Webber, 2012

Figure 2.

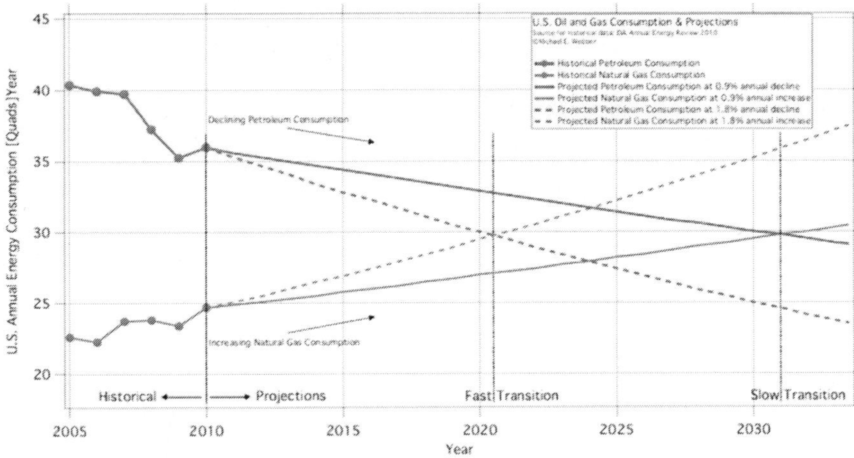

Source: EIA, M Webber, 2012

Figure 3.

The majority of contracts are long term take-or-pay contracts which have a price formula as an index linked to a basket of refined products (the "substitution fuels"). Historically, long-term contracts have played an important role in the development of the European gas market by providing a risk sharing arrangement between producers and buyers, enabling important new investment into production and infrastructure projects to be undertaken. The Eurozone realised that their growing gas needs, the bulk of which are met with Russian gas, can only be adequately supplied if Russia is able to invest in new gas fields and pipeline construction. They took a position that if gas is supplied exclusively through spot transactions, gas suppliers, Gazprom included, will not be willing to shoulder the risks associated with multi-billion dollar investments and substantial quantity risks. Corporate strategy aside, it would be impossible to access the international capital markets without guaranteed offtake contracts being in place. [1] - Thus, contracts of 20 years or more have been a normal occurrence in the European continent.

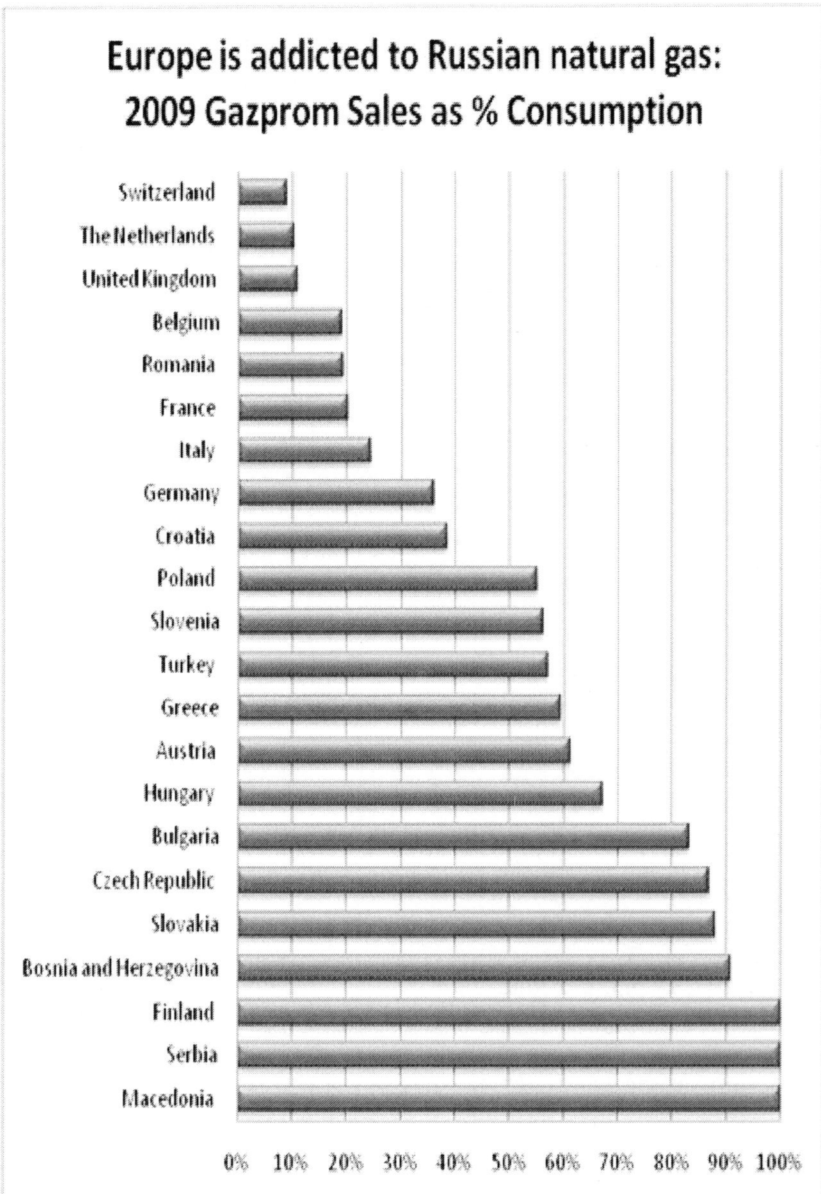

Source: IEA, 2010

Figure 4.

Major contracts concluded by Gazprom or affiliated companies with the EU consumers in 2006 and 2007

country	contract term	volume of annual supplies	description	additional agreements
Austria	2012–2027	7 billion m^3	contract signed by Gazprom with OMV, EconGas as well as with GWH and Centrex Europe Energy & Gas AG, affiliated to Gazprom	The rules for gas reception have changed: - change of gas recipient from OMV to EconGas; - GWH and Centrex, which are affiliated with Gazprom, were granted the right to sell gas (nearly 1.75 billion m^3 annually) to Austrian end-users
Germany (VNG)	2014–2031	5,25 billion m^3 (possibility of incre-asing supplies)	contract signed by WIEH (a joint venture of Gazprom and Wintershall) and Verbundnetzgas (VNG). Gazprom sells gas via WIEH in the markets of other EU member states	
Germany (E.ON)	2011–2036	4 billion m^3 (Nord Stream)	contract signed by Gazprom and E.ON on supplies via NS gas pipeline	
Germany (E.ON)	2020–2035	20 billion m^3	contract signed by Gazprom and E.ON, being an extention of the currently binding contracts (to expire in 2020)	
Italy	2017–2035	22 billion m^3	contract signed by Gazprom and ENI which constitutes a part of the new strategic partnership agreement; extention of the terms of the currently binding contracts	- companies affiliated with Gazprom (GMT Italia, Centrex) have gained access since 2007 to the Italian domestic market, and sales of up to 3 billion m^3 annually in 2010 is allowed - ENI has been promised the right to buy assets in Russia
France	2012–2030	12 billion m^3 + 2,5 billion m^3 (Nord Stream)	contract signed by Gazprom and Gaz de France ; the currently binding contract is prolonged, and an agreement for supplies via the Nord Stream gas pipeline is concluded	- Gazprom's subsidiary, GMT France, has been granted access to the French domestic market, and is allowed to sell up to a maximum of 1.5 billion m^3 of gas annually from 2007
Czech Republic (RWE)	2014–2035	9 billion m^3	contract signed by Gazprom and RWE Transgas; prolongation of the previous agreement on gas supplies to, and transit through, the Czech Republic	
Czech Republic (Vemex)	2008–2012	0,5 billion m^3	contract concerns supplies of gas to the Gazprom-controlled company which supplies gas to some Czech industrial customers	
Bulgaria	2011–2030	3 billion m^3	Gazprom and Bulgargaz have signed a contract prolonging the previously binding contract	
Romania (WIEH)	2012–2030	4,5 billion m^3	the previously binding contract has been prolonged	The contract's prolongation is made dependent on its renegotiation, and switching to cash payments for gas transit. The agreement provides for the increase of the gas transit through Bulgaria in exchange for co-operation in the implementation of Russian gas pipeline projects.
Romania (Conef)	2010–2030	2,5 billion m^3	Gazprom and Conef Energy SRL have signed a contract; supplies for Alro Stalina aluminium plant	
Denmark	2011–2031	1 billion m^3	Gazprom and DONG Energy have signed a contract for supplies from NS	

Based on *Gazprom in Europe: Faster expansion in 2006*, Ewa Paszyc, Centre for Eastern Studies, Warsaw, February 2007; data quoted from the companies' websites and news agencies

Source: Ewa Paszyc, Centre for Eastern Studies, 2008

Figure 5.

A number of drivers have begun to put significant pressure on the traditional model. Working in tandem, the economic growth on the continent, together with significant global environmental concerns and directives, has delivered a growing demand for natural gas. At the same time, as indigenous supplies begin to plateau and decline[2] - and governments become more reliant on imports, Security of Supply issues begin to make their way up national policy agendas. From a security of supply standpoint, there have traditionally been three pillars of national strategy for policymakers – development of indigenous supplies, diversification of suppliers and reduction in consumption.[3] - From the three pillars, the fastest route is evidently the diversification of suppliers as consumption reduction and indigenous suppliers requires significant lead-times. Certainly it is difficult to diversify in a timely manner if transit is to take place via pipelines, however, as the market of liquefied natural gas became more mature, it allowed for an efficient way to introduce new suppliers. Regasification terminals are significantly less complex than liquefaction terminals, and began to appear in a host of European coastlines.

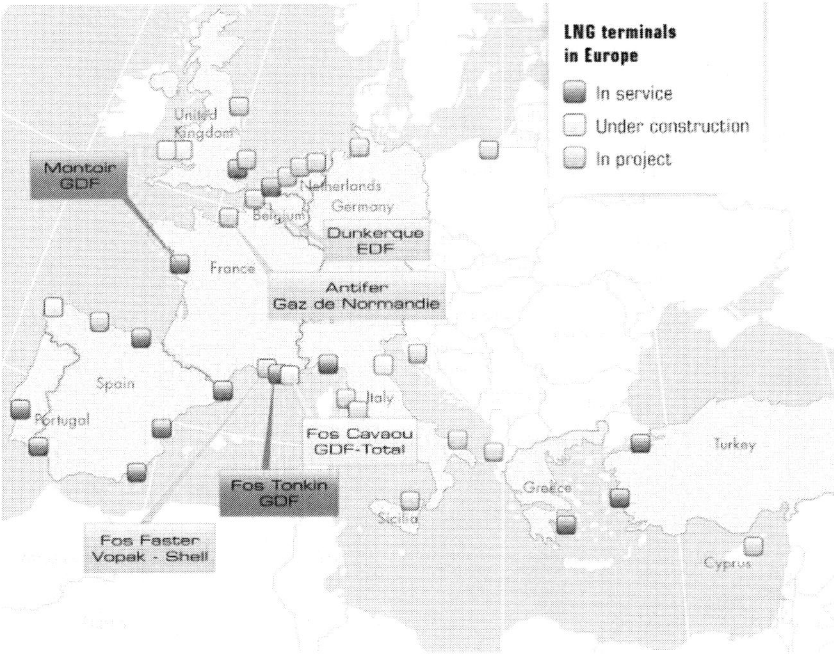

Source: GdF, 2011

Figure 6.

European competition rules have created somewhat of a stumbling block for these initially, but investment arrived in sufficient quantity to allow for an emerging spot market in the European gas hubs. The net effect of this has been an evolution of long-term contracts with certain traditional terms being re-examined and renegotiated. Some of the centrally important clauses such as duration/period are seeing a decrease from the frequently encountered fifteen to twenty-five years to perhaps eight to twelve years in length. This is, in part, due to the contract volumes also decreasing with new project supplying between three and ten BCM (Billion Cubic Metres) annually as opposed to the traditional ten to twenty BCM. Take-or-pay obligations are also become less stringent, with increasing "carry-forward" and "make-up" rights. Index pricing is being replaced in highly competitive markets by daily pricing derived from a liquid short term market, such as the UK National Balancing Point. Certainly this trend will apply to some of the new export contracts yet others, which intend to supply large volumes and require substantial infrastructure investment, will be done under traditional terms.[4] -

What cannot go unmentioned is the shale gas development. The flurry of exploration activity has seen significant results in adding major volumes to reserves in the US, and has become a game-changer in the US domestic market. The effect on global markets has not yet been so dramatic, although exploration activity for shale gas in the Eurozone has excited many a journalist and energy observer. Thus far, however, the UK has enforced a temporary ban on shale gas fracking and Poland's estimate of reserves has so far been cut by a factor of ten. How this develops could have a profound effect on the industry.

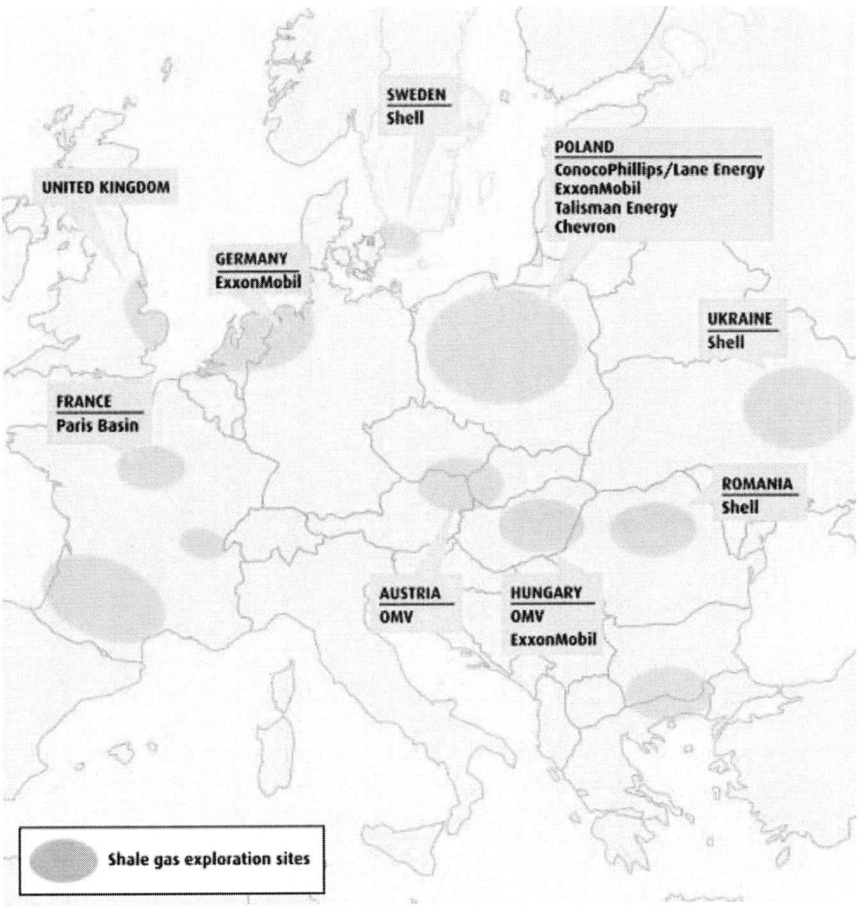

Source: Energy Tribute, 2011

Figure 7.

3. TRADITIONAL GASIFICATION USES

Where gas has been discovered in abandoned supply, the stakeholders had a very clear picture of how this asset can be monetised. The most straightforward solution has been the construction of a pipeline from the supply centre to the consumption centre, where gas would be used for heat and power generation, industry and grid supply. Pipelines can run for many thousands of kilometres, over different and difficult terrain, across borders, through mountains and under water.

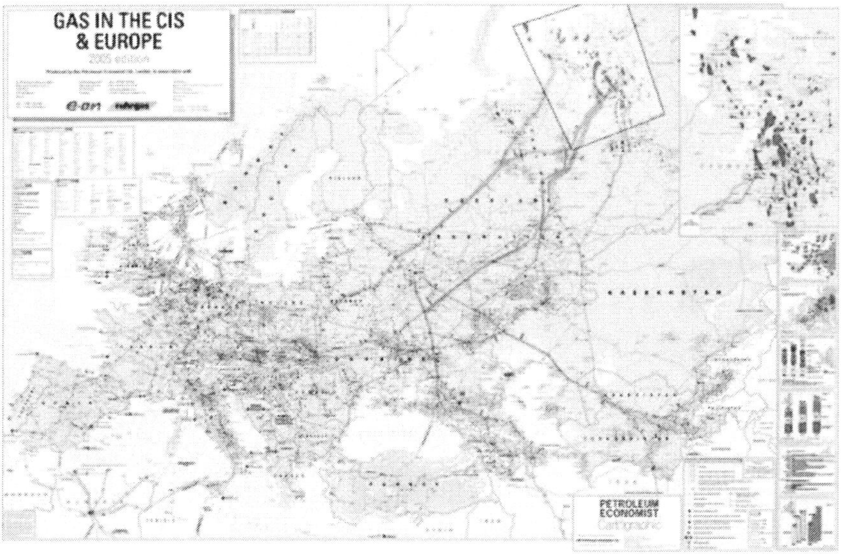

Source: Petroleum Economist, 2006

Figure 8.

If the consumption centres were satisfactorily supplied, or if the cost of the pipeline was prohibitively expensive, gas was either left in the ground, or converted into a final product that could be transported as a liquid or solid to other distant markets.

These conversion routes are what are known as the "Gas-to- " technologies and are specifically Gas-to-Chemicals, Gas-to-Liquids and Gas-to-Power. Recent advances in technology have allowed these processes to become available as economic methods not only to utilise stranded gas but to take advantage of pipeline gas that may be limited in its transport options.[5] -

Gas-to-Liquids (GTL) is a process that was initially discovered by Fisher and Tropsch during the World War II and has seen various applications thereafter. In essence, it is a petrochemical process that converts methane (major component of natural gas) into a synthetic diesel fuel that is environmentally clean as it contains no sulphur and is aromatics free. The first major commercial GTL facility was built in South Africa by Sasol, using coal gasification to produce the feedstock and manufacturing diesel oil. It is generally accepted that due to economies of scale, GTL facilities become economic with large output capacities and extremely low feedstock cost. As such, new plants are expected to be in excess of 100,000 barrels per day (bpd) of product, and located in the Middle East or Africa[6] - . Multi-billion dollar projects such as Shell's Pearl GTL in Qatar and Oryx GTL (Qatar Petroleum and Sasol) are leading examples of this technology in application. Another 200,000 bpd plant has been proposed in Australia.

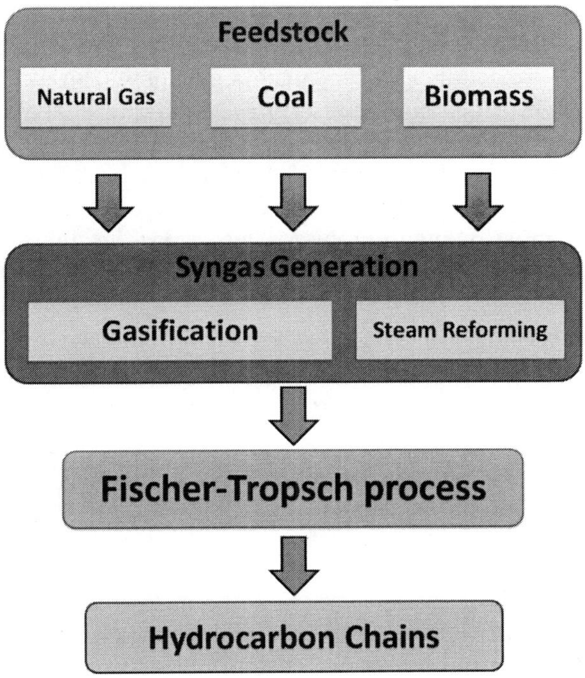

Source: Stamford University, 2010

Figure 9.

The Gas-to-Chemicals (GTC) process is a very mature process. This involves the conversion of methane to a chemical product, either an intermediate or final stage. Indeed, more value is captured the further down the process chain that one is able to proceed. The most common product is Ammonia which is used in the production of fertilizers. The high oil price has been somewhat of a double-edged sword for the price of fertilizers since the increase in the feedstock (where natural gas is still tied to the oil price) and also the increase in demand driven by the biofuels surge as a means to find alternative energy solutions. Because natural gas makes up about 70% of the cost of production, European based producers can no longer compete with producers with a low cost base such as Russia, or even Ukraine (due to special relationships with Russia[7] -). There is a clear link between financially stable fertiliser producers and a low gas source. As more producers are forced to shut down or relocate, and food scarcity continues to haunt developing countries, fertiliser production will remain a highly lucrative option for GTC processing. Another common product is methanol, and whilst a very price volatile product, it can itself be used as an intermediate to produce more valuable products. The methanol to olefins (MTO) process chain is a lucrative way to capture added value. Given the recent worldwide rise in the use of polymers, this particular process has spurred a myriad of activity. The process can be tuned to produce polypropylene and polyethylene. Given the issues outlined above, it makes sense to commission boutique-plants with capacities not exceeding 150,000 tonnes per year. Certainly scale economies are also achieved in this process, but given factors such as political risk and competitive pressures from new producers, it seems prudent to seek a short project pay-back period. Thus, given that the project is Capex sensitive, it is advisable to seek new, low cost technology that has become available in China and has half the cost of similar European technology.

Other polymers that may be of interest in the GTC segment are PBT (Polybutylene terephthalate) and PET (Polyethylene terephthalate). These are thermoplastic polymers that are used in the production of electrical insulators or plastic bottles and synthetic fibres respectively. PBT is a less widespread product, however its versatile nature sees the market grow at 7% annually. One route to its production would be via 1,4-Butanediol which can be produced from propylene, itself produced from methanol. Like many of these processes, the rights are protected by international patents and it is necessary to approach the patent holders to implement them. The manufacturing process of PBT via 1,4-Butanediol is patented by Zimmer, now part of Lurgi AG. In most cases, patent holders are willing to grant user licenses to return research and development costs. This is not the case with PET however, as the necessary intermediate is Acetic Acid, produced from methanol via only two economical routes. These routes are patented by BP

Chemicals and Celanese, which between them control the acetic acid market. These companies do not grant licenses to third parties and as such, gas owners would need to yield a majority stake to the licensors. Nevertheless, acetic acid, and subsequent PET production is an extremely lucrative method to monetise stranded or semi-stranded gas. These projects are indeed capital intensive and require considerable upfront investment. A 500,000 tonne acetic acid plant with a PET production line could cost in excess of $800+ million.

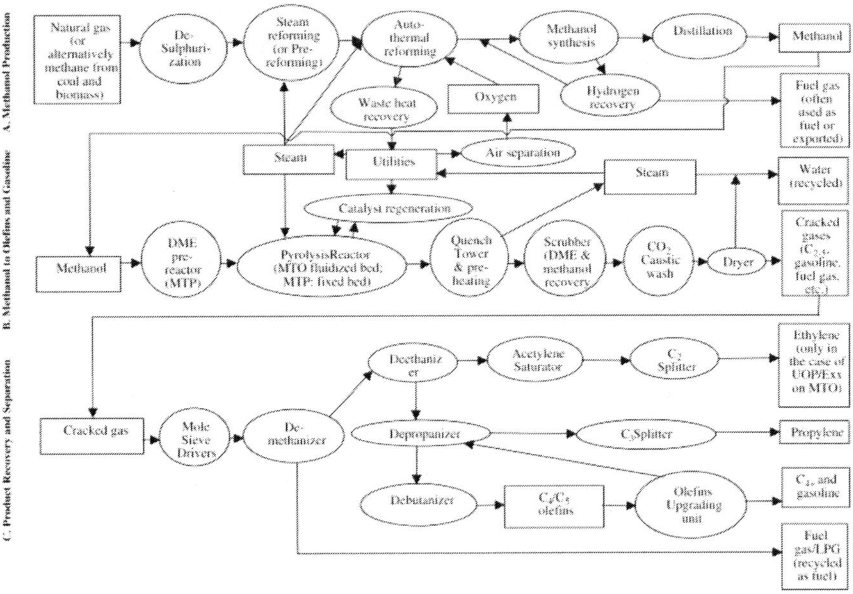

Source: Energy, Volume 33, Issue 5, 2008

Figure 10.

Gas-to-Power (GTP) involves the conversion of natural gas to electricity and normally implies in-situ generation. GTP has become a viable option since the introduction of the combined cycle gas turbine (CCGT), a system where the gas turbine is driven to produce electricity whilst the heat is used to manufacture steam to generate additional electricity through a steam turbine. These plants are much more efficient than their traditional counterparts[8] - and are more compact is size. Furthermore, the construction lead times usually do not exceed two years, which is a significant improvement on traditional power plants. In situ GTP is particularly applicable when gas is found in undeveloped urban centres with a low degree of residential and commercial gasification. Gasification refers specifically to the level of development and infiltration of the low pressure distribution networks that supply gas to local residents or small commercial users. Africa and India, both of which have discovered gas near populated areas, would see great benefit from such technology. Nevertheless, Nigeria, which holds Africa's largest natural gas resources, flares more gas than any other country, after Russia.

In fact, there has been some discussion about applying old jet engines as temporary gas turbines for local power production. Because gas is considered a clean fuel, and due to the CHP Directive[9] - in the EU urging the construction of such plants, CCGTs are likely to take a dominant role in the addition of new generating capacities, on the demand centre side. The major draw-back is the necessity to be located next to high-voltage electrical infrastructure, which makes it highly likely that gas transport pipelines will be found in the vicinity, in such a case yielding preference to the GTC process. In Russia's case, if a CCGT plant may be located near a European border, then, receiving access to the grid, it may be possible to export electricity to Russia's neighbours. However, as most of Russia's gas is located thousands of kilometres from the borders and large distances from major residential or industrial areas, CCGT is not a viable option. Instead, it becomes as viable option only at the receiving end of a gas chain. As there is significant delay and uncertainty surrounding the nuclear power route in Europe and the UK, CCGT will play an ever increasing role.

The status quo of the industry has thus far been a dominating position of major conglomerates and IOCs that have been controlling the entire value chain. Although one of the main barriers to entry for new players has been the extremely capital intensive nature of such projects, the technical complexity of these large scale undertakings has also been limiting the ability of niche operators to enter the market. Nevertheless, even if these issues were to be overcome, the proprietary technology required to efficiently run these processes sit with a handful of licence holders. As such, companies like Shell, Sasol, ExxonMobil and Statoil control the GTL process, for example. Independent producers that have access to natural gas, have engineering expertise and access to capital (such as in Russia or Latin America) must work with these license holders to implement GTL projects. This often adds a difficult commercial angle to an otherwise difficult technological process.

Who Are the Top Gas Flarers?				
	OFFICIAL DATA for 2004		WHAT IMAGERY SHOWS for 2004	
	Country	Billion Cu. M.'s	Country	Billion Cu. M's
1	Nigeria	24.1	Russia	50.7
2	Russia	14.9	Nigeria	23.0
3	Iran	13.3	Iran	11.4
4	Iraq	8.6	Iraq	8.1
5	Angola	6.8	Kazakhstan	5.8
6	Venezuela	5.4	Algeria	5.5
7	Qatar	4.5	Angola	5.2
8	Algeria	4.3	Libya	4.2
9	Indonesia	3.7	Qatar	3.2
10	Eq. Guinea	3.6	Saudi Arabia	3.0
11	USA	2.8	China	2.9
12	Kuwait	2.7	Indonesia	2.9
13	Kazakhstan	2.7	Kuwait	2.6
14	Libya	2.5	Gabon	2.5
15	Azerbaijan	2.5	Oman	2.5
16	Mexico	1.5	North Sea	2.4
17	U.K.	1.6	Venezuela	2.1
18	Brazil	1.5	Uzbekistan	2.1
19	Gabon	1.4	Malaysia	1.7
20	Congo	1.2	Egypt	1.7

Source: World Bank, 2005

Figure 11.

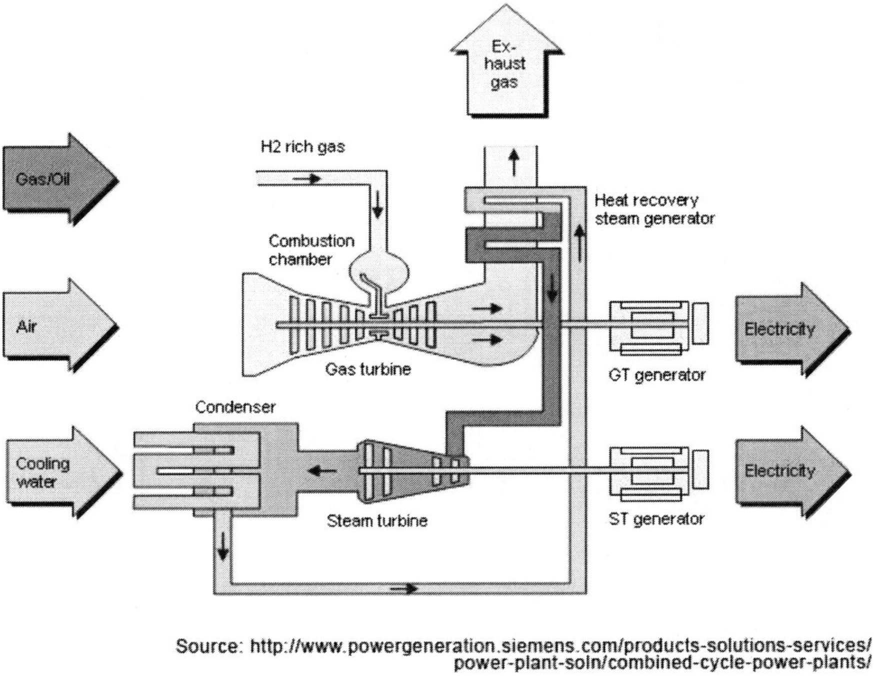

Source: Siemens, 2012

Figure 12.

4. NEW GASIFICATION MARKETS

One of the key ways in which the traditional model outlined above is changing is in the shift to boutique production – small and medium sized projects in which natural gas is used in situ to produce final, value added products. In the last five years, there has been a significant amount of research in reducing the size of gas conversion technology, from micro LNG developed in Australia to micro GTL being developed in America and Asia. This has been, in part, as a result of technological advancement in the field of materials, processing, catalyst and engineering. Two American firms have made significant process in showing the commercial viability of GTL processes without recourse to the proprietary technology of the majors. Rentech and Syntroleum have both developed technology which has seen application outside of the laboratory conditions. As a specific example,

Rentech, a medium sized US listed technology company, originally developed GTL technology as part of Texaco, and after a successful spin-out, remained as an independent developer. Whereas traditional GTL technology processes employ the use of cobalt catalysts with fixed bed reactors, Rentech has developed a way to use an iron based catalyst, which is seen as cheaper, and more efficient with a slurry bed reactor. Once capex costs are reduced, the economy of scale element becomes a secondary metric to reach required project rate of returns. Ultra Low Sulphur Diesel (ULSD) manufactured from Rentech production facilities in America has been used in vehicles and aircraft over the last ten years. The US Air Force has used Rentech GTL derived A1 Jet fuel in its aircraft, as part of its security of supply policy.

Boutique application means that stranded or semi-stranded gas reserves of a much smaller size can be successfully monetised. By decreasing required output from 100,000 bpd to 10,000 bpd or less for commercial production, fields of 5 BCM of recoverable reserves open up opportunity for GTL production. A huge market can be identified as CBM (Coal Bed Methane), where large coal deposits in areas such as China, Australia, Indonesia, Mongolia, Ukraine, can begin to utilise gas otherwise unable to reach a value generating market. The United States have significant CBM potential.

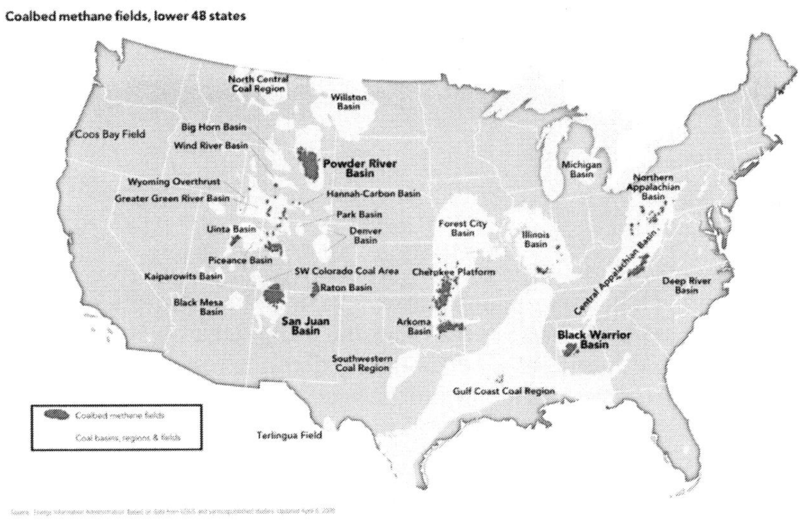

Source: Oil and Gas Journal, 2012

Figure 13.

In normal circumstances, small sized fields that are away from gas infrastructure means that they are stranded, meaning that investment required into pipeline construction render field development uneconomical. Small reactors allow for the production of liquid products that can be stored in canisters and transported by road or rail. For solid products, such as urea, it is possible to build up volumes in any port storage areas and then loaded on to larger vessels (typically 50,000+ tonnes).[10] - Urea market is fairly liquid, however, the price volatility means that feedstock costs must be fairly low in order to avoid risks of prolonged loss making.

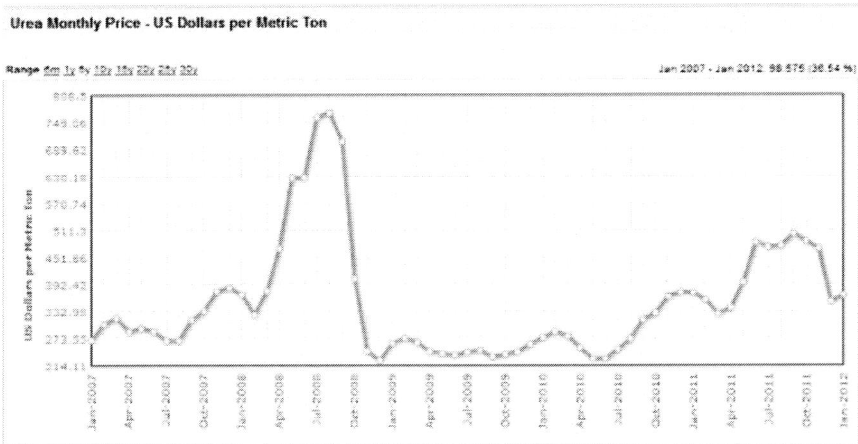

Source: Bloomberg, 2012

Figure 12.

One area which has seen significant attention is the gasification of biomass. This can be done on a micro-scale, meaning the most obvious applications are those of the municipal waste bodies and utilities or operators with large volumes of biomass waste products such as sawmills or sunflower oil producers. Some companies are claiming that they are able to achieve ULSD production in the volumes of 2,200 litres per day from a 10 tonne per day feedstock requirement. Woody biomass is gasified in a two-stage gasifier to produce Syngas with a 2:1 ration of hydrogen to carbon monoxide. This is then processed in a Fischer –Tropsch reactor to produce synthetic diesel fuel. Such small scale, modular application can reroute waste resources traditionally used to produce solid biomass fuels (pellets, briquettes, torrefied biomass etc) that can only be used in power or heat generation to liquid fuels that can be used in the transportation sector, either as blended additives, or for direct internal combustion. The key for a quick uptake of this technology is to reduce capex costs to a level where a 3 – 4 year simple payback can be achieved.

The advantage of the Biomass gasification is that it does not compete with food sources for feedstock, unlike traditional bio-fuels and hence is not party to significant pressure from political commentators and various pressure groups. By-products of agricultural production cycles, or forestry operations can increase efficiency and reduce transport / operations fleet carbon output.

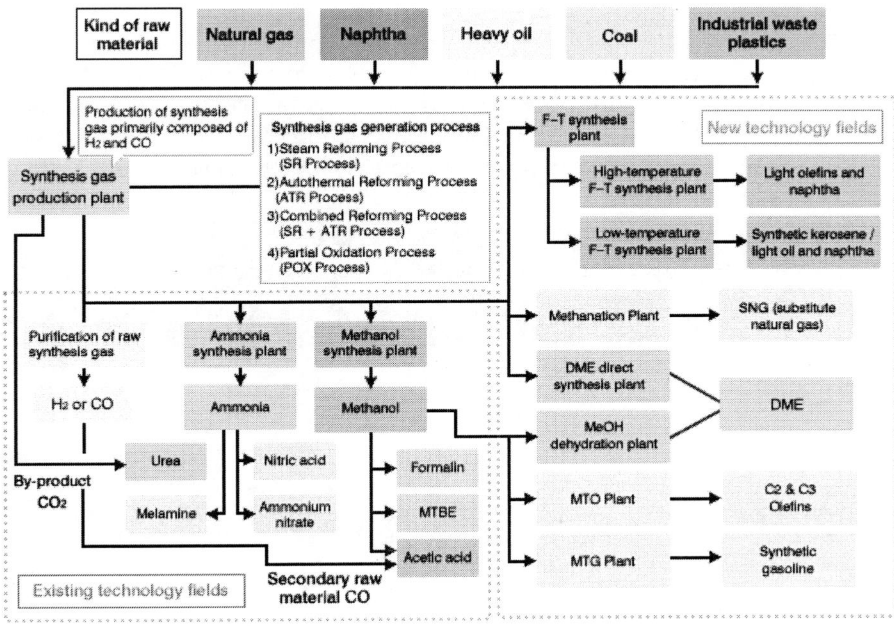

Source: Chiyoda, 2012

Figure 13.

5. ECONOMICS OF SMALL AND MID-SIZED GASIFICATION

The economics of boutique synthetic fuels production have recently shown similar parameters to those of major projects undertaken by IOCs / NOCs.

The author was directly involved in a feasibility study undertaken for a medium-sized GTL project, based on non-stranded gas in the Republic of Kazakhstan. The project economics were accepted by the contracted engineers and major, global investment banks focusing on natural resources.

The project entailed a facility with production of 120,000 metric tonnes per year of synthetic fuel (70% ULSD, 20% Naphtha, 10% kerosene), with a feedstock requirement of only 200 MMSCM of natural gas (dry, pipeline quality, high pressure) annually. This can be gas that is received from the gathering system of flared gas collections system, and directed to a processing facility or direct production or even pipeline gas. The price of gas was taken as US $2 per MMBTU.

Capital costs were considered at $150 million (which equates to c. 50,000 /bbl /day), with operational costs estimated at $7 / bbl. Although major operators are able to achieve a lower throughput costs, due to the super premium nature of ULSD and high conversion ratio, project profitability is more sensitive to capital costs. In this case, the project had a 4 year pay back and a 38% IRR.

Investment Analysis												
Discount Rate												*15.00%*
Transaction	Years											
	0	1	2	3	4	5	6	7	8	9	10	11
Cashflow from Operations	8,571	2,143	57,337	60,509	60,598	57,789	57,856	57,571	74,872	74,765	74,996	74,889
Investments	-75,000											
Net Cashflow	-66,429	2,143	57,337	60,509	60,598	57,789	57,856	57,571	74,872	74,765	74,996	74,889
Cumulative Cashflow	-66,429	-57,072	265	60,774	121,373	179,162	237,017	294,588	369,460	444,225	519,221	594,110
Discounted Cashflow	-66,429	1,863	43,355	39,786	34,647	28,732	25,013	21,643	24,476	21,253	18,538	16,097
Net Present Value	-66,429	-57,320	-13,965	25,821	60,468	89,199	114,212	135,855	160,331	181,584	200,122	216,218
Internal Rate of Return												37%

Source: Author, Midstream Energy Ltd, 2010

Figure 14.

The key for success of this project, and indeed any boutique application of gas conversion, is the ability to avoid "green-field" development. By placing new facilities on existing infrastructure, such as working refineries or old and abandoned heavy facilities, the capex figure can be kept to a manageable level to give satisfactory project returns. The Former Soviet Union (FSU), for example, has a large number of old chemical facilities that have ceased to operate and with a low cost of domestic gas, become good candidates for boutique GTL or GTC processing. One major advantage is the existence of transport infrastructure, both for the gas via pipeline and product via rail.

6. IMPACT ON POLICY

Ever since the major discoveries in the US of shale gas, new opportunities have arisen for application of "gas to" technologies. Observers have predicted that the US will become self-sufficient with respect to natural gas, and may become an exporter in the next decade. This has been further compounded by the recent permissions granted to Cheniere Energy for an LNG export terminal. It is incorrect to say that the US will become a net exporter, as it likely there will be imports of natural gas from Canada and some volumes of LNG from further afield. However, Shell has already announced that it is evaluating a large GTL project in the US. The key for such projects is the differential in price between natural gas and high-end products, a situation which reflects the current market in the US very well. Recent prices in the US (Henry Hub Futures) have been hovering at around the US $2 / MMBtu, whilst low sulphur diesel is currently trading at between USD $800 and $1000. If the price of crude oil continues to stay at or about $90+ per bbl, GTL projects become economically viable. This will also have a positive effect on Supply Security concerns, as the more transport fuels can be derived from domestic natural gas, the less dependence there is on oil imports.

When looking at other regions, there are similar advantages for China, as there are large opportunities in the near term for CBM gasification, and in the mid-term for shale gas development. China has announced significant finds of shale gas, and this can help to reduce dependence on oil imports. In fact, China is aware of the strategic disadvantage of having the bulk of the oil imports from the Middle East being shipped via the Malacca Strait. A well planned military operation can block this channel, effectively cutting China off from its oil flow.

African states, especially mature oil development areas such as Nigeria, have been unable to capitalise on the associated gas production, with various methods being undertaken to reduce gas flaring. In situ gas conversion, certainly in the first instance to power, and subsequently to fertiliser production, would be a coherent road map to develop the country's resources.

In Europe, there is less scope for this application, simply because due to liberalised markets, gas prices do not allow for economic production of other products, except for power generation and commercial and residential sectors. Furthermore, there is simply no spare capacity in the system to divert supplies from power and other sectors to gas processing. Economically, it makes more sense to produce in areas of low cost feedstock and deliver final products to the EU market.

7. FUTURE APPLICATIONS

One of the most advantages characteristics of synthetic fuels or more traditional gas processing products is the ability to utilise these in existing infrastructure without the need for a stock change. The biggest future growth will come from GTL, BTL and CTL processes and environmental concerns will play a role to increase the uptake of these fuels. As more stringent regulation places greater standards on reduced sulphur content in transportation fuels, more ULSD will be used as a blending fuel. Once the technological costs come down the cost curve, and producers will be incentivised to invest in direct GTL technology versus traditional deep refining, pressure will applied to the aviation industry to use synthetic fuels. Aviation is responsible for a major share of Green House Gas (GHG) emissions, and as such is a great potential consumer of synthetic fuels will come from this sector.

8. CONCLUSION

Natural gas is a versatile raw material that has traditionally been characterised by large complex infrastructure products, requiring full value chain integration. When not used as a fuel for power generation, natural gas has been an invaluable element in many household items and industrial chemicals. Due to the fact that supply and consumption centres have traditionally been separated by large distances, most natural gas projects required capital intensive pipeline construction. The financing of these required the mitigation of risks via long term offtake contracts. This was not the case in the Former Soviet Union, as government central planning directed investment and energy flows according to internal economic planning.

As a result, only large gas bearing basins were developed, with small fields either ignored or considered uneconomic for development. Oil reservoirs that contained a high gas-oil ratio were considered cumbersome in production areas where flaring was unacceptable, and in others where flaring was acceptable, natural gas remained as a nuisance.

With various advancements in technology, reduction in costs and improvements in technical knowhow, as well as economic and environmental conditions, there has been a focus on natural gas as the fuel of choice, ahead of crude oil, in most of the applications. This is likely to drive a trend where the growth in the consumption of gas will overtake oil in the long run, and perhaps become a major contributor to power, transportation and chemical sectors.

NOTES

1) The dynamics of funding such projects are very complex and are out of the scope of this article; however it is worth mentioning that domestic markets of major producers lack the hard currency to finance national champions whilst international capital markets generally shun away from risks associated with emerging market domestic consumption.

2) From the major producers, the UK North Sea production is in decline, Dutch production is capped, and Norwegian fields are in plateau (although there is heavy E&P activity).

3) Temporary relief was seen during the 2007-2011 financial crisis, as demand destruction allowed for a temporary shift from "sellers market" to "buyers market" and attention of the ministers was diverted to dealing with the financial economy and the failing UN Framework Convention on Climate Change.

4) Nord Stream and South Stream, for example.

5) Russian independent gas producers are prohibited from exporting natural gas by law (Federal Law "On Gas Export", 2006).

6) SassolChevron is in the process of building a 34,000 bpd plant in Nigeria.

7) This does not refer to inter-governemental price agreements but to private agreements between Gazpromexport and Ukrainian fertiliser producers.

8) A CCGT plant shows to have a conversion efficiency of 65%, as compared to a traditional gas turbine of 33%.

9) Directive on the promotion of cogeneration based on a useful heat demand in the internal energy market and amending Directive 92/62/EEC

10) This operation must be carried out in a fairly timely fashion, as urea has a tendency to degrade over time.

REFERENCES

1. Y. Grigoryev, 2007The Russian Gas Industry, Its Legal Structure, And Its Influence On World MarketsEnergy Law Journal28125 EOF146 EOF

2. Y. Grigoryev, 2007Today or not today: Deregulating the Russian gas sectorEnergy Policy3530363045

3. Z. Kang, T. Yundong, 2009The Status of World Shale Gas Resources Potential and Production Status as well as Development Prospect of China's Shale GasPetroleum & Petrochemical Today. 20093

4. Y. Grigoryev, 2012Future Of European Energy Security And Its Impact On Global Demand And Supply Centres. Proceedings World Gas Conference 2012, Kuala Lumpur.

5. Y. Grigoryev, 2012Challenges of the Russian regulatory framework and the monetisation of gas resources, Proceedings 2012 Russia & CIS Executive Summit, Downstream Oil and Gas, Dubai.

5

CHAPTER

SIMULATION OF OLIVE KERNEL GASIFICATION IN A BUBBLING FLUIDIZED BED PILOT SCALE REACTOR

Stavros Michailos, Anastasia Zabaniotou*

Department of Chemical Engineering, Aristotle University of Thessaloniki, Thessaloniki, Greece

ABSTRACT

The main aim of this study is to develop a comprehensive process model for biomass gasification in a pilot scale bubbling fluidized bed gasifier using the ASPEN PLUS simulator. A drawback in using ASPEN PLUS is the lack of a library model to simulate fluidized bed unit operation. However, it is possible for users to input their own models, using FORTRAN codes nested within the ASPEN PLUS input file, to simulate operation of a fluidized bed. The products of homogeneous reactions are defined by Gibbs equilibrium and reaction rate kinetics are used to determine the products of char gasification. Governing hydrodynamic equations for a bubbling bed and kinetic expressions for the char combustion were adopted from the literature. Different sets of gasification results for the operation conditions (temperature and air equivalence ratio (ER)) obtained from the our pilot-scale gasifier having a capacity of 1 kg/hr of olive kernel as feeding biomass, were used to demonstrate the validation of the model. The simulation results received from the application of the model were compared with the above experimental results and showed good agreement.

Keywords: Fluidised Bed Gasification; Biomass; Olive Kernel; Simulation; Aspen Plus

1. INTRODUCTION

The traditional approach necessary to establish comercial plant technology is based on comprehensive experimental investigations, progressing from a laboratory scale test unit to a pilot scale plant, before building a full-scale commercial demonstration plant. For process optimisation, an extensive investigation of the plant behaviour depending on various operating parameters is required for each scale up step. To support this optimisation procedure, mathematical models are helpful to reduce the temporal and financial efforts.

Pre-condition is a reliable simulation tool, which includes the mathematical formulation of all important chemical and physical processes by describing their dependency on operating parameters and their interdependencies [1]. However, only limited research has been performed to model those complex systems including also the hydrodynamics of biomass particles.

The development of numerical models for fluidizedbed gasification (FBG) documented in the literature [2] are devoted mainly to coal and less to biomass. Even though, biomass in comparison to coal is made up not only of lignin but also of cellulose and hemi cellulose, each one having its own thermal behaviour and making the modelling more difficult, the modelling approaches used for coal can serve for biomass gasification modelling as well. Ma et al. [3] developed a steady-state model which involved instantaneous devolatilization of coal at the top of the gasifier (freeboard region) and char combustion and gasification in the fluidized bed. Chejne and Hernandez [4] developed a comprehensive mathematical model to predict the behaviour of coal combustion and gasification on stacks in non-stationary operation. Sett and Bhattacharya [5] developed a mathematical model for the behaviour of a fluidised-bed charcoal gasifier, taking into account fluidised-bed hydrodynamic conditions, chemical reaction rates, mass diffusion rates, and the conservation of mass and energy. The model was solved numerically using an IBM 3083 mainframe computer. Ross et al. [6] modified their one-dimensional numerical isothermal model of a fluidised-bed coal gasifier in order to simulate the performance of a laboratory-scale gasifier.

A new numerical model based on the two-fluid model (TFM) including the kinetic theory of granular flow (KTGF) and complicated reactions has been developed to simulate coal gasification in a bubbling fluidized bed gasifier (BFBG) [7]. The collision between particles was described by KTGF. The coal gasification rates were determined by combining Arrhenius rate and diffusion rate for heterogeneous reactions or turbulent mixing rate for homogeneous reactions. The flow behaviours of gas and solid phases in the bed and freeboard could be predicted, which were not easy to be measured through the experiments. Hamel and Krumm [2] have developed a mathematical model for simulation of gasification processes of solid fuels in atmospheric or pressurised bubbling fluidised beds incorporating bed and freeboard hydrodynamics, fuel drying and devolatilization, and chemical reaction kinetics is presented. The model has been used to simulate four bubbling fluidised bed gasifiers, described in literature, of different scales from atmospheric laboratory scale up to pressurised commercial scale, processing brown coal, peat and sawdust. The gasifiers have been operated within a wide range of parameters using air, air steam or oxygen steam as gasification agent, operating with or without recirculation of fines at operating pressures up to 2.5 MPa. The simulation results for overall carbon conversion, temperature and concentrations of gaseous species agreed sufficiently well with published experimental data.

Corella and Sanz [8,9], at the University of Zaragoza and Madrid (Spain) started to study the modelling of fluidized bed biomass gasifiers in the mid-1980s. More recently Corella et al. [10], discussed the reaction network existing in a CFB biomass gasifier and the problems associated with the accuracy of the kinetic equations needed for the existing complex reaction network. Furthermore, he presented a model for bubbling fluidized bed (BFB) biomass gasifiers, gasifying with pure steam [10]. That model identified the four main, for modelling purposes, chemical reactions among the reaction network existing in the gasification process. With only four kinetic parameters, the model predicted quite well the BFB gasifier. More recently, Corella and Sanz [8,9] have presented a whole model for CFBBGs. Such model is 1-dimensional for steady state. The model has a semi rigorous character because of the assumptions that had to be introduced by lack of accurate knowledge of some parts of the modelling.

De Souza-Santos [11] developed a comprehensive mathematical model and commercially available computer program performing a comprehensive simulation of fluidized-bed equipment (CSFB Version 3.5), to use as a tool for engineering design and operation optimisation, by predicting the behaviour of a real unit during steadystate operation. His model is regarded as complete and it includes the conservation equations for the emulsion phase and bubbles, empirical equations for hydrodynamics, and it is also includes a through mass balance which considers that both drying and volatilisation are not instantaneous. Jiang and Morey [12] developed one-dimensional, steady state, numerical model for a fluidized bed biomass gasifier. The gasifier model consisted of a fuel pyrolysis model, an oxidation model, a gasification model and a freeboard model which were validated with experimental results.

Haggerty and Pulsifer [13] used the reaction model together with three different reactor models (a) plug-flow, (b) complete-mixing and (c) bubble-assemblage, where the bubble-assemblage model represents a valid alternative when modelling fluidized-bed gasifiers. Marias et al. [14] developed a mathematical model for the fluidizedbed incineration process using the waste composed of wood, cardboard and polyvinyl chloride. Aarsen's model treated an isothermal fluidized bed as two separate zones, the oxidation zone at the bottom and the gasification zone at the top [15]. Fuel pyrolysis took place at such a fast rate that only pyrolysis yields, which were assumed to vary with bed temperature, were needed. Mukosei et al. [16] examined the problem of the mathematical simulation of heterogeneous processes in a fluidized-bed reactor. These works and other more recent ones such as the work by He and Rudolph [17] proposed a new approach to the modelling of gross gas-solids flow through the riser in a circulating fluidized-bed system. This approach differs from the previous ones, which are found to be theoretically incorrect based on a fundamental analysis of the riser process hydrodynamics.

Panopoulos et al. [18] used the simple way of approaching the biomass gasification modelling by predicting thermodynamic equilibrium composition through Gibbs free energy minimisation calculations for the C, H, and O atoms of the fuel and the gasification agent mixture. Samuilov et al. [19] developed a mathematical model for the gasification in carbon dioxide of a single carbon particle. The porous structure of the particle, diffusion processes, the gasification processes on the pore surface according to the Langmuir-Hinshelwood model, and reactions on active carbon sites were taken into account. Deen et al. [20] reviewed the use of discrete particle models (DPMs) for the study of the flow phenomena prevailing in fluidized beds.

The ASPEN PLUS process simulator has been used by different investigators to simulate coal conversion; examples include methanol synthesis [21,22], indirect coal liquefaction processes [23], integrated coal gasification combined cycle (IGCC) power plants [24], atmospheric fluidized bed combustor processes [25], compartmented fluidized bed coal gasifiers [26], coal hydrogasification processes [27], and coal gasification simulation [28].

However, the work that has been done on biomass gasification is limited. Mansaray et al. [29] used ASPEN PLUS to simulate rice husk gasification based on material balance, energy balance, and chemical equilibrium relations. Because of the high amount of volatile material in biomass and the complexity of biomass reaction rate kinetics in fluidized beds, they ignored the char gasification and simulated the gasification process by the assumption that biomass gasification follows Gibbs equilibrium. In a typical atmospheric fluidized bed gasifier, feed, together with bed material, are fluidized by the gasifying agents, such as air and/or steam, entering at the bottom of the bed. The product gas resulting from the gasification process is fed to a gas-solid separator (i.e., cyclone) to separate solid particles carried by exhaust gas.

2. MODELLING APPROACH

Mathematical modelling is an alternative means to study fluidisation and fluidised bed biomass gasification. Modelling can not only account for the fundamental hydrodynamic behaviour of fluidisation and the coal gasification process, but also serve as a predictive tool to assist with the design, optimisation and scale-up of fluidised bed gasifiers. Therefore, in recent years mathematical modelling of fluidised beds has increasingly attracted more attention.

2.1. Assumptions

The following assumptions were considered in modelling the gasification process:

- Process is steady state and isothermal.
- Biomass devolatilization takes place instantaneously and volatile products mainly consist of H_2, CO, CO_2, CH_4, and H_2O.
- All the gases are uniformly distributed within the emulsion phase.
- Particles are spherical and of uniform size and the average diameter remains constant during the gasification ,based on the shrinking core model
- Char only contains carbon and ash.
- Reactions of N and S have not been taken account
- The reactions reached chemical equilibrium

2.2. Reactions

2.2.1. Combustion Zone

The combustible substance of a solid fuel is usually composed of elements carbon, hydrogen and oxygen. In complete combustion carbon dioxide is obtained from carbon in fuel and water is obtained from the hydrogen, usually as steam. The combustion reaction is exothermic and yields a theoretical oxidation temperature of 1450°C. The main reactions, therefore, are:

$$C + O_2 = CO_2\,(+393\ \text{MJ/kg mole}) \tag{1}$$

$$2H_2 + O_2 = 2H_2O\,(-242\ \text{MJ/kg mole}) \tag{2}$$

2.2.2. Reduction Zone

The products of partial combustion (water, carbon dioxide and uncombusted partially cracked pyrolysis products) now pass through a red-hot charcoal bed where the following reduction reactions take place.

$$C + CO_2 = 2CO\,(-164.9\ \text{MJ/kg mole}) \tag{3}$$

$$C + H_2O = CO + H_2\,(-122.6\ \text{MJ/kg mole}) \tag{4}$$

$$CO + H_2O = CO_2 + H_2\,(+42\ \text{MJ/kg mole}) \tag{5}$$

$$C + 2H_2 = CH_4\,(+75\ \text{MJ/kg mole}) \tag{6}$$

$$CO_2 + H_2 = CO + H_2O\,(-42.3\ \text{MJ/kg mole}) \tag{7}$$

Reactions (3) and (4) are main reduction reactions and being endothermic have the capability of reducing gas temperature. Consequently the temperatures in the reduction zone are normally 800°C - 1000°C. Lower the reduction zone temperature (~ 700°C - 800°C), lower is the calorific value of gas.

2.2.3. Pyrolysis Zone

Wood pyrolysis is an intricate process that is still not completely understood 14. The products depend upon temperature, pressure, residence time and heat losses. However following general remarks can be made about them. Up to the temperature of 200°C only water is driven off. Between 200°C to 280°C carbon dioxide, acetic acid and water are given off. The real pyrolysis, which takes place between 280°C to 500°C, produces large quantities of tar and gases containing carbon dioxide. Besides light tars, some methyl alcohol is also formed. Between 500°C to 700°C the gas production is small and contains hydrogen. Thus it is easy to see that updraft gasifier will produce much more tar than downdraft one. In downdraft gasifier the tars have to go through combustion and reduction zone and are partially broken down.

2.3. Hydrodynamic Parameters

The following assumptions were made in simulating the hydrodynamics:

- Fluidized bed reactors divided into two regions bed and freeboard
- The fluidization state in the bed is maintained in the bubbling regime
- The volume fraction of solids decreases as height increases, corresponding to the coalescence of bubbles in the bed and the returning of solid particles to the bed in the TDH zone
- Volumetric flow rate of gas increases along with height, corresponding to the production of gaseous products
- The mixing of solid particles, consisting of ash, char particles, and bed material, is perfect
- The reactor is divided into a finite number of equal elements with constant hydrodynamic parameters
- The fluidized bed is one-dimensional; any variations in conditions are considered to occur only in the axial direction.

2.4. Global Model of Fluidized Bed Gasifier

The proposed gasification dynamic model represents two stages: instantaneous devolatilization of straw and combustion of the char at the bottom of the gasifier and the gasification in the fluidized bed. Thus, a two-phase representation of the fluidized bed incorporates the phenomena of jetting, bubbling, slugging and dynamic mass and energy balances. Figure 1 shows the fluidized bed domain divided into three regions: jet, bubbles and slugs. The following assumptions were made regarding to the gas low divisions and the low patterns of gas and solids:

1) the fluidized bed consists of a dilute phase (jets, bubbles and = or slugs) and an emulsion phase2) the emulsion phase is divided into an interstitial gas phase and a solid phase 3) mass and heat exchange take place between

the dilute phase and interstitial gas and between the interstitial gas and the solids4) the fluidizing gas enters the bed through nozzles in a jet form. The jets degenerate into bubbles, which rise through the bed and grow by coalescence with other bubbles to form slugs5) slugging occurs if the bubble diameter becomes larger than one third of the reactor diameter, (Slugging occurs in improper fluidization 6) the dilute gas and the interstitial gas are in plug low and the dilute gas is free of solids7) the gas behaves ideally, and 8) the produced gas consist of CO, CO_2, H_2, H_2O, CH_4 and N_2.

2.5. Mass and Energy Balances

The olive kernel feed flow is calculated by the following equation [30]:

$$n = E_{th} / (F_b \times HHV) \tag{8}$$

For the calculation of the air that is demanded for the stoichiometric combustion of biomass is used [30]:

$$CH_aO_bN_cS_d + eO_2 \rightarrow CO_2 + a/2\,H_2O + cNO + dSO_2 \tag{9}$$

where the coefficients a, b, c, d represent the molecular fractions H/C, O/C, N/C and S/C . Therefore we can calculate the air flow for several values of equivalence ratio (ER) according to previous equation [30]

With the assumption that gasification char is consisted of fixed carbon and ash, is calculated the solid mass flow from the equation [30]:

$$F_s = (\%C_f + \%ash) \times F_{b,dry} \tag{10}$$

Then the overall mass balance is formed for the gasification reactor and the mass flow for the gas gasification is estimated as following [30]:

$$F_{th} = F_b + F_{air,gasif} - F_s \tag{11}$$

Correspondingly the overall energy balance is described as below [30]:

$$E_b + E_{air,gasif} + E_{ext} = E_{th} + E_{loss} \tag{12}$$

Es represents the thermal content of biomass feed flow which is estimated from the following equation [30]:

$$E_s = E_{Fb} \times LHV / 3600 \tag{13}$$

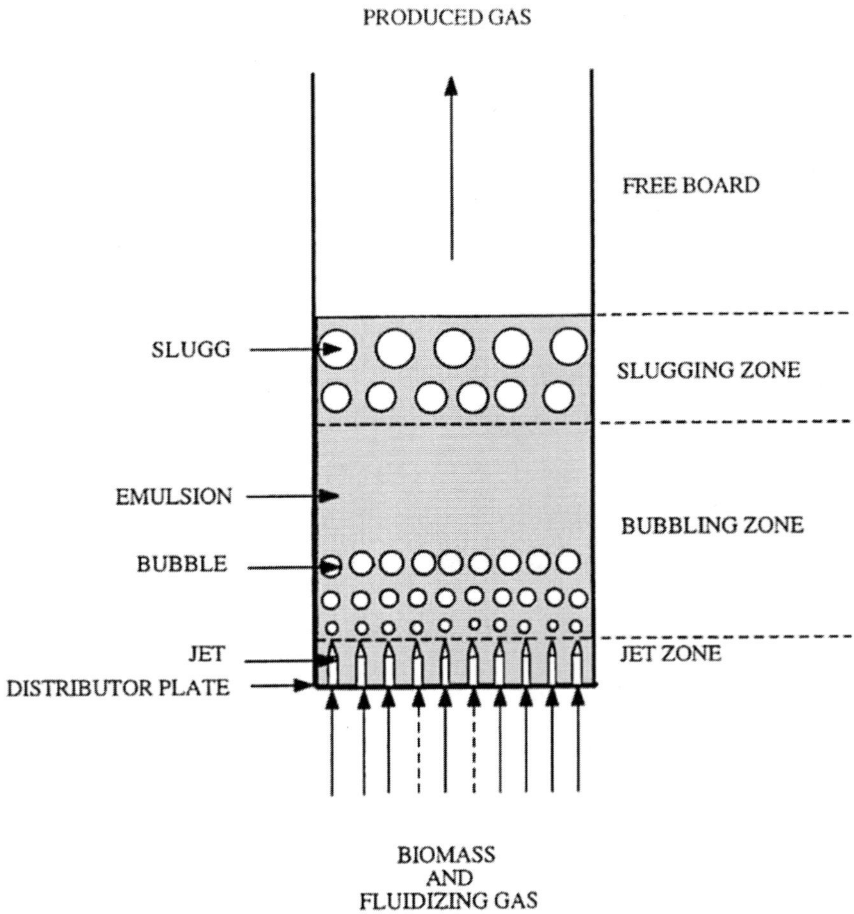

Figure 1. The fluidized bed gasifier zones.

For the air it is assumed that enters the reactor at environmental temperature, so its enthalpy content is considered zero (Eair, gasif = 0) .

Eth represents the thermal energy and the enthalpy of gas product and it is given by the following equation [30]:

$$E_{th} = E_u + E_s \tag{14}$$

where

$$E_u = \left(m_{gas} \times LHV_{gas}\right) / \left(3.6 \times \rho_{gas}\right) \tag{15}$$

The following equation estimates the enthalpy of each component of gas gasification in gasification temperature

$$E_s = \left[m_{gas} \times \sum(y_i h_i)\right] / \left[3600 \times \sum(y_i MB_i)\right] \tag{16}$$

Finally Eloss represents the energy losses from the system due to ash and char removal and the losses to environment from the walls of the gasifier. Therefore:

$$E_{loss} = E_w + E_{char} \tag{17}$$

It is assumed that the energy losses from the gasifier walls constitute the 10% of the overall losses. The energy losses due to char gasification removal is calculated as below [30]:

$$E_{loss} = (\%C_{fixed}) \times m_{char} \times (LHV_{gas} + h_{char}) / 3600 \tag{18}$$

2.6. Air Equivalence Ratio

Equivalence ratio (ER) constitutes a significant factor of biomass gasification. The wise choice of it discourages the propulsion of the oxidation reactions. It is defined by the following equation [31]:

$$ER = A_\pi \,(\text{kg/kg biomass}) / A_\theta \,(\text{kg/kg biomass}) \tag{19}$$

A_π represents the mass ratio between air and biomass that has been used in laboratory and A_θ represents the stoichiometrically demanded mass ratio of air for the total combustion of the same quantity of biomass.

The alteration of air flow effects the process since it leads it to the two terminal situations:

1) High air equivalence ratio ($ER > 1$) approaches the process of combustion giving as gasification product mainly CO_2

2) Low air equivalence ratio ($ER < 0.25$) benefits the process of pyrolysis giving primarily as product syngas ($CO + H_2$).

2.7. Producer Gas LHV

Temperature and ER also significantly affect the heating value of producer gas. Main research target is the production of a producer gas enriched in

CO, H_2, and CH_4. The presence of these combustible species contributes to the production of a medium to high heating value gas, suitable for further exploitation in internal combustion engines (ICE) and turbines for power production. LHV calculation was made using the following equation [32]:

$$LHV_{gas} = \frac{(30CO + 25,7H_2 + 85,4CH_4) \times 4,2}{1000} \, MJ/Nm^3 \quad (20)$$

where, CO, H_2 and CH_4 are the molar ratios of the species in the producer gas as measured in our gasification experiments.

3. EXPERIMENTAL

3.1. Biomass Characteristics

In the laboratory olive kernel was used as biomass feeding and its characteristics are illustrated in the Table 1. Olive kernel is the final solid residue emerging from olive oil production industries. Olives are processed in an oil extracting plant to recover the oil content. After a first residue drying, residual oil is hexane extracted generating a residual solid product with a moisture content of around 10 - 12 wt% called olive kernel. Traditionally, such kind of residue is sold as fuel for combustion in small boilers and specially designed furnaces due to its significant calorific value (HHV ~21 MJ/kg).

Nowadays, there are several boilers fuelled with olive kernels destined for residential and greenhouse heating or in some cases to generate steam for electricity production in conventional steam turbines. However, a more efficient and environmentally friendly alternative with respect to CO_2 and CH_4 mitigation to the atmosphere, is olive kernel gasification; aim is to produce a high calorific value gas to fuel a gas engine or even a gas turbine in a combined cycle of heat/power production.

3.2. Gasifier Characteristics

3.2.1. The Bench Scale FBG Unit

A 5 kW_{th}, bench scale, bubbling fluidized bed gasifier (FBG) was designed and manufactured for biogenic material gasification. It is an atmospheric air blown FBG unit that consists of four sections: 1) the biomass feeding section, 2) the air supply, control and preheating section, 3) the gasification facility, 4) the gas sampling and off line analysis section.

Producer gas was sampled using airtight gas sampling bags and analyzed with a GC system for quantitative determination of CO, CO_2, H_2, CH_4, C_2H_4 and C_2H_6.

3.2.2. The Olive Kernel Feeding Section

The FBG unit is equipped with a feeding hopper suitable for feeding of low density biomass. Feeding takes place through a system of a constant speed screw feeder and a variable speed screw feeder (motorized via inverter), separated by a rotary valve in order to avoid the hot gas backflow that could pyrolyse olive kernels and cause serious tar fouling problems. Additionally, a small proportion of fluidization air (~1 l/min) is supplied for backflow prevention. The motorized screw feeder is able to control the biomass feeding, based on a series of calibration curves and the maximum mass flow rate of olive kernel could reach 120 g/min. Olive kernel pass through the metal to metal rotary valve and is introduced into the reactor by the second screw feeder, into the hot zone of the reactor, 120 mm above the air distributor.

Table 1. Olive kernel characteristics.

Proximate analysis (wt%)	
Moisture content	12.30
Fixed carbon	17.18
Volatile matter	79.90
Ash	3.63
Ultimate analysis (wt%)	
C	48.59
H	5.73
N	1.57
O	40.65
HHV (MJ/kg)	20.46
LHV (MJ/kg)	19.20
Flow rate (kg/hr)	1.00

3.2.3. The Air Supply, Control and Preheating Section

Laboratory's air supply unit provides the FBG unit with the primary and secondary air, necessary for the fluidization and gasification process, always in hypo-stoichiometric amounts obtaining the desired ER ratio. It consists of a central compressor, a main valve to control the air flow rate and an air flowmeter with maximum flow capacity of 200 l/min. The primary air for gasification process is introduced from the bottom of the FBG and a uniformly perforated plate (151 holes of 1mm diameter) is used in order to retain bed fluidization. Primary air is preheated up to 350°C, by passing through a stainless steel tube that is kept in touch with the hot zone of the reactor inside the furnace. In the beginning of each operation, while the reactor is still cold, air preheating is obtained via an electric resistance placed at the bottom of the FBG. Secondary air contributes in producer gas temperature increase, which results in further tar cracking improving carbon conversion efficiency.

3.2.4. The 5 kW_{th} FBG Test Unit

Figure 2 shows the bench scale FBG reactor with nominal capacity of 5 kW_{th}. Bed and freeboard sections are made of stainless steel with diameters of 60 mm and 90 mm respectively; the total reactor height is 1400 mm. A more detailed analysis about the reactor design and construction can be found in a previous published work [33]. FBG design enables the easy dismantling and thorough cleaning, in order to avoid not only memory effects, but also possible tar fouling under the hot flow experimental conditions. The gas cleaning system consists of a cyclone of 10 μm cut point. Limited experimental work has done using high temperature ceramic filters and various types of metallic foam for particulate abatement and tar elimination.

3.2.5. Bed Material

The bed material is of great importance in a fluidised bed reactor as it plays a dual role acting as heat transfer medium and, if it is not an inert material, as an internal catalyst for extended tar cracking.

A major problem of gasification applicability is tar formation and its minimisation is seen as one of the greatest technical challenges to overcome for the successful development of commercially attractive gasification technologies. According to the literature, tar production in an air steam gasification process could be as low as ~20 g/Nm^3 [31].

Tar treatment technologies are categorised either as upstream, when the tar problem is tackled inside the gasifier (named also primary or internal methods) or downstream, when tar is cleaned in separate hot gas devices (named also secondary or external methods). Although the secondary methods are proven to be effective and fully controlled, primary treatment technologies are gaining more interest due to their economic benefits. In primary methods operating parameters such as temperature, gasifying agent, bed material, equivalence ratio and residence time can play an important role on the formation and decomposition of tars. Cao et al. [31] concluded that the utilization of some catalysts in the gasifier and the concepts of two stage gasification are of prime importance. They worked on a laboratory scale, two-region, combined fluidised bed reactor for air gasification of sawdust producing a H_2 rich gas, with an LHV of 5 MJ/Nm^3 and tar concentration below 10 mg/Nm^3.

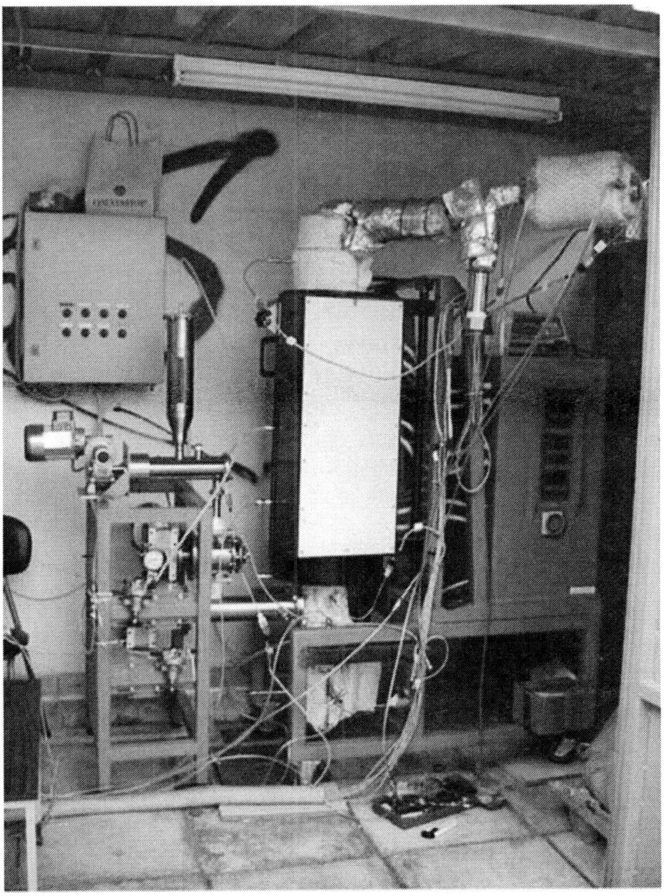

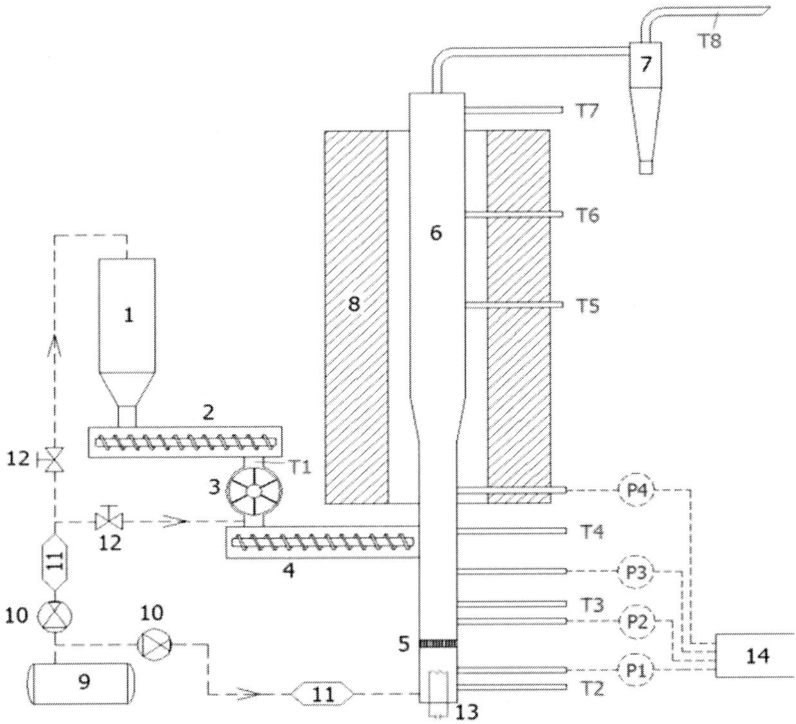

Figure 2. The 5 kWth, bench scale FBG unit, 1-biomass hopper; 2-screw feeder (inverter); 3-rotary valve; 4-screw feeder; 5-air distributor; 6-fluidized bed reactor; 7-cyclone; 8- electric furnace; 9-air compressor; 10-pressure regulator; 11-air flow meter; 12-isolation valve; 13-electrical heater; 14-differential pressure transmitter; P1 to P4 pressure measurement; T1 to T8 temperature measurement.

It is stated that several in-gasifier materials have catalytic action and eliminate in-bed tar by activating reforming reactions. Such materials modify the kinetic constants of most catalytic reactions involved in gasification. The in-bed additives are classified into two categories: 1) laboratory scale catalysts, which are very active but expensive; those catalysts are deactivated by coking when used at high (above 1000 kg/hm^2 cross sectional area) biomass throughputs (realistic for commercial applications), 2) natural occurring solids such as dolomite, limestone, olivine, high iron-content solids, which have been proved to have only a relatively small tar-elimination activity.

Nordgreen et al. [34] used metallic iron and iron oxides to catalytically crack tars in a secondary reactor, downstream the atmospheric fluidized bed gasifier fuelled with Swedish birch. They worked at a temperature range of 700°C - 900°C and low equivalence ratio almost near to pyrolysis conditions ($0 < ER < 0.2$), concluding that catalytic breakdown of the tar reached 100%.

The first set of experiments was carried out with quartz sand (Table 2) as bed material, in the size range: 500 - 425 μm. Although quartz sand was an easy and cheap to find material, it caused severe defluidization due to inevitable agglomeration and tar formation tendency at relatively low gasification temperature ($T < 800$°C). Thus, and taking into account that an iron containing bed material might act as a catalyst for tar destruction, quartz sand was replaced with olivine (Table 2). A 500 - 425 μm size fraction of olivine was also used and it was found to be durable for a prolongued operating period and in temperature range 800°C - 950°C.

3.2.6. Producer Gas Sampling and Analysis

The sampling line is kept at a constant temperature of ~300°C in order to avoid tar condensation and sampling line fouling. Tar condensation and particle removal takes place in the gas cleaning section (Figure 3) consisting of a water washing, moisture trap, impinger bottles with isopropanole, a fiber filter and a silica gel filter.

When the appropriate gasification conditions are achieved producer gas is sampled, using a membrane pump and airtight gas sampling bags and analyzed in laboratory's GC system (Model 6890N, Agilent Technologies equipped with FID and TCD detectors connected in series). The Gas Chromatograph is fitted with two columns HP-Plot Q with helium as carrier gas. GC's temperature profile was an isothermal at 50°C and the retention time of the analysis process was 38 min. The standard gas mixture used.

GC calibration composed from CO, CO_2, H_2, CH_4, C_2H_4 and C_2H_6 1 vv% of each one of the above components balanced in helium.

Table 2. Comparison of physicochemical characteristics of commercial quartz sand and olivine.

Physical characteristics		
Bed material	Silica sand	Olivine
Bed material solid density	2650 kg/m^3	3300 kg/m^3
Bed material bulk density	1700 kg/m^3	1900 kg/m^3
Chemical characteristics		
SiO_2 [%wt]	99.5	43.5%
Fe_2O_3 [%wt]	0.10	7.5% - 8.2%
Al_2O_3 [%wt]	0.10	0.3%
CaO [%wt]	0.03	0.4%
MgO [%wt]	*n.a.*	46% - 48%

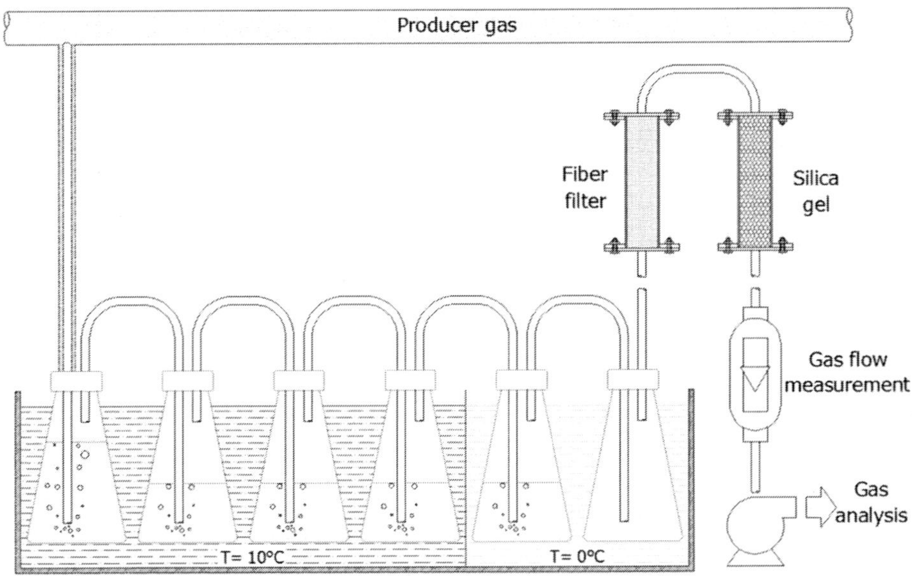

Figure 3. Producer gas sampling section.

4. SIMULATION WITH ASPEN PLUS

4.1. Simulation Software

ASPEN Plus was selected for modelling the gasifier. It is a steady state chemical process simulator, which was developed to evaluate synthetic fuel technologies. It uses unit operation blocks, which are models of specific process operations (reactors heaters, pumps etc.). The user places these blocks on a flowsheet, specifying material and energy streams. An extensive built in physical properties database is used for the simulation calculations. ASPEN Plus has the capability to incorporate Fortran code into the model. This feature is utilised for the definition of non-conventional fuels, e.g. biomass, specific coals and for ensuring the system operates within user defined limits and constraints. The development of a model in ASPEN Plus involves the following steps:

1) Stream class specification and property method selection.
2) System component specification (from databank).
3) Defining the process flowsheet (unit operation blocks and connecting material and energy streams).
4) Specifying feed streams (flow rate, composition, and thermodynamic condition).
5) Specifying unit operation blocks (thermodynamic condition, chemical reactions etc.).

4.2. Flow Sheet Development

ASPEN PLUS blocks that have been used in order to simulate the gasification process are described in Table 3. Figure 4 shows the flowsheet for the gasification process.

4.2.1. Drying

The ASPEN PLUS stoichiometric reactor, RSTOIC, was used to simulate the process of biomass drying. In this step biomass is separated from moisture. A SEPARATION COLUMN model was used before the RYIELD reactor to separate the moisture from dried biomass. Within the ASPEN PLUS environment, the separation column is the most appropriate unit operation to achieve this goal. The amount of volatile material can be specified from the biomass approximate analysis.

Table 3. Reactor blocks description utilized in the simulation.

Reactor block	Description
RYIELD	Models a reactor by specifying reaction yields of each component. This model is useful when reaction stoichiometry and kinetics are unknown and yield distribution data or correlations are available
RGIBBS	Models single-phase chemical equilibrium, or simultaneous phase and chemical equilibrium by minimizing Gibbs free energy, subject to atom balance constraints. This model is useful when temperature and pressure are known and reaction stoichiometry is unknown
RPFR	Models a plug flow reactor. This model is useful when reaction kinetics is known. This model is useful when solids, such as char, are participating in the reactions
RSTOIC	Models stoichiometric reactor with specified reaction extent or conversion. Reactors where reaction kinetics are unknown or unimportant but stoichiometry and extent of reaction are known.

4.2.2. Biomass Decomposition

The ASPEN PLUS yield reactor, RYIELD, was used to simulate the decomposition of the feed. In this step, biomass is converted into its constituting components including carbon, hydrogen, oxygen, sulfur, nitrogen, and ash, by specifying the yield distribution according to the biomass ultimate analysis.

4.2.3. Volatile Reactions

The ASPEN PLUS Gibbs reactor, RGIBBS, was used for volatile combustion, in conformity with the assumption that volatile reactions follow the Gibbs equilibrium. Biomass consists of mainly C, H, N, O, S, and ash. Carbon will partly constitute the gas phase, which takes part in devolatilization, and the remaining carbon comprises part of the solid phase (char) and subsequently results in char gasification. A COMPONENT SEPARATOR model was used before the RGIBBS reactor to separate S and N within the ASPEN PLUS environment, the component separator is the most appropriate unit operation to achieve this goal. The amount of volatile material can be specified from the biomass approximate analysis. The MIXER block was used before RGIBBS so as to add air to gas mixture.

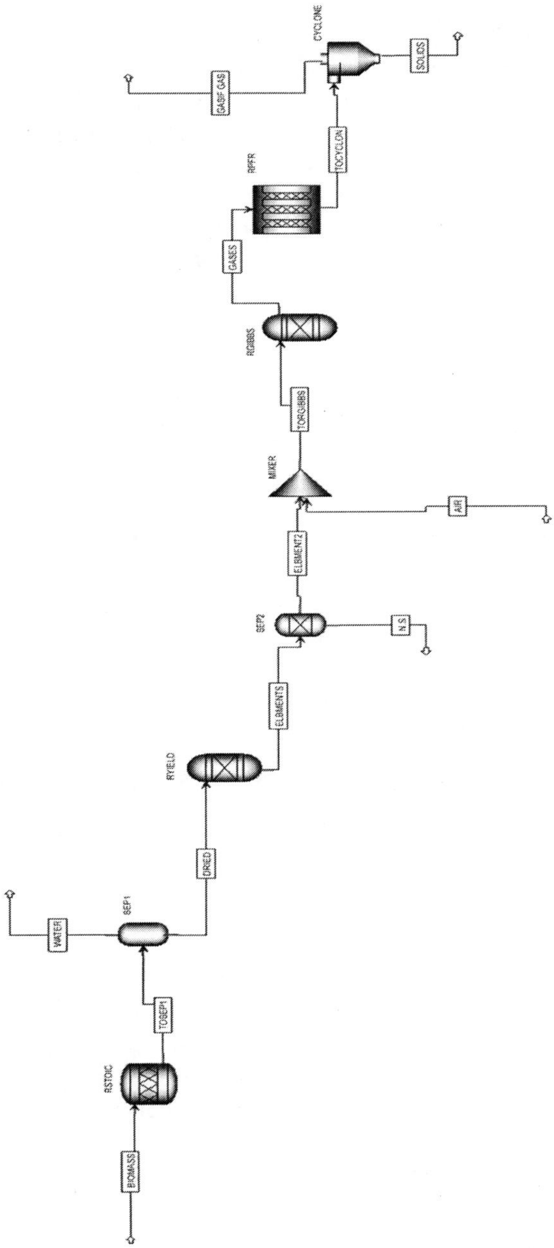

Figure 4. Comprehensive simulation diagram for the fluidized bed gasification process.

4.2.4. Char Gasification

The ASPEN PLUS PFR reactor, RPFR, performs char gasification by including gasification reactions and chemical kinetics, as mentioned previously. The hydrodynamic and kinetic parameters, such as superficial velocity and voidage are constant in these reactors. Also considering the assumption that char contains only carbon and ash, the amount of carbon in the volatile portion can be calculated by deducting the total amount of carbon in char from the total carbon in biomass.

4.2.5. Solid Separation

A CYCLONE SEPARATOR block was used in this model in order to distinguish carbon solid from the gas mixture that it is received as final product. A CYCLONE SEPARATOR block in ASPEN PLUS environment separates solids from gas using gas vortex in a cyclone and it is used for rating and sizing cyclones.

5. RESULTS

5.1. Model Validation

In order to validate the simulation results, experimental data from gasification of olive kernel in a pilot-scale fluidized bed gasifier was used. Biomass produces more tar and unburned hydrocarbon in lower temperatures, which decreases hydrogen production. The error related to the prediction of hydrogen is the result of ignoring tar production in the simulation.

A model was developed for the gasification of biomass in a bubbling fluidized bed gasifier using the ASPEN PLUS simulator. To provide the model, several ASPEN PLUS unit operation blocks were combined and, where necessary, kinetic expressions and hydrodynamic models were developed using data and models from the literature. The model was used to predict the results of pilot-scale gasification of olive kernel with air. The simulation results for the product gas composition versus temperature and air equivalence ratio were compared with experimental results.

Sensitivity analysis is performed to monitor the dependence of different parameters on the composition of the produced gas from the reactor. In this work, the temperature and air equivalence ratio (ER) are analysed in order to evaluate their effects on the composition of the produced gas and on the

LHV of gasification gas. The optimum operating conditions according to simulation results is ER = 0.2 and T = 750°C. During the sensitivity analysis the model input data was kept the same as for model validation with one parameter being varied at any given time Simulation results were compared with all sets of experimental data. The analysis of data for product gases in the optimum conditions is shown in **Tables 4** and **5**. Carbon monoxide and methane show the highest and lowest error, respectively, in all sets of experiments.

5.1.1. Effect of Temperature

Figures 5-10 show the simulation results compared with experimental data for product gas composition versus three different temperatures in the range of 750°C - 9850°C.

Table 4. Analysis of data (gasification gas composition).

	Gas composition (v/v%)		ERROR %
	EXP	SIM	
CO	15.5	17.3	–4.85
H_2	24.1	25.4	–5.39
CO_2	19.5	20.2	–3.59
CH_4	3.8	4.2	–9.58
$CO + H_2$	39.6	41.7	–4.77

Table 5. Analysis of data (LHV).

	EXP	SIM	ERROR %
LHV (MJ/Nm^3)	6.54	6.77	–3.51

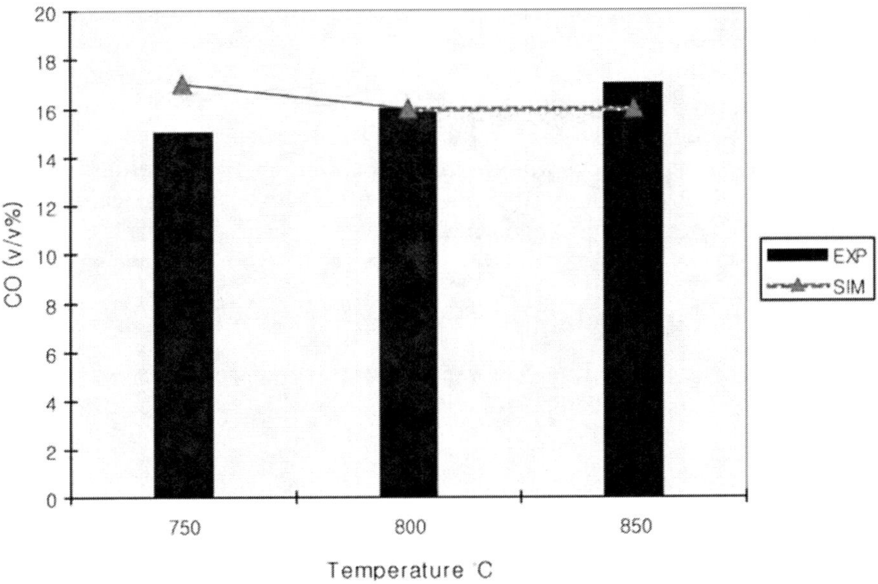

Figure 5. Effect of temperature on CO (ER = 0.2).

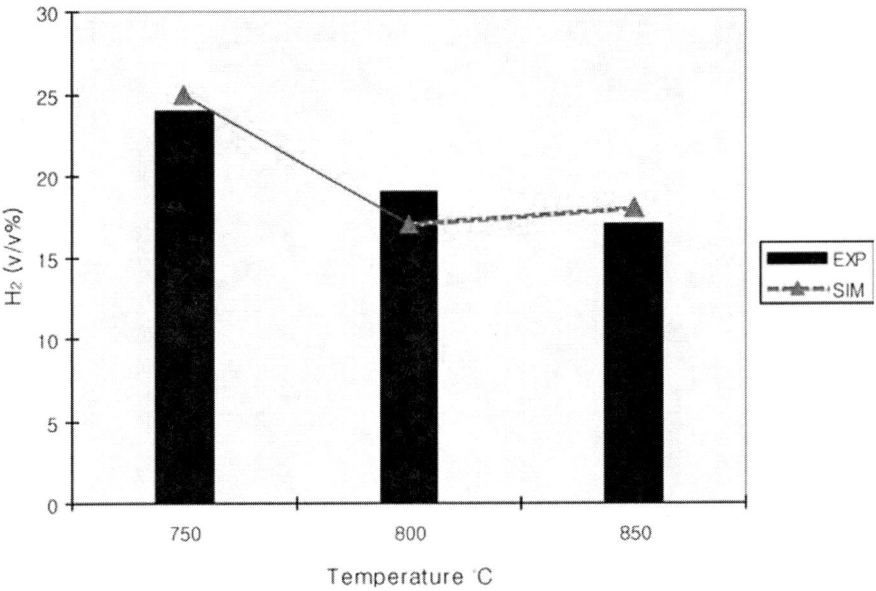

Figure 6. Effect of temperature on H_2 (ER = 0.2).

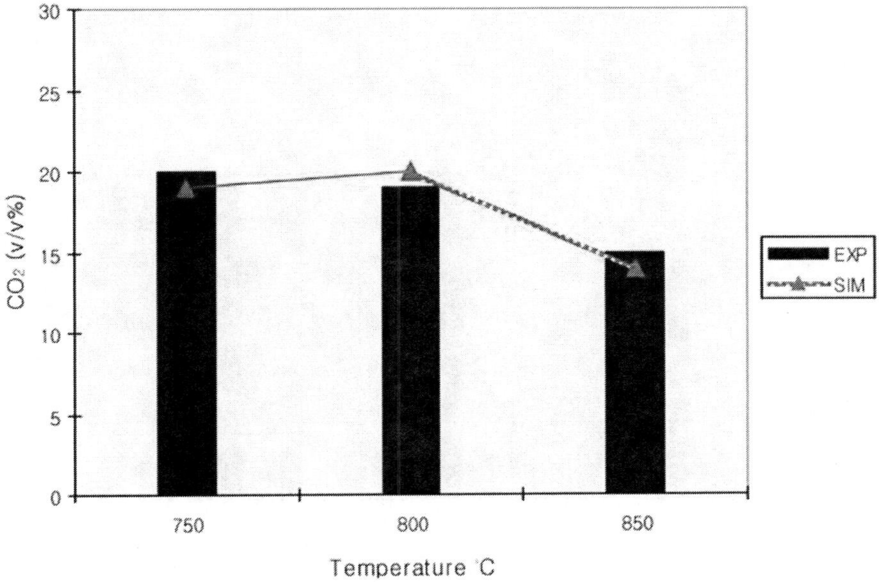

Figure 7. Effect of temperature on CO_2 (ER = 0.2).

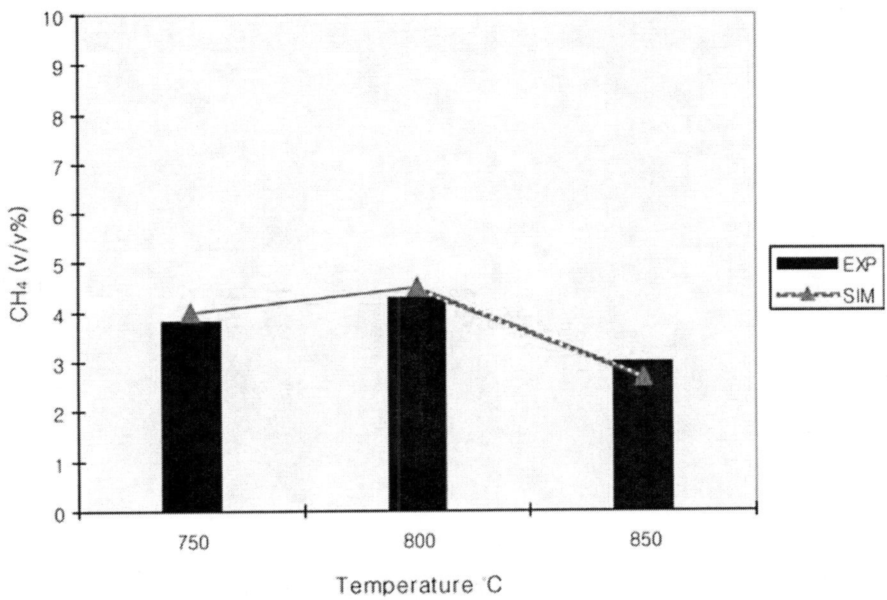

Figure 8. Effect of temperature on CH_4 (ER = 0.2).

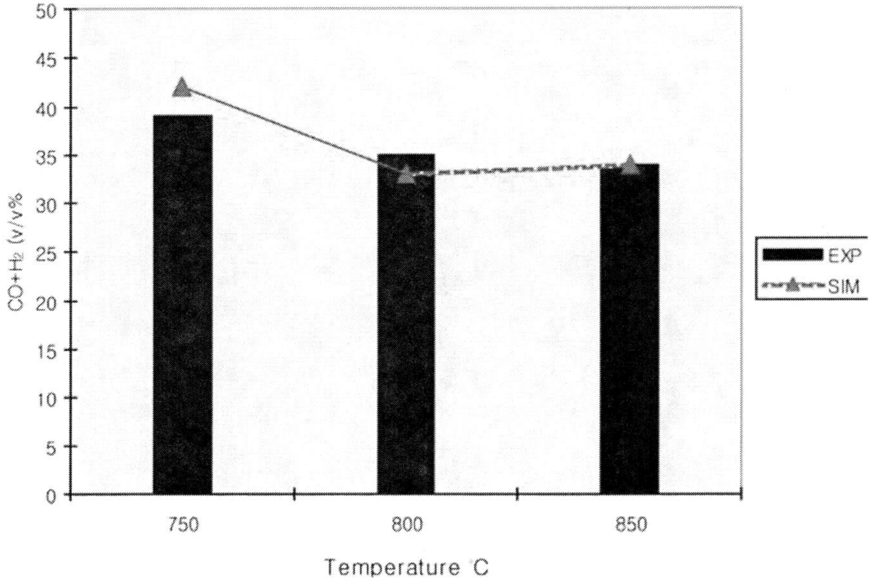

Figure 9. Effect of temperature on syngas (ER = 0.2).

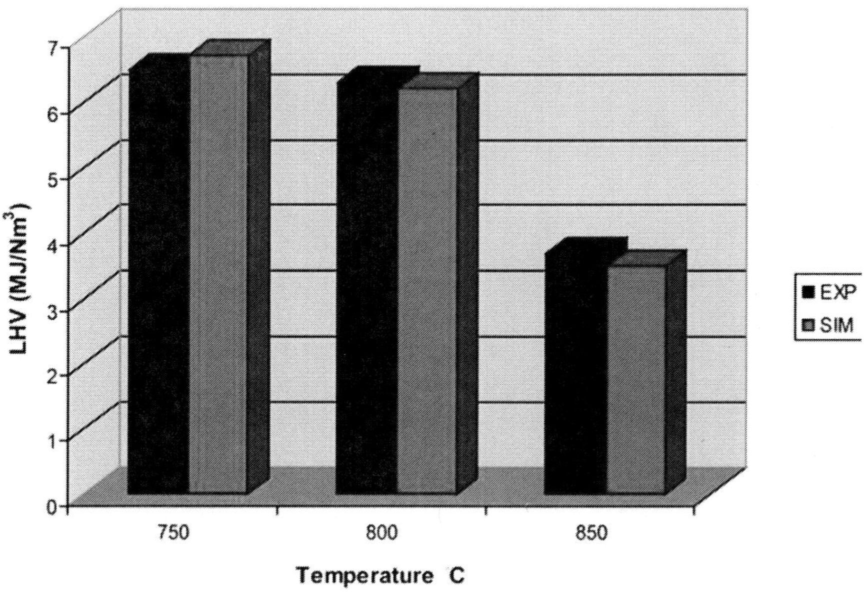

Figure 10. Effect of temperature on LHV (ER = 0.2).

5.1.2. Effect of Air Equivalence Ratio

Simulation results and experimental data for gas composition versus three different air equivalence ratios in the range of 0.2 - 0.4 are shown in **Figures 11-16**.

5.2. Modelling Predictions

Since the reliability of proposed model has been proved it can be now used in order to predict gasification gas composition and LHV for several values of temperature and air equivalence ratio.

Figures 17-19 presents model predictions versus temperature and **Figures 20-22** the predictions versus air equivalence ratio while Figure 23 shows LHV behaviour versus temperature and air equivalence ratio.

5.3. H_2/CO Ratio in Producer Gas

High yields of syngas (H_2 and CO) were obtained from olive kernel gasification with air and this behaviour could be attributed to the increased content of cellulose and hemicellulose in the raw material. Such information was known from previous experimentation in a laboratory scale, fixed bed gasification unit [35].

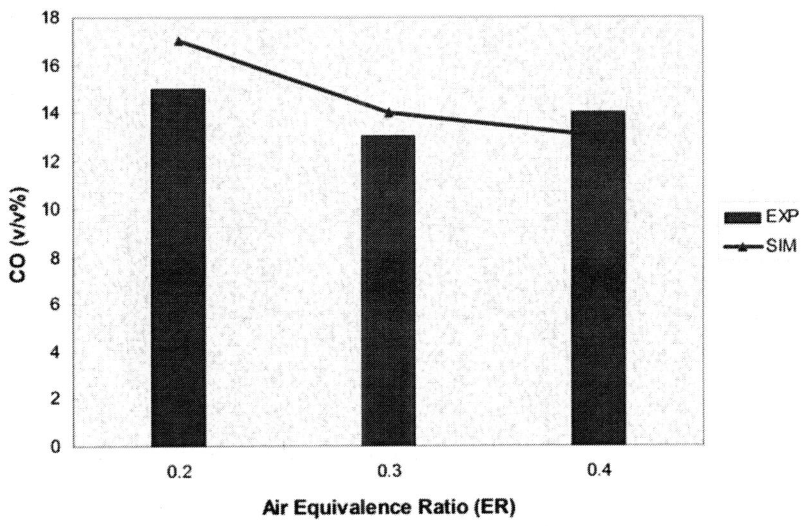

Figure 11. Effect of air equivalence ratio on CO (T = 750°C).

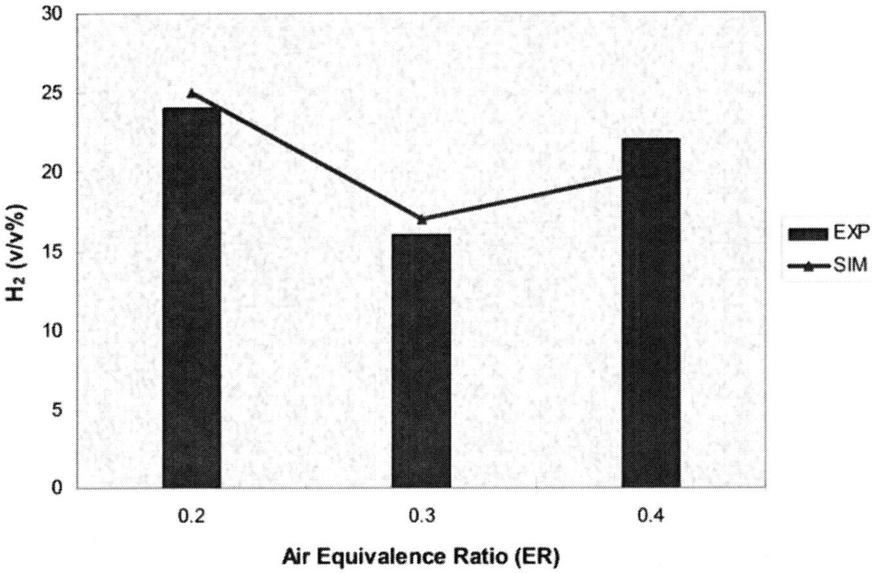

Figure 12. Effect of air equivalence ratio on H_2 (T = 750°C).

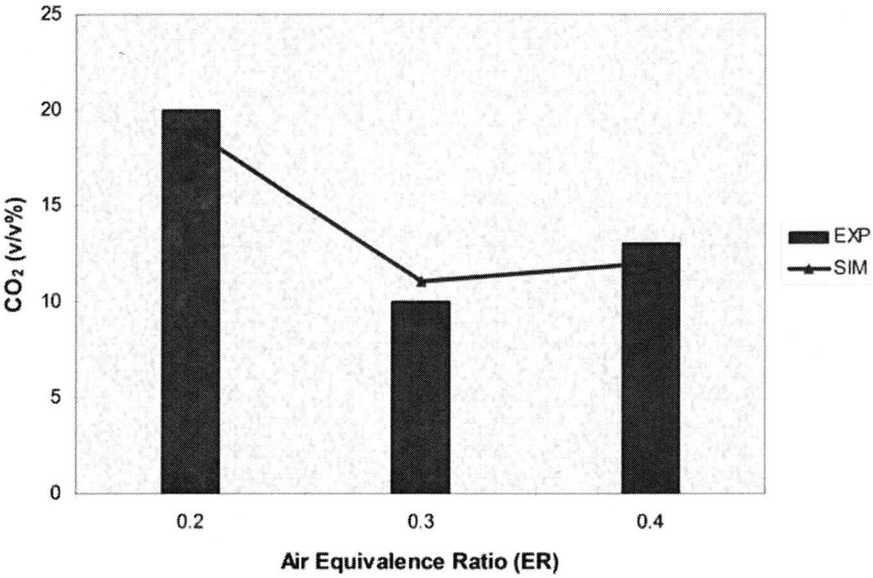

Figure 13. Effect of air equivalence ratio on CO_2 (T = 750°C).

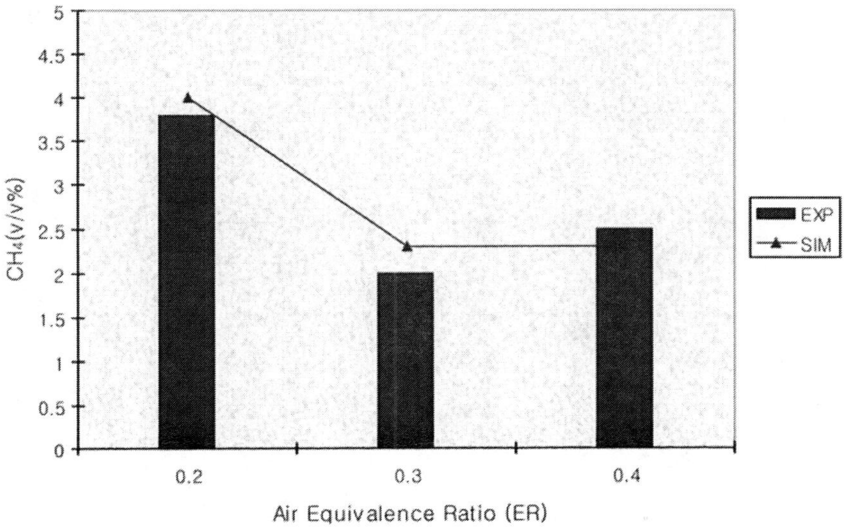

Figure 14. Effect of air equivalence ratio on CH_4 (T = 750 °C).

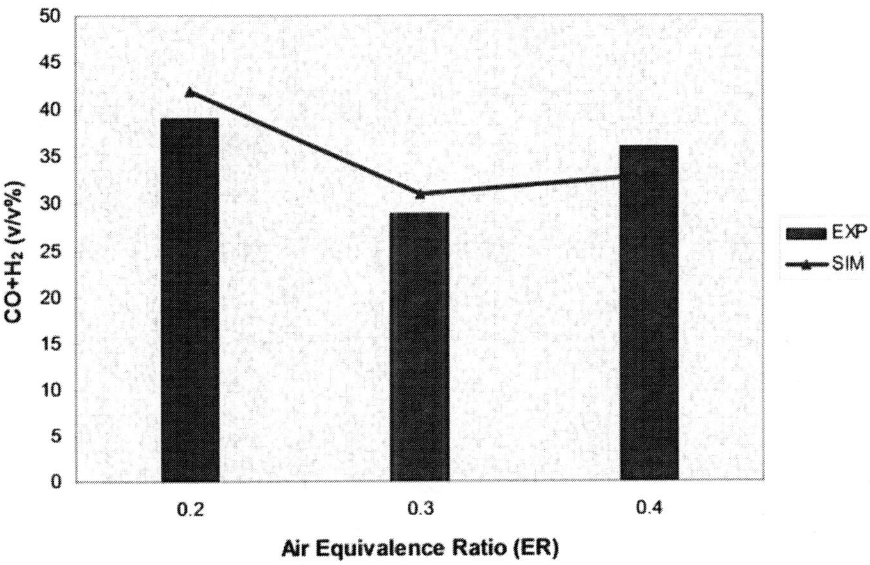

Figure 15. Effect of air equivalence ratio on syngas (T = 750°C).

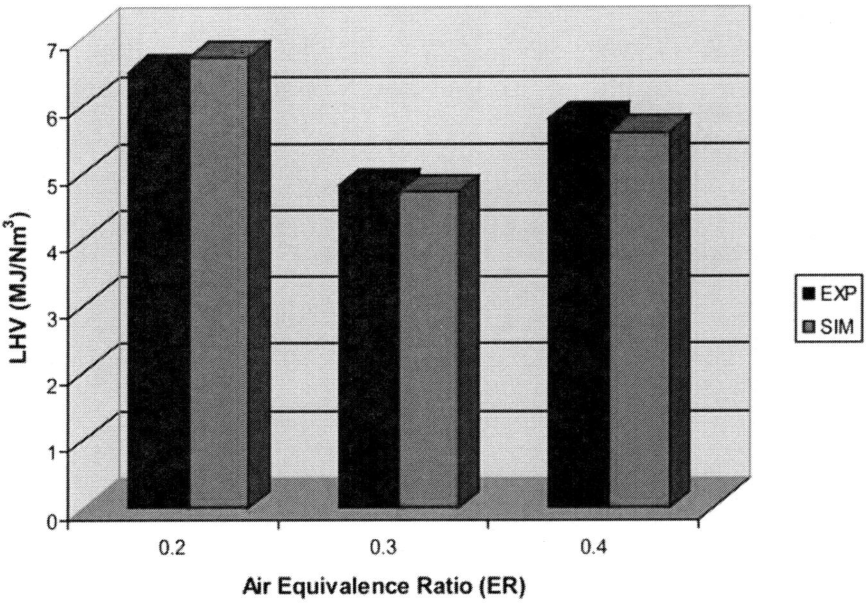

Figure 16. Effect of ER on LHV (T = 750°C).

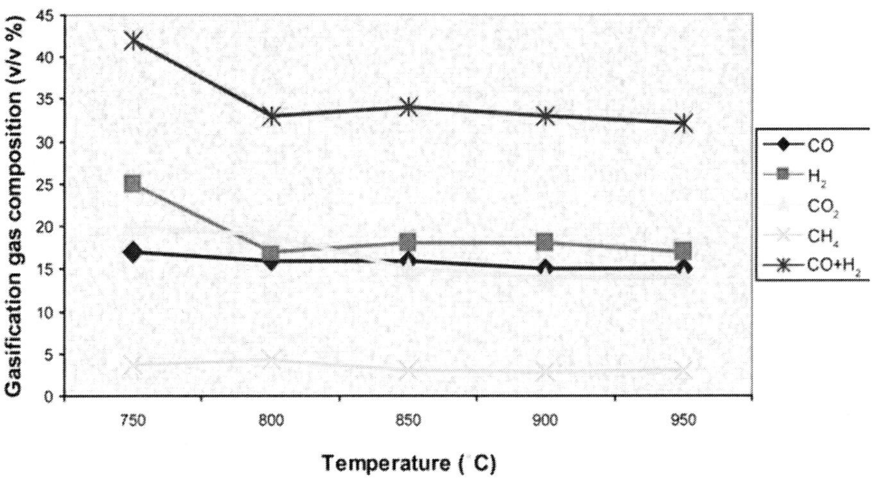

Figure 17. Effect of temperature on gas composition (ER = 0.2).

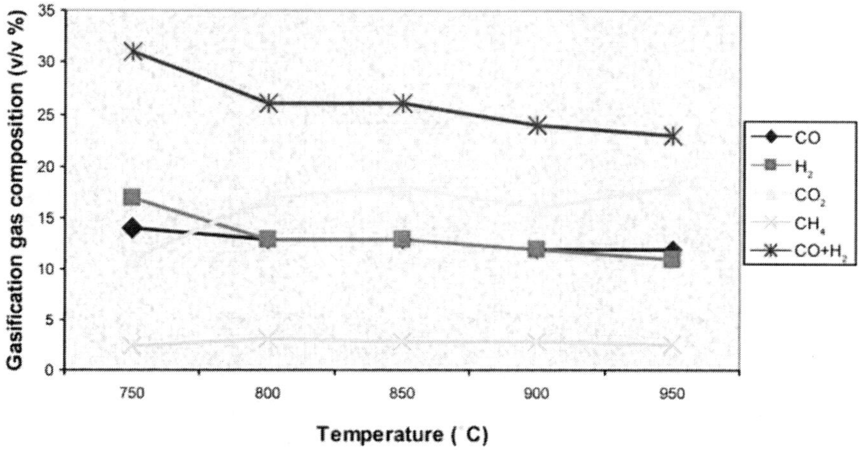

Figure 18. Effect of temperature on gas composition (ER = 0.3).

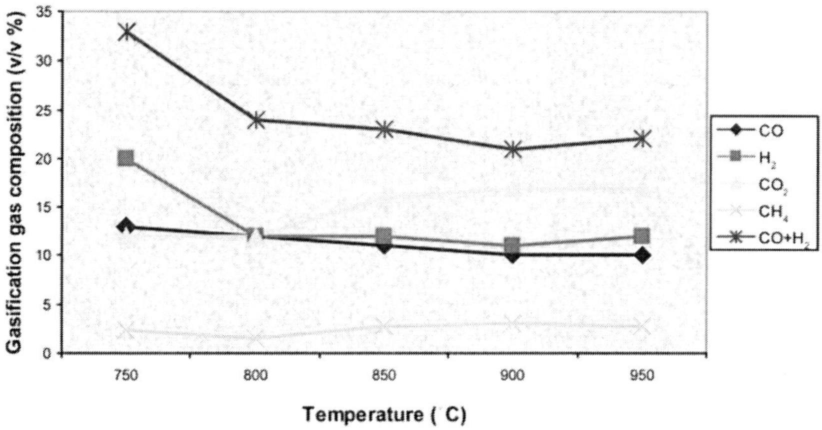

Figure 19. Effect of temperature on gas composition (ER = 0.4).

Uses of syngas could be distinguished according to the H_2/CO molar ratio. Producer gas with H_2/CO ratio between 1 to 2 can be used for the production of 2nd generation biofuels via Fischer Tropsch synthesis (H_2/CO = 2) and high added value chemical products like methanol. The present study indicates that producer gas of olive kernels gasification at 750°C and ER = 0.2 with a molar ratio of H_2/CO = 1.75 could be useful for Fisher Tropsch synthesis. The effect of temperature on H_2/CO molar ratio is presented in Figure 24.

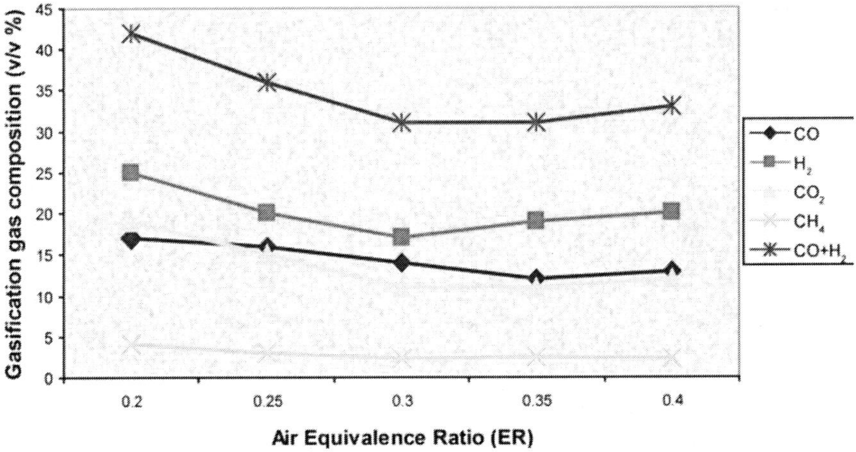

Figure 20. Effect of air ratio on gas composition (T = 750°C).

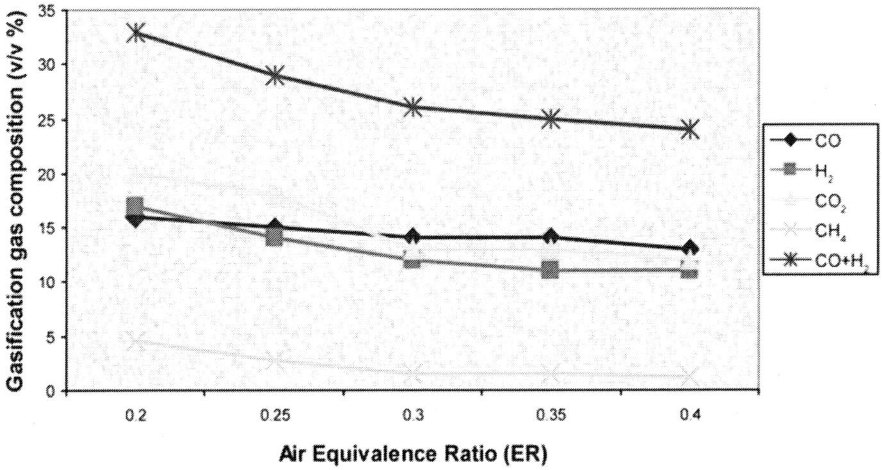

Figure 21. Effect of air ratio on gas composition (T = 800°C).

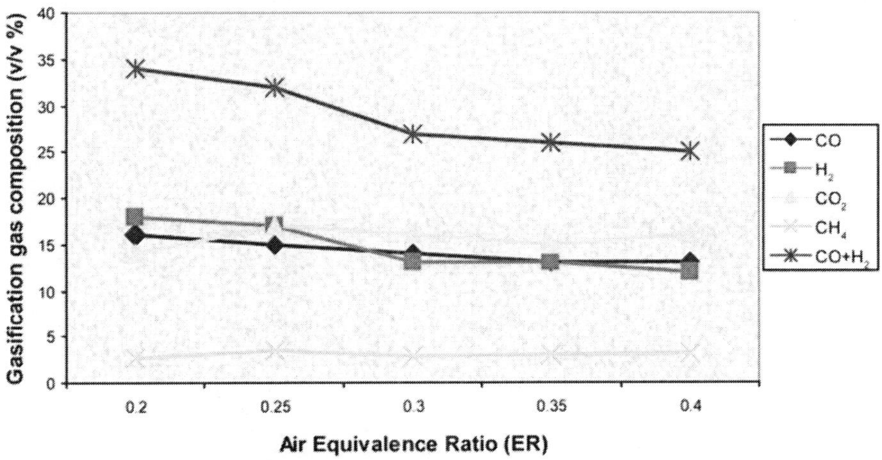

Figure 22. Effect of air ratio on gas composition (T = 850°C).

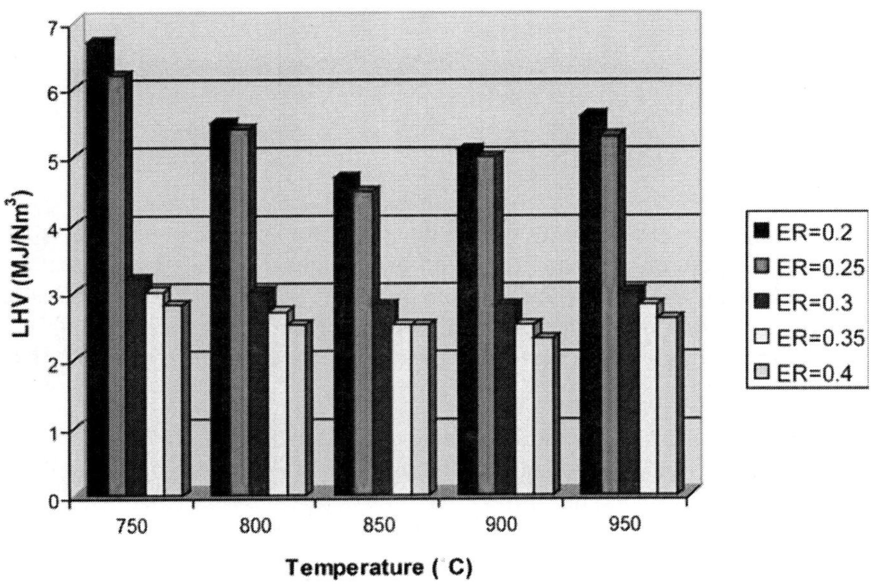

Figure 23. Effect of air ratio and temperature on LHV.

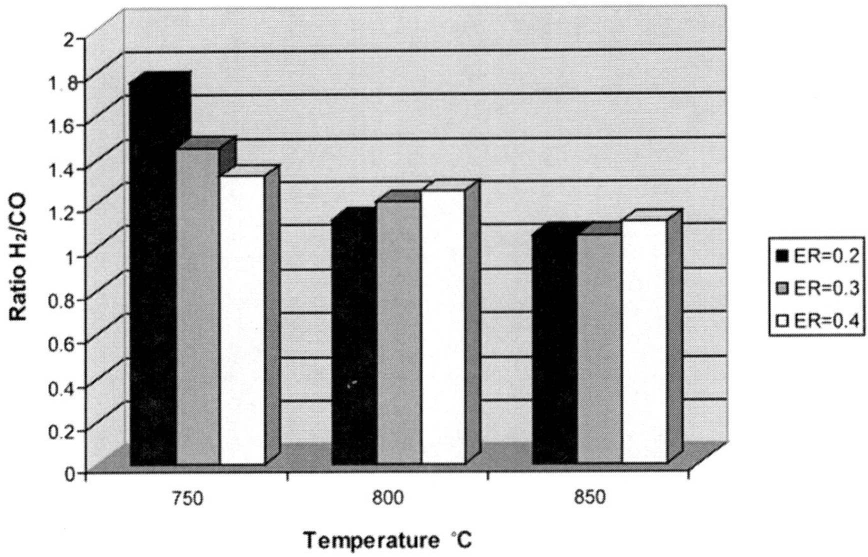

Figure 24. Effect of temperature and ER on H_2/CO molar ratio.

6. CONCLUSIONS

The objective of this study was to develop simulation capable of predicting the steady-state performance of an atmospheric fluidized bed gasifier by considering the hydrodynamic and reaction rate kinetics simultaneously. The products of homogeneous reactions were defined by Gibbs equilibrium, and reaction rate kinetics were used to determine the products of char gasification.

A BFB biomass gasifier model was developed using ASPEN Plus. The results obtained from the sensitivity analyses are in good agreement with experimental work. Therefore, the model is capable of predicting accurately gasifier performance over a wide range of operating conditions.

The influence of ER and temperature were investigated and the main conclusion is that the optimum operating conditions are ER = 0.2 and temperature 750°C something that agrees with experimental data. At these conditions the concentration of syngas ($CO + H_2$) in gasification gas was maximized (41.7 v/v%) and LHV took its highest value (6.77 MJ/Nm^3).

By increasing the ER the concentrations of CO and H_2 were decreasing in the favour of CO_2 production. As a result of this the LHV of gasification gas was also decreasing which are in accordance with the experimental results Based on the results obtained in the present study, the following conclusions are drawn:

1) The proposed combined transport and kinetic model is successfully validated with the experimental data reported in the literature.
2) The model predicted composition of producer gas matches very well with the experimental data.
3) The molar fractions of CO and CH_4 increase and that of water vapor decreases with time in the pyrolysis zone of the gasifier due to a subsequent increase in the temperature of pyrolysis zone.
4) The molar fraction of water vapor in the gaseous phase is the highest in the pyrolysis zone and decreases as it travels through the oxidation and reduction zones of the gasifier.
5) The molar fractions of CO and CH_4 in the gaseous phase increase in the pyrolysis zone as gas travels through it. However, the molar fractions of CO, H_2 and CH_4 decrease in the oxidation zone due to high temperature oxidation and hence the molar fraction of CO_2 increases.

Good qualitative agreement between model prediction and experimental data was achieved. However, to improve the simulation results, some modifications should be considered. Those modifications concern tar formation and mass transfer inside the particles:

1) In this study simulation results of parametric study of temperature and air ratio on gas composition (H_2, CO, CO_2, and CH_4) without considering tar formation were achieved. However, tar formation consideration in the model might improve the predicted results. For this reason, detailed experimental data about the influence of operating conditions on the formation of tar along with the kinetics studies is needed in order to obtain a thorough evaluation.
2) For the future experimental work it is proposed to use the chemical formula $C_xH_yO_z$ as that of tar. The parameters (x, y, z) are temperature and heating rate dependent. Tar production can be taking in account by defining nonequilibrium products in the RGIBBS.
3) Mass transfer inside solid particles is also an important parameter in gas–solid reactions and heat transfer inside particles. Heat transfer is also important between phases, particles. This is another feature that should be included in the model in order to achieve better simulation prediction. Radial dispersion inside the reactor will also help to see wall effects on the hydrodynamics of the fluidized bed reactor.

7. ACKNOWLEDGEMENTS

The study is funded by EU LIFE PLUS project "SMARtCHP Demonstration of a Small scale Mobile Agricultural Residue gasification unit for decentralized Combined Heat and Power production": http://smartchp.eng.auth.gr/en.

REFERENCES

1. S. Hamel and W. Krumm, "Mathematical Modelling and Simulation of Bubbling Fluidised Bed Gasifiers," Powder Technology, Vol. 120, No. 1-2, 2001, pp. 105-112.
2. S. Hamel and W. Krumm, "Modelling and Simulation of Bubbling Fluidized Bed Gasification Reactors," Fuel Energy Abstracts, Vol. 43, No. 4, 2002, p. 250.
3. R. P. Ma, R. M. Felder and J. K. Ferrell, "Modelling a Pilot-Scale Fluidized Bed Coal Gasification Reactor," Fuel Processing Technology, Vol. 19, No. 3, 1988, pp. 265-290.
4. F. Chejne and J. P. Hernandez, "Modelling and Simulation of Coal Gasification Process in Fluidised Bed," Fuel, Vol. 81, No. 13, 2002, pp. 1687-1702.
5. A. Sett and S. C. Bhattacharya, "Mathematical Modelling of a Fluidised-Bed Charcoal Gasifier," Applied Energy, Vol. 30, No. 3, 1988, pp. 161-186.
6. D. P. Ross, H. M. Yan and D. K. Zhang, "Modelling of a Laboratory-Scale Bubbling Fluidised-Bed Gasifier with Feeds of Both Char and Propane," Fuel, Vol. 83, No. 14-15, 2004, pp. 1979-1990.
7. L. Yu, J. Lu, X. Zhang and S. Zhang, "Numerical Simulation of the Bubbling Fluidized Bed Coal Gasification by the Kinetic Theory of Granular Flow (KTGF)," Fuel, Vol. 86, No. 5-6, 2007, pp. 722-734.
8. J. Corella and A. Sanz, "Modelling Circulating Fluidized Bed Biomass Gasifiers. A Pseudo-Rigorous Model for Stationary State," Fuel Processing Technology, Vol. 86, No. 9, 2005, pp. 1021-1053.
9. A. Sanz and J. Corella, "Modelling Circulating Fluidized Bed Biomass Gasifiers. Results from a Pseudo-Rigorous 1-Dimensional Model for Stationary State," Fuel Processing Technology, Vol. 87, No. 3, 2006, pp. 247-258.

10. J. Corella, J. Herguido, J. M. Toledo and J. I. Gomez-Cı- vicos, "Modeling Fluidized Bed Biomass Gasifiers. Part II: Gasification with Steam in a Bubbling Fluidized Bed," Proceedings of First World Conference on Biomass for Energy and Industry, Sevilla, Vol. 2, 2001, pp. 1971-1975.

11. M. L. De Souza-Santos, "Comprehensive Modelling and Simulation of Fluidized Bed Boilers and Gasifiers," Fuel, Vol. 68, No. 12, 1989, pp. 1507-1521.

12. H. Jiang and R. V. Morey, "A Numerical Model of a Fluidized Bed Biomass Gasifier," Biomass and Bioenergy, Vol. 3, No. 6, 1992, pp. 431-447.

13. J. F. Haggerty and A. H. Pulsifer, "Modelling Coal Char Gasification in a Fluidized Bed," Fuel, Vol. 51, No. 4, 1972, pp. 304-307.

14. F. Marias, J. R. Puiggali and G. Flamant, "Modelling for Simulation of Fluidized Bed Incineration Process," American Institute of Chemical Engineers Journal, Vol. 47, No. 6, 2001, pp. 1438-1460.

15. F. G. van den Aarsen, A. A. C. M. Beenackers, W. P. M. van Swaaij, "Modelling of a Fluidized Bed wood Gasifier," In: K. Ostergaard and A. Serensen, Eds., Fluidization V, Proceedings of the 5th Engineering Foundation, Engineering Foundation, New York, 1986.

16. V. I. Mukosei, L. I. Kheifets and R. V. Dzhagatspanyan, "The Mathematical Simulation of a Fluidised-Bed Reactor," Inzhenerno-Fizicheskii Zhournal, Vol. 12, No. 4, 1967, pp. 508-514.

17. Y. He and V. Rudolph, "Gas-Solids Flow in the Riser of a Circulating Fluidized Bed," Chemical Engineering Science, Vol. 50, No. 21, 1995, pp. 3443-3453.

18. K. D. Panopoulos, L. E. Fryda, J. Karl, S. Poulou and E. Kakaras, "High Temperature Solid Oxide Fuel Cell Integrated with Novel Allothermal Biomass Gasification Part I: Modelling and Feasibility Study," Journal of Power Sources, Vol. 159, No. 1, 2006, pp. 570-585.

19. E. V. Samuilov, M. F. Faminskaya and E. S. Golovina, "Model and Calculation of the Gasification of a Single Carbon Particle," Combustion, Explosion Shock Waves, Vol. 40, No. 1, 2004, pp. 77-84.

20. N. G. Deen, M. V. S. Annaland, M. A. Van der Hoef and J. A. M. Kuipers, "Review of Discrete Particle Modeling of Fluidized Beds," Chemical Engineering Science, Vol. 62, No. 1-2, 2007, pp. 28-44.

21. R. A. Kundsen, T. Bailey and L. A. Fabiano, "Experience with ASPEN While Simulating a New Methanol Plant," In: R. S. H. Mah and G. V. Reklaitis, Eds., Selected topics on computer-aided process design and analysis, American Institute of Chemical Engineers Symposium Series No. 214, 1982.

22. K. T. Schwint, "Great Plains ASPEN Model Development, Methanol Synthesis Flowsheet. Final Topical Report," National Technical Information Service, USA, 1985.

23. R. E. Barker, J. M. Begovich, J. H. Clinton and P. J. Johnson, "ASPEN Modelling of the Tristate Indirect Lique Faction Process," USA Oak Ridge National Laboratory, Oak Ridge, 1983.

24. J. N. Phillips, M. R. Erbes and R. H. Eustis, "Study of the Off-Design Performance of Integrated Coal Gasification," Proceedings Winter annual meeting of the American Society of Mechanical Engineers, Anaheim, 7-12 December 1986.

25. P. L. Douglas and B. E. Young, "Modelling and Simulation of an AFBC Steam Heating Plant Using ASPEN/SP," Fuel, Vol. 70, No. 2, 1990, pp. 145-154.

26. H. M. Yan and V. Rudolph, "Modeling a Compartmented Fluidized Bed Coal Gasifier Process Using ASPEN PLUS," Chemical Engineering Communication, Vol. 183, No. 1, 2000, pp. 1-38.

27. L. Backham, E. Croiset and P. L. Douglas, "Simulation of a Coal Hydrogasification Process with Integrated CO_2 Capture," Greenhouse Gas Control Technologies 7, Vol. 2, No. 2, 2005, pp. 1941-1945.

28. H. G. Lee, K. M. Chung, C. Kim, S. H. Han and H. T. Kim, "Coal Gasification Simulation Using ASPEN PLUS," US-Korea Joint Workshop on Coal Utilization Technology, Pittsburgh, 1992, pp. 447-474.

29. K. G. Mansaray, A. M. Al-Taweel, A. E. Ghaly, F. Hamdullahpur and V. I. Ugursal, "Mathematical Modelling of a Fluidized Bed Rice Husk Gasifier," Energy Sources, Vol. 22, No. 1, 2000, pp. 83-98.

30. P. Basu, "Combustion and Gasification in Fluidized Beds," Taylor Francis, Halifax, 2006.

31. Y. Cao, Y. Wang, J. T. Riley and W. P. Pan, "A Novel Biomass Air Gasification Process of Producing Tar-Free Higher Heating Value Fuel Gas," Fuel Processing Technology, Vol. 87, No. 4, 2006, pp. 343-353.

32. P. M. Lv, Z. H. Xiong, J. Chang, C. Z. Wu, Y. Chen and J. X. Zhu, "An Experimental Study on Biomass Air-Steam Gasification in a Fluidized Bed," Bioresource Technology, Vol. 95, No. 1, 2004, pp. 95-101.

33. A. Zabaniotou, V. Skoulou, G. Koufodimos and Z. Samaras, "Conceptual Desing and Preliminary Hydrodynamic Study of an Agro Biomass Bench Scale Gasification Fluidized Bed Reactor," International Journal of Chemical Reactor Engineering, Vol. 6, No. 1, 2008, pp. 1-17.

34. T. Nordgreen, T. Liliedahl and K. Sjöström, "Metallic Iron as a Tar Breakdown Catalyst Related to Atmospheric, Fluidised Bed Gasification of Biomass," Fuel, Vol. 85, No. 5-6, 2006, pp. 689-694.

35. P. Ollero, A. Serrera, R. Arjona and S. Alcantarilla, "The CO_2 Gasification Kinetics of Olive Residue," Biomass and Bioenergy, Vol. 24, No. 2, 2003, pp. 151-161.

6

CHAPTER

THE CONCEPT, DESIGN AND PERFORMANCE OF A NOVEL ROTARY KILN TYPE AIR-STAGED BIOMASS GASIFIER

Huiyuan Shi [1,*,†], Wen Si [2,†] and Xi Li [1]

[1] *Department of Environmental Science and Engineering, Fudan University, Shanghai 200433, China*

[2] *College of Information and Computer Science, Shanghai Business School, Shanghai 200235, China*

ABSTRACT

Tar formation is the main bottleneck for biomass gasification technology. A novel rotary kiln type biomass gasification process was proposed. The concept design was based on air staging and process separation. This concept was demonstrated on a pilot scale rotary kiln reactor under ambient pressure and autothermic conditions. The pilot scale gasifier was divided into three different reaction regions, which were oxidative degradation, partial oxidation and char gasification. A series of tests was conducted to investigate the effect of key parameters. The results indicate that under optimum operating conditions, a fuel gas with high heat value of about 5500 kJ/Nm^3 and gas production rate of 2.32 Nm^3/kg could be produced. Tar concentration in the fuel gas could be reduced to 108 mg/Nm^3 (at the gasifier outlet) and 38 mg/Nm^3 (after gas conditioning). The cold gas efficiency and carbon conversion rate reached 75% and 78%, respectively. The performance of this gasification system shows considerable potential for implementation in distributed electricity and heat supply projects.

Keywords: biomass gasification; air staged; rotary kiln; partial oxidation; energy balance

1. INTRODUCTION

Thermal conversion of biomass is a promising way to produce gaseous and liquid fuels [1]. Among all the techniques, gasification has received extensive attention because it produces valuable gaseous products, which can be used as a fuel for internal combustion engines, gas turbine, boilers, and cooking [2]. Extensive studies have been conducted on biomass gasification processes to investigate how to increase the energy transformation efficiency, reduce pollution during the gasification process and improve the device reliability [3,4,5]. Among all the obstacles that impede the development of gasification technology, tar formation is the most cumbersome one [6]. Tar is a complex mixture of hydrocarbons, which includes aromatic compounds, along with other oxygen-containing hydrocarbons and complex polycyclic aromatic hydrocarbons (PAHs). Tar is defined by EU/IEA/US-DOE meeting as all organic contaminants with a molecular weight larger than benzene [7]. Tar easily causes fouling in ducts, valves, and engines, and polymerizes to form more complex structures during combustion, thus decreasing the reliability of equipment and increasing pollutant emissions [8,9]. Tar limits the usage of syngas in application processes. The minimum allowable limit for tar is dependent on the end user application. As tabulated by Milne and Evans [10], the general tolerance limits for engines, turbines and other combustion devices range from 5 to 100 mg/Nm^3.

There are two types of tar control technologies, which are syngas conditioning and inside gasifier treatment. The first method is proved to be effective but it is costly and may cause secondary pollution. The second method is gaining much more attention due to its economic benefits.

As reviewed by Milne and Evans [7], the formation and conversion of tar is seriously affected by the material composition, temperature, gasification agent(s), oxidation processes and residence time. These are all important operation parameters for a biomass gasifier. Plenty of works are available on tar formation and destruction behavior under the effect of the parameters mentioned above. It is generally agreed that tar is produced during biomass pyrolysis around 300 to 500 °C. It can be divided into primary, secondary and tertiary tar, according to its reaction history. The primary tar is characterized by cellulose-derived products such as levoglucosan, hydroxyacetaldehyde and furfurals. The secondary tar species are phenolics and olefins, while tertiary tar includes methyl derivatives of aromatics, such as methyl acenaphthylene, methylnaphthalene, toluene, and indene. Tar composition varies with the temperature program used during pyrolysis. High heating rates lead to a high tar production and a high proportion of primary and secondary tar, hence, flush pyrolysis technology is applied for bio-oil production [11]. Low heating rates result in a low tar production and high tertiary tar proportion, that is the reason why slow pyrolysis is used for torrefaction and gasification. Tars produced by pyrolysis may change under certain conditions. Chen et al. [12] reported that biomass pyrolysis tar decomposed by more than 90% at 1000 °C. Besides decomposition, tar also experiences polymerization reactions above 800 °C. Their study also revealed that primary and secondary tar could be converted into highly polymerized and highly aromatic components, such as naphthalene and indene. Houben [13] reported that tar decomposition reactions can be accelerated under an oxidative environment. Pyrolysis tar conversion rates under a partially oxidative atmosphere are higher than under inert conditions. Moreover, oxygen-containing compounds are favored, which are mainly phenol and its homologous series.

As reviewed above, oxygen supply is a key issue for biomass gasifier design. Several gasifier concepts have been developed in the past 20 years [14]. The majority of them have a single inlet for air, no matter whether it is an updraft, downdraft or fluidized gasifier. Few of them has more than one air inlet. Pan et al. [15] reported a 88.7 wt % of tar reduction by injecting secondary air above the biomass feeding point in a fluidized bed configuration. Narv et al. [16] performed secondary air injection in the freeboard of a fluidized bed gasifier and observed a tar reduction of nearly 50 wt % with a temperature rise of about 70 °C. The Asian Institute of Technology (AIT) developed a two stage biomass gasifier [7]. Experimental results showed that the tar concentration in flue gas is about 40 times less than in a single-stage reactor under similar operating conditions. This concept involves a downdraft gasifier with two air inlets. Tar produced in first pyrolysis stage will pass through a high temperature char bed and decompose at elevated temperatures. Yoshikawa [17] combined a fixed-bed pyrolyzer with a high temperature reformer using a high temperature steam/air mixture and demonstrated that the injection of high temperature steam/air mixture into the pyrolysis gas effectively decomposes tar and soot components in the pyrolysis gas into gaseous products. Henriksen et al. [18] reported successful working and operating experiences with a two-stage gasifier. Although it has only one oxygen inlet, the pyrolysis and gasification process are separated by a partial oxidation zone, where the produced pyrolysis tar can be decomposed to a great extent.

From the extensive literature search, it can be observed that the temperature profile and oxygen injection are key design issues that influence tar production during gasification. A few unique gasifiers have been developed. Some of them were based on an "air staging" concept and some separated gasification into many processes through structure modification. In the current work, a novel gasifier system with a thermal capacity of 1.5 MW_{th} was proposed. The conceptual configuration of the gasification reactor is illustrated in Figure 1. The gasifier is a rotary kiln with a 6 degree angle of inclination. Unlike a traditional rotary kiln, it is divided into three regions, which are oxidative degradation, partial oxidation and char gasification. Biomass is fed into the reactor from the left end and passes through the gasifier by rotating the reactor. Air used for gasification is injected into three different regions by a blower.

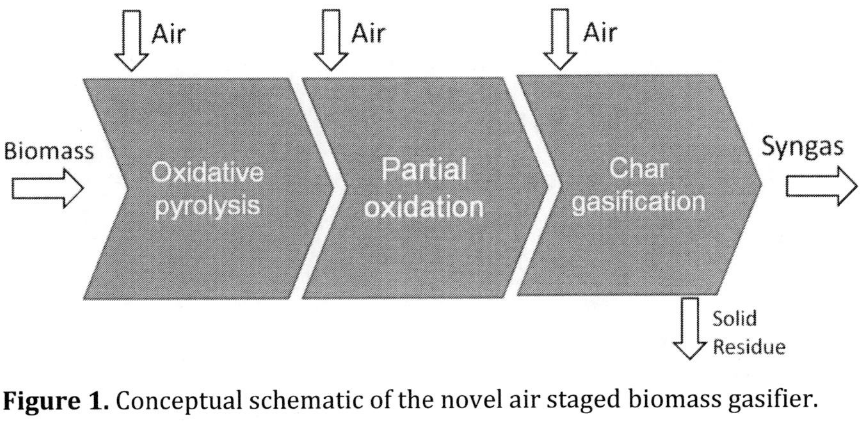

Figure 1. Conceptual schematic of the novel air staged biomass gasifier.

The pyrolysis reaction takes place in the first region under an oxidative atmosphere. Then, the pyrolytic gas with tar species is drawn into the second region by an induced fan. In this stage, homogeneous partial oxidation reactions happen. Tar components decompose into small molecule gaseous products. After the pyrolytic and partial oxidation process, the volatiles in the biomass solid fuel are released completely and this leaves a char with high carbon content. In the third stage, the char gasification reaction takes place. Besides, biomass char also acts as absorbent to adsorb tar components [19].

The air staging strategy does not just reduce tar production; it also produces an energy self-sufficient system. DTU developed a two-stage gasifier [18], where the oxidation and pyrolysis processes are separated. However, its pyrolysis stage is heated externally by exhaust gas from a gas engine. In the present gasifier, every single stage is auto-thermal. Heat for pyrolysis is provided by partial oxidation of the solid biomass fuel in an oxidative environment. Besides, radiation heat from the second region also helps maintain the temperature in the pyrolysis region. In the present study, the optimal operation parameters and main factors that influence tar production, low heat value (LHV) of the fuel gas, gas production rate, cold gas efficiency and carbon conversion rate were investigated. In addition, the heat balance was calculated and discussed.

2. MATERIALS AND METHODS

2.1. Material

Mixed wood chips with an average size of 1 × 2 × 5 cm were used as feedstock. The corresponding proximate and ultimate analysis results are shown in Table 1. The volatile content is as high as 60%, while the fixed carbon content is about 17%. Since the biomass used in this test is a mixture of waste wood without drying, its water and ash content is a bit higher than those of other wood materials. Air was used as gasification agent in all tests.

Table 1. Proximate and ultimate analysis of wood chips.

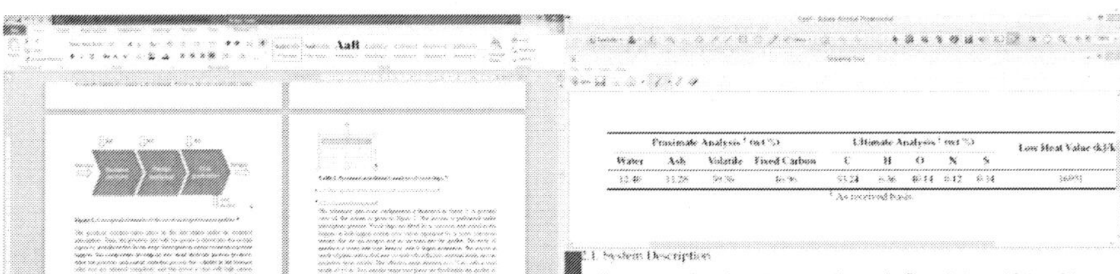

2.2. Description of the System and Experiment Setup

2.2.1. System Description

The schematic pilot-scale configuration is illustrated in Figure 2. A pictorial view of the system is given in Figure 3. The process is performed under atmospheric pressure. Wood chips are lifted by a conveyor and stored in the hopper. A loch hopper system (two valves separated by a screw conveyer) ensures that no gas escapes and no air leaks into the gasifier. The body of gasifier is a rotary kiln type furnace, with 6 degree inclination. The reactor is made of plain carbon steel and covered with refractory material inside and an insulation layer outside. The effective inner diameter is 0.75 m, with a total length of 15 m. Two annular shape steel plates are fixed inside the gasifier to divide the inner space into three different regions. The length of each region is 5 m, 2 m and 8 m, respectively. A variable frequency motor is used to drive and control the rotation speed.

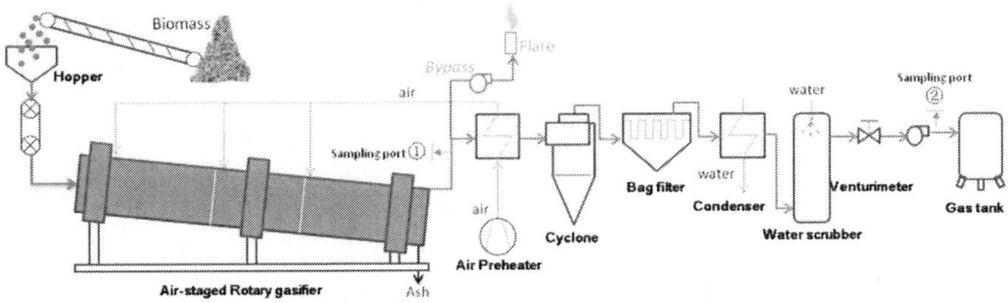

Figure 2. Flow diagram of the gasification plant.

Figure 3. A view of the1.5 MW_{th} biomass gasification system.

Gasification agent is provided by an air blower. Air is preheated by the hot product gas and divided into three intakes. Three flow meters are used to control each air injection separately. The air feed system for the rotary kiln type furnace is a patented technology, which ensures no air and fuel gas leakage from the rotating furnace. First air is fed into the pyrolysis region of the reactor through tuyeres placed radically around the circumference. Secondary air is injected into the secondary part of the reactor and distributed in gaseous phase area, so that partial oxidation of the pyrolytic gas would take place and oxidation of biomass char would be prevented at the bottom of the reactor. Third air is injected into the third part of reactor through nozzles located at the bottom of the reactor. This is to ensure a fine mixture of air and hot char, and hence promote the gasification reaction.

A cyclone is used as dedusting unit for fuel gas cleaning which is heat-insulated to prevention any condensation of trace tar. An air preheater is placed after the dust remover and followed by an indirect water cooler. The cooled producer gas is then allowed to pass through a washing tower in order to remove the moisture and trace tar content in the syngas. The ash remover, gas cooler and washing tower constitute of the gas conditioning system. After gas conditioning, the syngas goes to the tank through the blower. Just before the gas tank, a water bubbler is installed in the path for safety reasons to avoid backfire. A venturimeter is installed in the path to measure the gas flow rate. In a Venturi flow meter, the mass flow rate is proportional to the square root of the pressure difference across the venture inlet and the throat.

The fuel gas exits the gasifier and enters into the following de-dusting and cooling units only if the system has been brought to a steady state. Before that, a bypass vent is used to exhaust fuel gas produced during the startup period.

2.2.2. Preparation and Procedures

Biomass fuel is prepared first. Its size and moisture content are carefully controlled according to the needs (refer to Table 1). The system is then tested for leakage by sealing the gasifier at the fuel inlet and ash tank using a water seal and subjecting it to air at a pressure of 50–70 mm water gauge with the help of an air blower. Every joint was examined carefully with soap suds. After the leakage tests, the bypass fan is started. A 200 Pa negative pressure is maintained inside the gasifier through an interlock pressure control system. Before feeding the biomass, a startup burner fueled with natural gas is used to heat the reactor. When the desired temperature distribution is achieved in the center of first region (700–750 °C), second region (650–700 °C) and third region (500–550 °C), the feeder valve is opened to feed biomass into the gasifier. The reactor rotation driving motor and air blower are turned on simultaneously. Then, the startup burner is taken out. At the same time, the assisting fuel gas (natural gas) and secondary air are injected into the second region to ignite the pyrolytic gas produced in the first region. When the required temperature (>900 °C) is reached in the second region, the assisting fuel gas is turned off and the air injection system is continuously operated. After the temperature in the second region reaches a steady state, we turn on the air injection at the third region. After that, the feedstock/air ratio is carefully adjusted to reach steady state operation conditions. In general, it takes around 8 h to make the system stable with respect to the reactor temperature. After the system reaches a steady state, product gas is shifted to the gas conditioning system. All parameters are kept constant for about 2 h for sampling and data analysis.

2.2.3. Measuring and Sampling Instrument

To position thermocouples inside the rotary gasifier is a great challenge. Calibrated S-type (platinum-rhodium) thermocouples are used for temperature measurement. The wires are electrically separated from each other by ceramic beads. All the electric wires are placed inside a 1Cr18Ni9Ti stainless steel pipe coated with insulation layer. The pipe is fixed at the center axis of the reactor with two bearings at both ends to ensure an immobile position when the reactor is rotating (Figure 4). A digital data collector is used to read out the temperature.

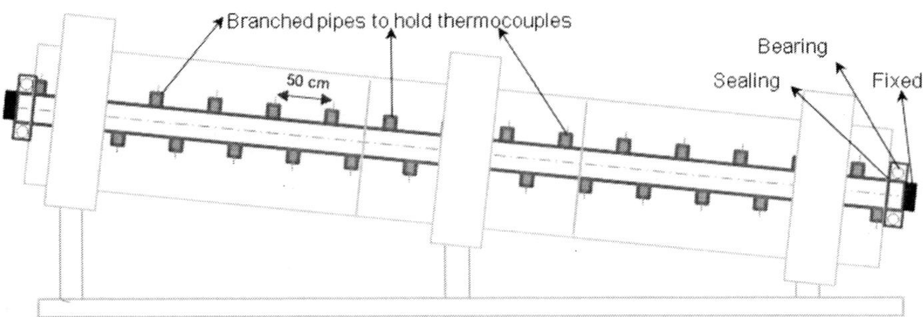

Figure 4. Position of thermocouples inside the gasifier.

A model Gasboard-3100 gas analyzer produced by Wuhan Cubic Optoelectronics Cp., Ltd. (Wuhan, China) is used to measure the composition of the producer gas. The resolution ratio and accuracy are 0.01% and 1%, respectively. Two sensors are used for gas composition measurement, which are a Non-Dispersive Infra-Red (DNIR) and a Thermal Conduction Detector (TCD). The volume concentrations of CO_2, CO, CH_4, H_2 and O_2 can be measured simultaneously. The measuring range for CO_2, CO, CH_4, H_2 and O_2 are 0%–50%, 0%–40%, 0%–25%, 0%–40% and 0%–25% respectively. Water vapor, dust and tar are removed before gas analysis. The gas analyzer is coupled with the tar sampling system.

The tar sampling system operates as follows: syngas is induced by a pump with a controlled flow rate of 1 L/min. The syngas is first cleaned by a dust filter, and then washed by four cold gas washing bottles, which are immersed in an ice-water bath. The sampling line and ash filter are electrically heated to 350 °C to avoid condensation of tar as noted in [7]. There might be some components that has a condensation temperature lower than 350 °C. These species cannot be collected in the sampling bottle. The first two bottles are filled with glass balls, in order to increase the contact area. Large molecule tars will be trapped in the first two bottles. Then, gas is washed by a solution of methanol and trichloromethane with a volume ratio of 1:4. After this process, tar can be trapped in four bottles. The clean gas is pumped into the gas analyzer for composition measurement. After gas sampling, all the pipes and bottles are washed with methanol and chloroform solution. The washing liquids are sealed and kept in a refrigerator for tar composition and mass analysis. After GC analysis, the solvent was evaporated in a RE3000A type rotary evaporator for 30 min, and the rest was placed in a drying oven with a temperature of 35 °C for 1 h in an inert atmosphere for the final evaporation and then weighed as gravimetric tar and sealed in a container. The total tar concentration is calculated by the mass of sampled tar minus the sampled gas volume.

Tar composition is analyzed by gas chromatography/mass spectrometry (GC/MS) using an Agilent 6890 N gas chromatograph and an Agilent 5975 C mass spectrometer (Agilent, Santa Clara, CA, USA). The GC split ratio is 10:1. The sample inlet temperature is set at 280 °C. A HP-5MS column (30 m × 0.25 mm, 0.25 μm) is used at a flow rate of 2.4 mL/min. The carrier gas is helium with a flow rate of 30 mL/min. The column temperature increases from 45 to 180 °C at 5 °C/min and then to 300 °C at 45 °C/min, where it is held for 10 min. The sample injection volume is 1 μL. The GC/MS was quantified by the external standard method with 2, 5, 10, 20, 40, 60, 80 and 100 ppm solutions of eight typical tar components: benzene, toluene, styrene, phenol, naphthalene, ethylbenzene, indene, and benzaldehyde. The chromatographic system has limitations for measuring higher tars as the heavier components are retained and do not elute through the GC column [7].

2.2.4. Biomass Feed Rate Calibration

Biomass is gravity feed by a loch hopper system to control the feed rate and exert a sealing function. In the gasifier, which is like a helical vane drum with 6 degree inclination, biomass could be transmitted by rotating the gasifier. The average biomass feed rate was plotted against the gasifier rotation speed. The determination coefficient was 0.9811, which was due to the non-uniform size of the wood chips. The loch hopper system runs intermittently to match the gasifier rotation speed in order to achieve a certain biomass feed rate. As the rotation speed varied from 1/6 r/min to 1 r/min, the biomass residence time inside the gasifier ranged from 114 min to 33 min. In the current study, the gasifier rotation speed was fixed at 5/9 r/min. The corresponding biomass feed rate and residence time were 300 kg/h and 55 min, respectively.

3. RESULTS AND DISCUSSION

The aim of this study was to demonstrate the concept of a novel air-staged gasification process to produce tar-free fuel gas. A series of tests were conducted to investigate the impact of parameter variation such as air injection percentage (AIP) of the three regions and equivalent ratio (λ) on temperature distribution, gas composition, tar formation, gas production rate, cold gas efficiency, and carbon conversion rate.

The gasifier is designed for a nominal fuel power of 1.5 MW_{th}. This is equal to 300 kg/h biomass fuel fed into the gasifier with a low heat value of 17 MJ/kg.

The equivalent ratio (λ) is defined by Equation (1):

$$\text{Equivalent ratio } (\lambda) = \frac{\text{Air injected into the gasifier}}{\text{Stoichiometric air}} \tag{1}$$

The stoichiometric air V_0 is calculated by Equation (2):

$$V_0 = 0.0889C_{ar} + 0.265H_{ar} - 0.0333\,(O_{ar} - S_{ar}) \tag{2}$$

Here V_0 refers to the stoichiometric air required for fuel combustion (Nm^3/kg), Car, Har, Oar, Sar refer to the elemental mass compositions of C, H, O and S (wt %), which are available in Table 1.

Besides λ, AIP also plays an important role on the temperature profile inside the gasifier. According to the description of the conceptual design, the purpose of the three air injections is not identical. The first air is to supply the heat required for the pyrolysis reactions; the secondary air is used for partial oxidation of the pyrolytic gas, which results in a high temperature environment; the third air acts as an agent for char gasification. Hence, optimization of AIP is crucial for acquiring a desired temperature distribution.

A series of test runs were carried out with the 1.5 MW_{th} pilot scale gasifier to understand the behavior of this gasifier and improve its performance.

3.1. Optimization of AIP

We have made plenty of tests covering all the equivalence ratios (0.28 to 0.45). Three different AIP conditions were tested, which were condition 1 (1:7:2), condition 2 (2:6:2) and condition 3 (3:5:2). Results showed that an optimum ratio of (2:6:2) applied to all the equivalence ratios. In order to be concise, only the temperature curve when $\lambda = 0.36$ is drawn in Figure 5.

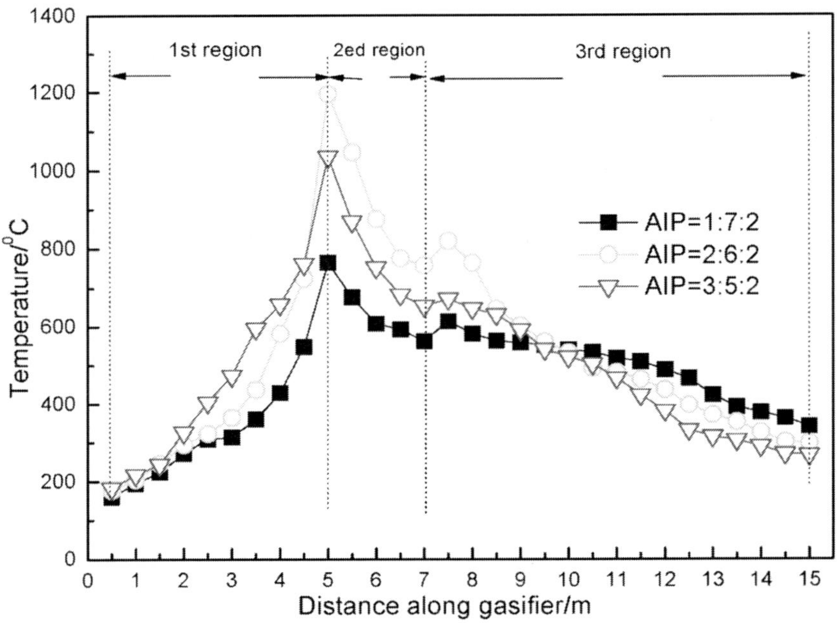

Figure 5. Temperature profile varied with AIP.

The temperature of the pyrolysis region rose up as the injected air increased. As described before, the heat required for the pyrolytic reactions is supplied by biomass and char oxidation. Hence, the more oxygen was injected, the higher the temperature rose. However, the second region behaved differently. In the second region, oxygen was injected into the gaseous phase, where it mixed with pyrolytic gas and resulted in a homogeneous partial oxidation reaction. Therefore, the environment temperature suddenly increased. In this experiment, although air injection in the second region decreased from 0.7 to 0.6, the peak temperature rose from 765 to 1196 °C. This indicated that more combustible gas was produced under condition 2. Under condition 1, the temperature in the first region was too low for the biomass decomposition reactions to take place. The pyrolytic gas produced was less than under condition 2. Therefore, even though the oxygen injection in the second region was higher; there was not enough gas for combustion, and as a result, the temperature decreased. Under condition 3, the temperature in the second region decreased. Although the pyrolysis reaction was enhanced due to the higher temperature, a large amount of injected oxygen may cause an oxidation of the pyrolytic products in the same time they are produced. In addition, a lack of oxygen in the second region restrains the secondary oxidation of the pyrolytic gas. For these reasons, the temperature in the second region was lower than under condition 2.

Heat required for char gasification is transferred by radiation and hot gas convection from the partial oxidation process. Therefore, the temperature profile in the third region was greatly affected by the second region, even though the air injection in the third region was identical for the three conditions.

Based on the above analysis, condition 2 is the most optimal. The highest temperature was achieved in the second region, which resulted in a maximum tar reduction efficiency and char gasification rate. Thus, in the following tests, AIP was fixed at 2:6:2 to obtain the best temperature distribution. Next the total air injected into the gasifier was varied to investigate the effect of λ on gasifier performance.

3.2. Temperature Profile vs. λ

We have conducted many experiments under different λ values (from 0.2 to 0.6). Results indicated that the auto thermal operation happens only if λ is larger than 0.28. Thus, we just present the results when $0.28 < \lambda < 0.45$. The temperature distribution along the axis of the gasifier is illustrated in Figure 6. It can be seen that the variation of temperature curves were alike for different λ values. In addition, the shape of the temperature curves was in accordance with the three regions inside the gasifier.

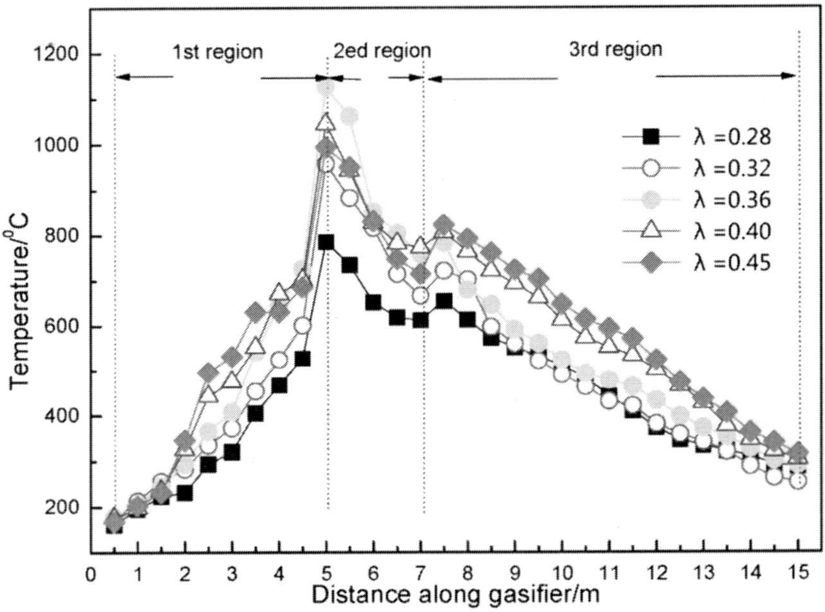

Figure 6. Temperature profile variation with λ.

When biomass fuel was fed into the gasifier, it was heated by radiation of the hot solid fuel and gas in the second region. The environment temperature at the inlet immediately rose up to around 200 °C. As the biomass was moved by rotating the gasifier, the temperature gradually increased. When the temperature reached 350 °C, the pyrolysis reactions started. It was reported that pyrolytic reactions could be enhanced under an oxidative environment [20]. At the end of the first region, the biomass decomposition reactions were almost finished. A slight increase of temperature could be observed due to hot gas radiation from the second region. Along the temperature curves in the first region, no obvious peaks were detected, which indicated that the oxidation reaction took place smoothly. Researchers have concluded that in such a temperature range (300–500 °C), the oxidation reactions almost take place between solid fuel and oxygen [21]. The combustion of pyrolytic gas requires a higher activation energy which cannot be reached within this temperature range [22,23]. Under the current temperature, a large amount of CO and CH_4 were produced, however they would not be burned in the first region and would experience partial oxidation in the next stage.

A step temperature increase appeared at the inlet of the second region. This is because of the combustion of pyrolytic gas. Oxygen injected at the first stage has all been consumed. The pyrolytic gas with high heat value was drawn to the partial oxidation area, where the gas met with oxygen and underwent an intense combustion reaction. Hence, the temperature exceeded 1000 °C.

At the inlet of the third region, another obvious peak was observed. The third injected air would cause combustion of the residual syngas that was not consumed in the second region. Besides, heterogeneous oxidation reactions would also occur when air contacts the hot biomass char. These homogeneous and heterogeneous oxidation reactions resulted in an increase of the environment temperature, while the heat released was instantly adsorbed by gasification reactions of biomass char. Hence, the temperature gradually decreased to around 300 °C at the outlet of the gasifier.

When λ increased, the temperature profile responded regularly. Firstly, the average temperature of the pyrolysis region increased. In the presented novel gasification process, pyrolysis reactions take place under an oxidative environment, and the oxygen concentration in the pyrolysis zone varied with λ. Since the heterogeneous oxidation reaction rate is slower than that of homogeneous oxidation, the heat release rate is slow and gentle, thus the temperature increased slightly. The oxidative reactions mainly happen on the surface of biomass or biomass semi-coke, where degradation of biomass large molecules occurs as well. The oxidation and degradation reactions almost take place at the same time and in the same location. Hence, it is an "in-situ," oxidative degradation process [24]. In this way, the heat released from oxidation could be effectively adsorbed by reduction reactions. Therefore, the overall temperature won't rise too much, while a high decomposition rate of biomass fuel can be achieved. This is the major advantage of this air-staged gasifier.

The steep growth at the inlet of second region can be attributed to severe combustion of the pyrolytic gas. As shown in Figure 6, when λ increased from 0.28 to 0.36, the peak temperature rose from 965 to 1196 °C. It is obvious that a higher λ promoted partial oxidation reactions, and hence released more heat to this region. The temperature didn't vary linearly with λ. When λ increased from 0.36 to 0.45, the temperature decreased from 1196 to 1124 °C. This is because that under higher λ conditions, biomass pyrolysis is already finished. Increased air injection will dilute the pyrolytic gas since a large amount of N_2 is injected into the system. Therefore, the partial oxidation rate of the pyrolytic gas is slowed down and the temperature of the second region decreased accordingly.

In the third region, a distinct peak was observed at the inlet as well. However, it behaved differently with λ than in the second region. The peak temperature increased with λ in all tests. No inflection point was detected. This is because in the third region, the combustible gas concentration is low, therefore, heterogeneous oxidation reactions prevail. Oxygen was inclined to react with the biomass char rather than the gases. After the previous two stages, biomass devolatilization was completed. Char produced in the present gasification process would be more active, since it was produced by oxidative degradation. Researchers have reported that biomass char produced by pyrolysis in certain oxygen concentration environments has a highly porous structure, and the specific surface area could reach a value of 321 m^2/g [19]. Moreover, oxidative char is rich in oxygen- containing functional groups, which may act as an "activation center" during gasification and oxidation [25]. In conclusion, biomass char produced in an oxidative environment is more reactive. For this reason, an increase in λ results in a promotion of heterogeneous oxidation reactions and heat release and therefore, the average temperature of the third region rose with λ. At the outlet of the third region, the temperature dropped down to around 300 °C. Some components of the biomass tar would condense under 350 °C. Thus, a lower temperature would be helpful for trapping tar. In addition, a low temperature syngas outflow also indicates a high energy efficiency.

According to the above analysis, the temperature features of this gasifier can be summarized as follows: the moderate temperature and oxidative environment in the first region promoted the pyrolytic reaction and increased the production of gaseous products. A high temperature area is formed by homogeneous partial oxidation of the pyrolytic gas in the second region. This resulted in a high tar destruction rate. Biomass char produced by oxidative degradation had a high reaction activity. Hence, the gasification reaction in the third region was favored.

3.3. Syngas Quality

The fuel gas mass yield variation with λ is shown in Figure 7, Figure 8 and Figure 9. The main components are CO, CO_2, H_2, CH_4 and trace amounts of C_xH_y. The unit of gas yield is kg/kg C, where C equals the mass of biomass plus the carbon mass content. The carbon mass content is listed in Table 1.

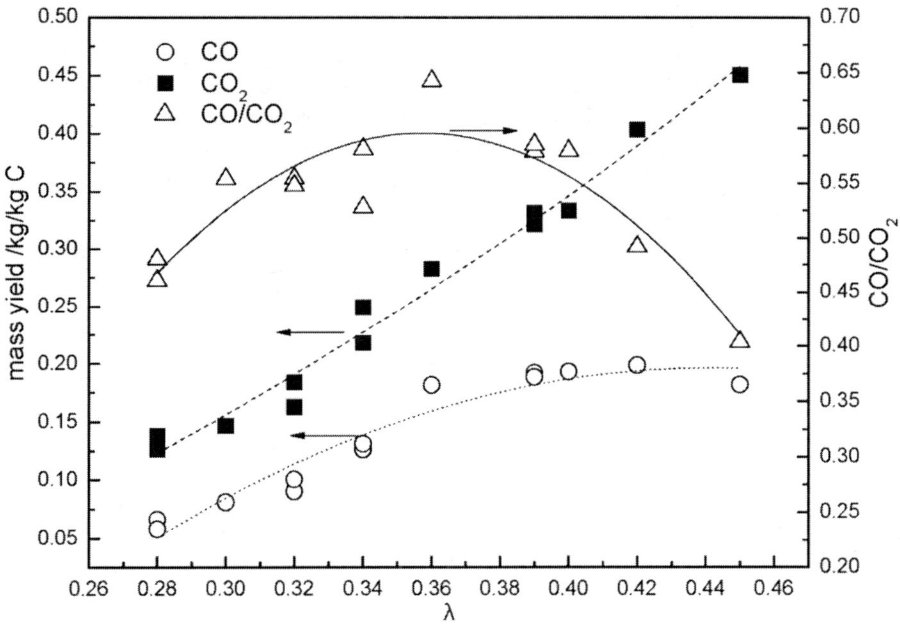

Figure 7. CO and CO_2 mass yield.

CO first increased then decreased as λ increased. As λ increased from 0.28 to 0.42, the yield of CO increased from 0.125 to 0.373 kg/kg C. After that, CO decreased to 0.341 kg/kg C. However, the mass yield of CO_2 continuously increases with λ. The mass of CO_2 is always larger than that of CO for all conditions.

During the biomass thermal chemical conversion process, the production of CO_2 is always accompanied by CO release. It seems that there is a rivalry between the formation of CO and CO_2. The proportion of CO and CO_2 is determined by the temperature and oxygen concentration in the environment. When the temperature was low, an increase of λ resulted in production of CO, while, CO_2 might be favored under a high temperature environment. Looking at the CO/CO_2 value in Figure 7, it is clear that when λ is below 0.36, the temperature is not high enough to initiate homogeneous oxidation. That is kinetically control reaction, thus CO keeps increasing as a result of the promoted pyrolysis reaction. Meanwhile, as λ surpasses 0.36, a significant increase of temperature in the second stage could be observed in Figure 6. Hence, CO would be oxidized into CO_2 and lead to a decrease of CO/CO_2. Under this condition, it seems that the diffusion rate dominates the oxidation reaction.

H_2 yield is shown in Figure 8. The peak value of 0.029 kg/kg C appears at λ = 0.36. The release of H_2 is mainly due to bonds between H radicals that are produced by the decomposition of larger molecules. Moreover, the polycondensation and dehydrocyclization of the biomass char structure would be another source of H_2 formation. Hence, H_2 production has a strong relationship with the environmental temperature. However, when λ exceeded 0.36, H_2 may be consumed through combustion.

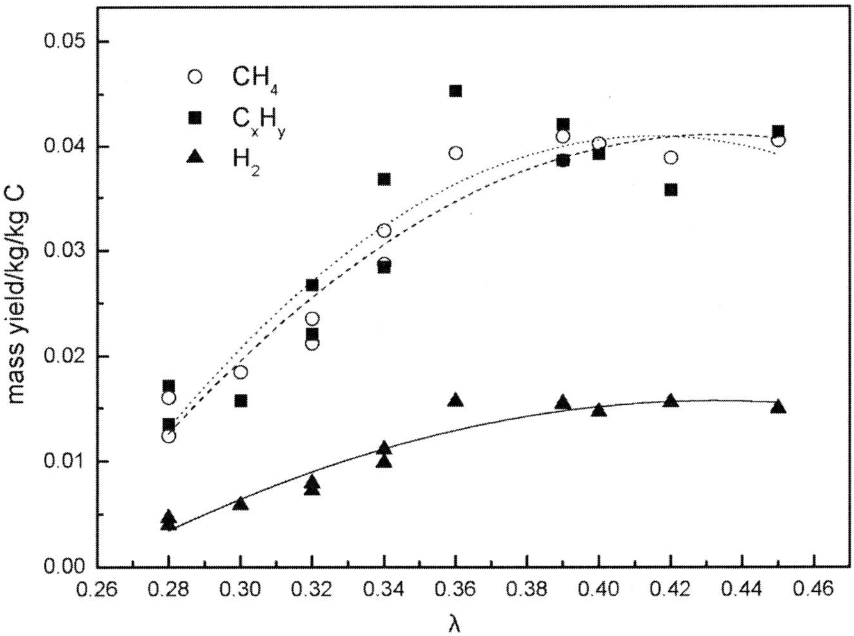

Figure 8. CH_4, C_xH_y and H_2 mass yield.

CH_4 and C_xH_y are shown in Figure 8. The formation of CH_4 is always considered to be the product of demethylation reactions. Therefore, the amount of methyls in the biomass structure and the environmental temperature would be important for CH_4 production. The peak values of CH_4 and C_xH_y were 0.074 kg/kg C and 0.085 kg/kg C at the point of λ = 0.36.

The syngas low heat value (LHV) is calculated by Equation (3):

$$LHV\,(kJ/Nm^3) = 126V_{CO}(\%) + 108V_{H2}(\%) + 359V_{CH4}(\%) + 595V_{CxHy}(\%) \quad (3)$$

where V_{CO}, V_{H2}, V_{CH4} and V_{CxHy} are the volume composition of CO, H_2, CH_4 and C_xH_y in the syngas.

The variation of LHV is shown in Figure 9. A higher λ resulted in a rise of the temperature inside the gasifier. However, the quality of syngas was not always increasing. When λ increased from 0.28 to 0.36, LHV grew from 4214.66 kJ/Nm³ to 5469.85 kJ/Nm³, whereas, LHV decreased to 4065.50 kJ/Nm³ when λ increased to 0.45. Thus, optimization of λ was critical in the present biomass gasification process.

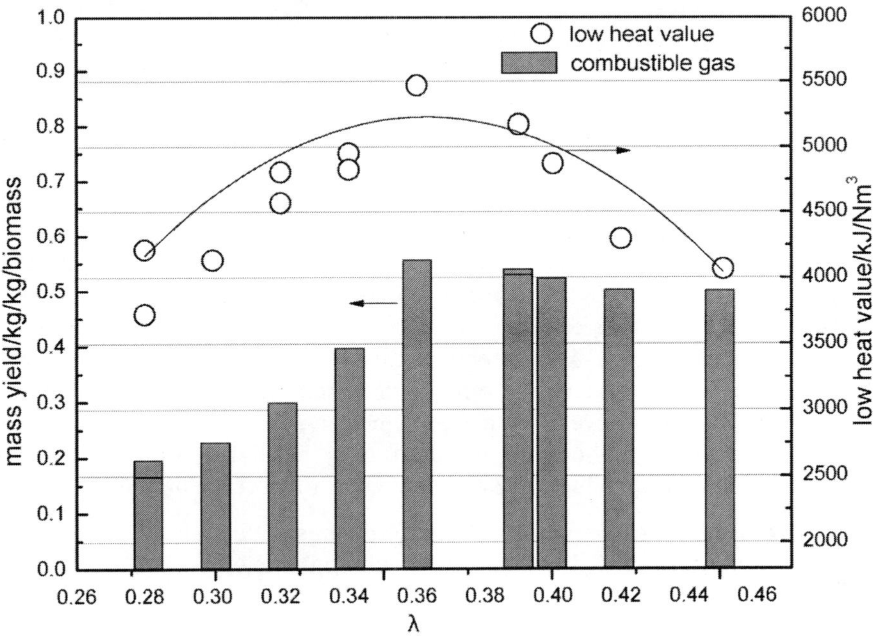

Figure 9. Mass of combustible gas and low heat value of syngas.

The total combustible gas and LHV in syngas reached a maximum value of 0.556 kg/kg biomass and 5469.85 kJ/Nm³ at λ = 0.36 in the current study. Under these conditions, the temperature in the second region was comparatively high. Since the oxygen concentration was not too high, the combustible gases would not be consumed by combustion. It is also found that when λ exceeded 0.36, the environmental temperature rose up to a high level. Therefore, the oxidation of produced fuel gas would reduce the heat value of the syngas. It is clear that λ = 0.36 is a threshold point for improving or deteriorating syngas quality in this study. Several factors contributed to this transformation, such as the high decomposition rate of biomass in an oxidative environment, higher tar destruction efficiency, and more reactive char for gasification.

3.4. Tar Yield and Composition

Biomass tar sampled from sample port 1 (before conditioning) and 2 (after conditioning) were analyzed carefully. The total tar yield variation with λ is shown in Figure 10. It was observed that when λ increased from 0.28 to 0.45, the tar yield was reduced from 404 mg/(kg biomass) to 238 mg/(kg biomass). In the present gasifier, biomass tar was mainly formed at the pyrolysis stage. After that, tar decomposition and conversion happened in the second and third regions. During the tar conversion process, large molecular species not only cracked but polymerized to form high molecular weight aromatics. When the oxygen concentration in the first region was increased, tar production was suppressed. It is reported in the literature that under an oxidative environment, a large number of free radicals are formed [26]. These high reactive radicals will attack the weak bonds between tar molecules, hence pyrolytic tar would easily decompose. Increasing λ resulted in a temperature rise in the second and third regions as shown in Figure 6. A high temperature environment is beneficial for tar decomposition. Moreover, biomass char can act as an adsorbent and catalyst for tar conversion as described before. This effect would be enhanced under higher temperatures. Looking at the above factors, the total biomass tar production was reduced by an increase of λ.

Figure 10 also shows that tar yield obviously decreased after the gas conditioning process. Although the scrubbing efficiency slightly decreased as the input tar concentration was reduced, the average efficiency could be kept at around 65%. This indicated that the gas conditioning system has good load regulation ability.

Figure 11 shows a GC chromatogram illustrating the composition of a typical tar sample from sampling port 1. Each remarkable peak was marked with the corresponding species. The highest peak corresponded to benzene, followed by toluene, styrene, phenol and naphthalene. Benzene, toluene and xylene have lower boiling points. These components are not problematic for further usage. In addition, these species could increase the syngas lower heating value and anti-knocking properties during combustion in a gas engine. However, these tar components easily condense below 300 °C, which could cause fouling in the gas duct and other process facilities.

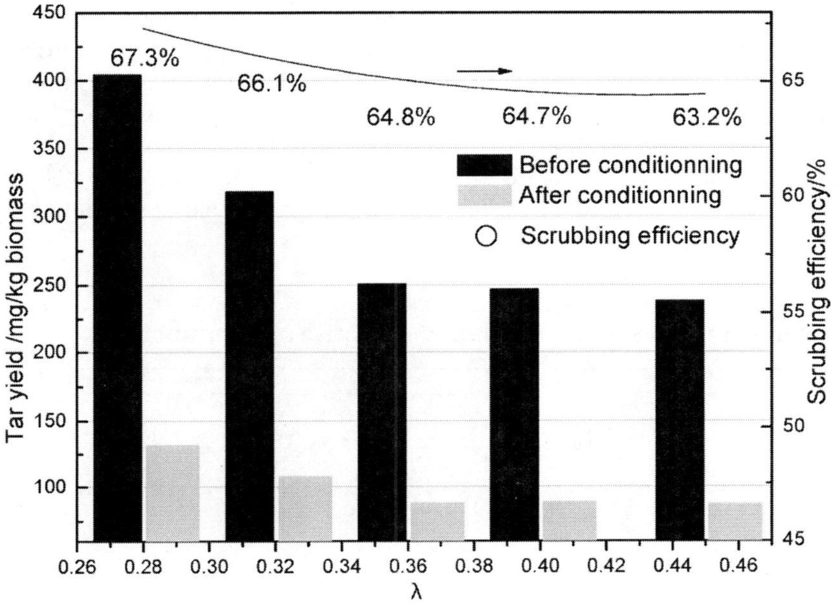

Figure 10. Tar concentration variation with λ.

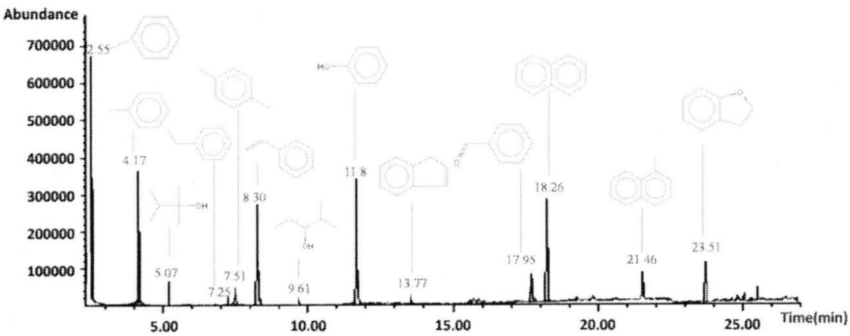

Figure 11. GC chromatogram of a typical sample from sampling port 1.

Tar composition and species relative concentrations are listed in Table 2. The relative concentrations are calculated by the corresponding peak area minus the sum of the areas of all the detected peaks. The listed 13 species represent about 85% of the total detected tar components. Among all the detected species, benzene was the most abundant one. Since benzene is the simplest aromatic with singe ring, any side chain breakage or decomposition of aromatics would produce benzene. It can be seen from Table 2 that when λ increased from 0.28 to 0.45, its relative concentration increased from 27.22% to 48.73%. This implies that tar decomposition and side chain breakage were favored under higher λ. Besides benzene, toluene and naphthalene are the other two most abundant components. When λ increased from 0.28 to 0.45, their concentration rose up from 9.76% and 6.75% to 15.32% and 13.52%, respectively. In comparison, toluene and naphthalene have more thermo-stable structures.

Table 2. Tar concentration and species relative concentration (%).

No.	Molecular Formula	Species Name	Equivalent Ratio (λ)					
			0.28	0.32	0.36 Before Conditioning	0.36 After Conditioning	0.40	0.45
1	C_6H_6	Benzene	37.22	39.21	43.25	41.35	45.64	48.73
2	C_7H_8	Toluene	9.76	9.86	12.46	12.06	14.63	15.32
3	$C_8H_{16}O_2$	2-Butanol, 2,3-dimethyl-	2.45	2.75	2.01	/	1.65	1.1
4	C_8H_{10}	Ethylbenzene	1.21	1.12	0.46	/	0.73	0.37
5	C_8H_{10}	*p*-Xylene	1.12	0.42	0.52	/	0.35	0.26
6	C_8H_8	Styrene	6.53	5.72	4.36	1.68	2.1	2.41
7	$C_6H_{14}O$	3-Pentanol, 2-methyl-	0.87	0.46	0.63	/	0.37	0.43
8	C_6H_6O	Phenol	9.24	6.26	2.68	1.72	2.74	2.35
9	C_9H_8	Indene	0.55	0.94	1.12	/	1.42	1.32
10	$C_7H_6O_3$	Benzaldehyde	2.13	0.96	0.86	/	0.87	0.91
11	C_8H_{10}	Naphthalene	6.75	10.14	12.56	31.32	13.22	13.52
12	$C_{11}H_{10}$	Naphthalene, 1-methyl-	2.31	2.41	2.33	3.17	2.42	2.51
13	C_8H_8O	Benzofuran, 2,3-dihydro-	4.31	4.14	0.85	/	0.67	0.72
Percentage of above species (with benzene) from all detected species %			84.45	84.39	84.09	91.3	86.81	89.95
Percentage of above species (without benzene) from all detected species %			47.23	45.18	40.84	49.95	41.17	41.22

The amounts of phenol and its derivatives were comparatively less. Since hydroxyls easily break away from benzene rings under high temperatures and in an oxidative environment, phenolic products could be converted to benzene easily. As seen from Table 2, as λ increased from 0.28 to 0.45, the phenol concentration decreased from 9.24% to 2.35%. Likewise, benzofuran decreased from 4.31% to 0.72%. Other oxygen-containing tar components presented a decreasing trend with increase of λ as well.

Tar composition variation before and after conditioning is illustrated in Table 2. It can be seen that after gas conditioning, the number of species was reduced remarkably. Six detected species accounted for up more than 90% of the total tar. The primary constituents are benzene, toluene, styrene, phenol, naphthalene and 1-methylnaphthalene.

3.5. Gas Production Rate and Carbon Conversion Rate

Gas production rate and carbon conversion rates are defined by Equations (4) and (5).

$$\text{gas production rate} = \frac{\text{syngas flow rate } \left(\text{Nm}^3/\text{h}\right)}{\text{biomass feeding rate } (\text{kg/h})} \quad (4)$$

Carbon conversion rate:

$$\text{carbon conversion rate} = \frac{\text{carbon element in syngas } (\text{kg})}{\text{carbon element in feedstock } (\text{kg})} \times 100\% \quad (5)$$

As shown in Figure 12, the gas production rate increased continuously with λ. This is because gas production rate takes no account of heat value. The increase of gas production rate after λ = 0.36 is mainly due to a rapid production of CO_2 (Figure 7). Similarly, the carbon conversion rate displays an increasing trend with λ as well. Under the conditions when λ was below 0.36, the carbon conversion rate increased rapidly. When λ was greater than 0.36, the rate of increased was slightly slowed down. As λ increased from 0.28 to 0.36, more air was injected; gasification and oxidation reactions were enhanced. As a result, a large number of combustible gases were produced. With further air being injected, the fuel gas would be combusted, since homogeneous oxidation reactions are faster than heterogeneous oxidation reactions under a high temperature [23]. However, homogeneous oxidation will not affect the carbon conversion rate, therefore, an increase of λ from 0.36 to 0.45 would have a weaker influence on carbon conversion rates than the condition when λ < 0.36.

3.6. Cold Gas Efficiency

Cold gas efficiency indicates the energy conversion efficiency from biomass feedstock to syngas. It is defined by Equation (6):

Cold gas efficiency:

$$\text{cold gas efficiency} = \frac{\text{syngas low heat value } \left(\text{kJ/Nm}^3\right) \times \text{ gas production rate } \left(\text{Nm}^3/\text{kg}\right)}{\text{biomass low heat value } (\text{kJ/kg})} \times 100\% \quad (6)$$

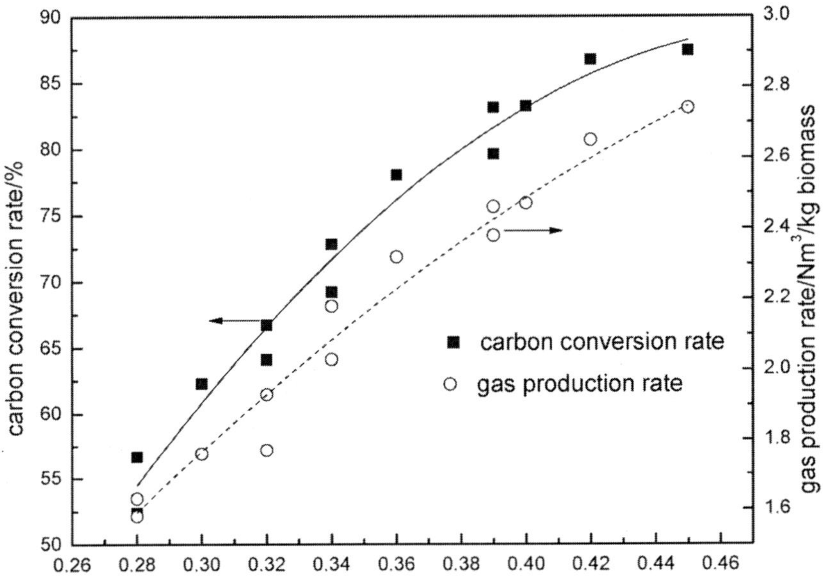

Figure 12. Carbon conversion rate and gas production rate.

A comparison of the theoretical and experimental cold gas efficiency results is plotted in Figure 13. The theoretical thermodynamic equilibrium calculations based on Gibbs free energy minimization is chosen to model the biomass gasification process. For the equilibrium calculations, the software program FactSage 5.4.1 has been used. FactSage is an integrated thermochemical databank system consisting of calculation modules and databases. It enables the user to access and manipulate pure substance and solution databases and perform thermochemical calculations such as multiphase chemical equilibria. It was seen that under equilibrium conditions there is a general increase in the cold gas efficiency with λ. The maximum value of 79% was obtained at λ = 0.32. This peak value is higher than the operational results and occurs at a lower λ. The highest cold gas efficiency of 75% can be achieved when λ was 0.36 under the operation conditions. Since the equilibrium model assumes an isothermal and long enough residence time to achieve equilibrium, the error may be attributed to the non-isothermal temperature inside the gasifier and the short residence time.

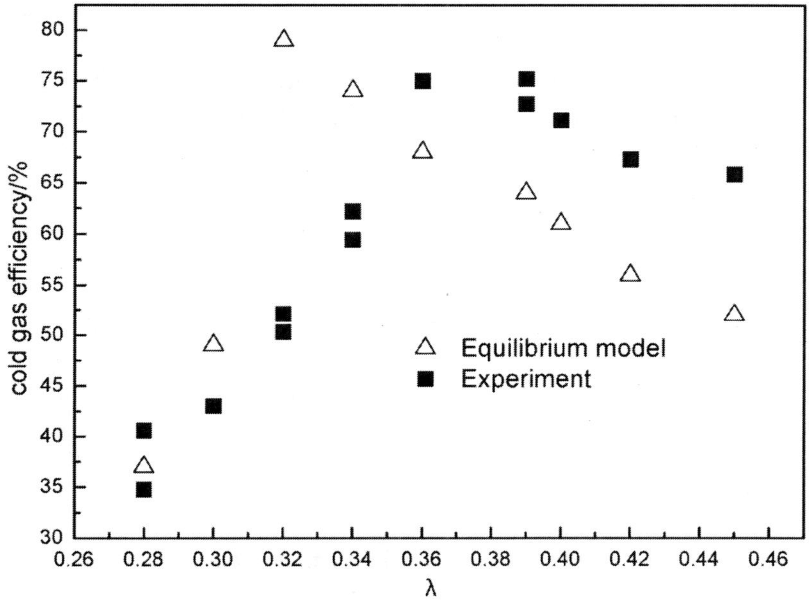

Figure 13. Cold gas efficiency of theoretical vs. experimental results.

3.7. Energy Balance Analysis

The total energy balance is calculated using Equations (7) to (10):

$$E_{input} = E_{biomass} + E_{air} + E_{motor} \tag{7}$$

$$E_{output} = E_{syngas} + E_{solid\ residue} + E_{condensation} \tag{8}$$

where E = mass × heat value:

$$\text{Energy conversion efficiency} = \frac{\text{Esyngas}}{\text{Einput}} \times 100\% \tag{9}$$

$$\text{Error} = \frac{\text{Einput} - \text{Eoutput}}{\text{Einput}} \times 100\% \tag{10}$$

Figure 14 shows the schematic diagram of the system with the energy input and output of products at each stage. Total energy input was determined by biomass feeding rate, total air injection and motor power. In the present test, the energy input was 1420.4 kW. The syngas energy was 1057.5 kW, so the energy conversion efficiency is 74.45%. This was due to heat losses from the system and also the heat reserved in the char residue. Besides, some of the energy is also lost in the gas conditioning system through the scrubbing and condensation processes. The characteristics of the solid residue are listed in Table 3.

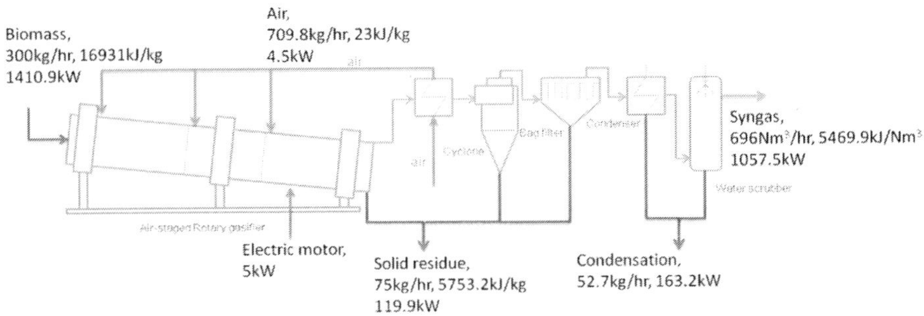

Figure 14. Energy input and output of the gasification system.

Table 3. Proximate and ultimate analysis of the solid residue.

Proximate Analysis [1] (wt %)				Ultimate Analysis [1] (wt %)					Low Heat Value (kJ/kg)
Water	Ash	Volatile	Fixed Carbon	C	H	O	N	S	
1.35	88.29	0	10.36	5.93	3.17	0.12	0.24	0.31	5753.2

[1] As received basis.

The heat value of the solid residue was determined by a bomb calorimeter test. The average heat value is 5753.2 kJ/kg. The solid residue ash production rate is 75 kg/h, which is measured by regular weighing of all the ash from the gasifier, cyclone and bag filter (shown in Figure 14). As syngas comes in contact with the cyclone internal walls, the temperature drops and tar components along with water vapor may condense. In this process, ash and char would contact with the condensates, causing the ash and char to become wet, decreasing its calorific value. The condensation heat is measured by monitoring the scrubbing and cooling water temperature variation as well as their flow rate. The heat inputs and outputs drawn in Figure 14 were not totally balanced. The error of 4.95% could be due to system heat dissipation and other possible heat losses that could not be measured.

3.8. Comparison of Performance of Biomass Gasifier

In Table 4 the experimental results are compared with those reported in the literature. Generally, fluidized bed gasifiers provide an intensive contact between gas and solid biomass which results in higher reaction rates and conversion efficiencies. Fixed bed gasifiers typically have a lower heat and mass transfer and often generate higher amounts of char. The peak temperature in a fixed bed system is also lways higher than in a fluidized bed gasifier, therefore, higher heat value syngas is produced. A recent review [14] reported that multi-stage gasifiers show an obvious advantage compared with other types. The separation of heating and drying, pyrolysis, oxidation and gasification processes makes it possible to control and optimize the different steps separately. Multi-stage gasifiers produce high purity gas with low levels of tar and high process efficiencies. In the present study, the rotary kiln gasifier combined both the advantages of fixed bed and fluidized bed gasifiers to form a new multi-stage gasifier concept, which has not been reported in any of the previous work. This gasifier provides a more intensive mixture of gas and solid fuel than a fixed bed one, meanwhile, its temperature is higher than that of a fluidized bed one. Table 4 shows that the present gasifier has a comparatively higher heat value of gas with low tar concentration.

Table 4. Comparison of the experimental results with published in literature.

Biomass Fuel	Low Heat Value of Syngas MJ/m^3	Tar Concentration (g/m^3)	Cold Gas Efficiency (%)	Gasifier Type	Reference
Biomass/waste	2.9–4.9	10~40/0.01–0.9/1~23	39.1–56.9/<70	fluidized bed	[14,27,28]
Agrol, willow, and DDGS, sawdust	2.7–4.3	5–12/1~30	50-70	Circulated fluidized bed	[14,28–30]
Sesame wood and rose wood/hazelnut shell	4.5–6.3/4~5	0.015–0.5/0.01~6	30–80.91	Downdraft fixed bed	[14,29,31–33]
Sewage sludge and woodpellets	4.9–5.3/5~6	30–150/10~150	20–60	Updraft fixed bed	[14,29,33–35]
Straw, manure fibres, sewage sludge	5.2~7/6.6	4.8/<0.015	87%~93%	Multi-stage	[14,36]
Waste wood	5.5	0.038	75	Rotary kiln	present study

4. CONCLUSIONS

A new gasification process concept was demonstrated in a pilot scale gasification system. This concept is based on air-staged injection with separated three thermo-chemical conversion regions inside the gasifier. Results showed that under optimum conditions, the gas production rate and syngas heat value could reach 2.32 Nm^3/kg and 5469.85 kJ/Nm^3, respectively. Tar concentration before and after gas conditioning could be reduced to 108 mg/Nm3 and 38 mg/Nm^3, respectively. Cold gas efficiency and carbon conversion efficiency could reach about 75% and 78%. A comparison of this study with literature data shows that a higher calorific value gas with lower tar concentration is produced in the present gasifier. The performance of this gasification system showed a considerable potential for implementing it in distributed electricity and heat supply projects.

ACKNOWLEDGMENTS

This work was carried out with financial supports from National Natural Science Foundation of China (Grant No. 61171008 and No. 21103024).

AUTHOR CONTRIBUTIONS

Huiyuan Shi and Wen Si conceived and designed the experiments; Huiyuan Shi performed the experiments; Wen Si and Xi Li analyzed the data; Huiyuan Shi wrote the paper.

CONFLICTS OF INTEREST

The authors declare no conflict of interest.

REFERENCES

1. Saxena, R.C.; Seal, D.; Kumar, S. Thermo-chemical routes for hydrogen rich gas from biomass: A Review. Renew. Sustain. Energy. Rev. **2008**, 12, 1909–1927.
2. Balat, M.; Balat, M.; Kırtay, E.; Balat, H. Main routes for the thermo-conversion of biomass into fuels and chemicals. Part 2: Gasification Systems. Energy Convers. Manag. **2009**, 50, 3158–3168.

3. Fredriksson, H.O.A.; Lancee, R.J.; Thüne, P.C.; Veringa, H.J.; Niemantsverdriet, J.W. Olivine as tar removal catalyst in biomass gasification: Catalyst dynamics under model conditions. Appl. Catal. B Environ. **2013**, 130–131, 168–177.

4. Virginie, M.; Adánez, J.; Courson, C.; de Diego, L.F.; García-Labiano, F.; Niznansky, D.; Kiennemann, A.; Gayán, P.; Abad, A. Effect of Fe-olivine on the tar content during biomass gasification in a dual fluidized bed. Appl. Catal. B Environ. **2012**, 121–122, 214–222.

5. Barisano, D.; Freda, C.; Nanna, F.; Fanelli, E.; Villone, A. Biomass gasification and in-bed contaminants removal: Performance of iron enriched Olivine and bauxite in a process of steam/O_2 gasification. Bioresour. Technol. **2012**, 118, 187–194.

6. Kirkels, A.F.; Verbong, G.P.J. Biomass gasification: Still promising? A 30-year global overview. Renew. Sustain. Energy Rev. **2011**, 15, 471–481.

7. Milne, T.A.; Evans, R.J. Biomass Gasifier"Tars": Their Nature, Formation, and Conversion; National Renewable Energy Laboratory: Golden, CO, USA, 1998.

8. Goyal, H.B.; Seal, D.; Saxena, R.C. Bio-fuels from thermochemical conversion of renewable resources: A Review. Renew. Sustain. Energy Rev. **2008**, 12, 504–517.

9. Kirubakaran, V.; Sivaramakrishnan, V.; Nalini, R.; Sekar, T.; Premalatha, M.; Subramanian, P. A review on gasification of biomass. Renew. Sustain. Energy Rev. **2009**, 13, 179–186.

10. Evans, R.J.; Milne, T.A. Molecular characterization of the pyrolysis of biomass. 1. Fundamentals. Energy Fuels **1987**, 1, 123–138.

11. Wei, L.; Xu, S.; Zhang, L.; Zhang, H.; Liu, C.; Zhu, H.; Liu, S. Characteristics of fast pyrolysis of biomass in a free fall reactor. Fuel Process. Technol. **2006**, 87, 863–871.

12. Chen, Y.; Luo, Y.; Wu, W.; Su, Y. Experimental investigation on tar formation and destruction in a lab-scale two-stage reactor. Energy Fuels **2009**, 23, 4659–4667.

13. Houben, M.P. Analysis of tar removal in a partial oxidation burner. Ph.D. Thesis, Eindhoven University of Technology, Eindhoven, The Netherlands, 2004.

14. Heidenreich, S.; Foscolo, P.U. New concepts in biomass gasification. Prog. Energy Combust. Sci. **2015**, 46, 72–95.

15. Pan, Y.G.; Roca, X.; Velo, E.; Puigjaner, L. Removal of tar by secondary air injection in fluidized bed gasification of residual biomass and coal. Fuel **1999**, 78, 1703–1709.

16. Narv, P.I.; Orpso, A.; Aznar, M.P.; Corella, J. Biomass gasification with air in an atmospheric bubbling fluidized bed. Effect of six operational variables on the quality of produced raw gas. Ind. Eng. Chem. Res. **1996**, 35, 2110–2120.

17. Yoshikawa, K. R&D (Research and Development) on distributed power generation from solid fuels. Energy **2006**, 31, 1656–1665.

18. Henriksen, U.; Ahrenfeldt, J.; Jensen, T.K.; Gobel, B.; Bentzen, J.D.; Hindsgaul, C.; Sorensen, L.H. The design, construction and operation of a 75 kW two-stage gasifier. Energy **2006**, 31, 1542–1553.

19. El-Hendawy, A.-N.A. Surface and adsorptive properties of carbons prepared from biomass. Appl. Surf. Sci. **2005**, 252, 287–295.

20. Liu, X.; Li, W.; Xu, H.; Chen, Y. A comparative study of non-oxidative pyrolysis and oxidative cracking of cyclohexane to light alkenes. Fuel Process. Technol. **2004**, 86, 151–167.

21. Wang, H.; Dlugogorski, B.Z.; Kennedy, E.M. Coal oxidation at low temperatures: Oxygen consumption, oxidation products, reaction mechanism and kinetic modeling. Prog. Energy Combust. Sci. **2003**, 29, 487–513.

22. Putzeys, O.; Bar-Ilan, A.; Rein, G.; Fernandez-Pello, A.C.; Urban, D.L. The role of secondary char oxidation in the transition from smoldering to flaming. Proc. Comb. Inst. **2007**, 31, 2669–2676.

23. Senneca, O.; Chirone, R.; Salatino, P.; Nappi, L. Patterns and kinetics of pyrolysis of tobacco under inert and oxidative conditions. J. Anal. Appl. Pyrolysis **2007**, 79, 227–233.

24. Yi, S.U. Study on Product Properties and Reaction front Propagation Charateristics during Oxidative Pyrolysis of Biomass. Ph.D. Thesis, Shanghai Jiao Tong University, Shanghai, China, 2013.

25. Wang, H.; Dlugogorski, B.Z.; Kennedy, E.M. Analysis of the mechanism of the low-temperature oxidation of coal. Comb. Flame **2003**, 134, 107–117.

26. Hayes, C.J.; Merle, J.K.; Hadad, C.M. The chemistry of reactive radical intermediates in combustion and the atmosphere. In Advances in Physical Orgnanic Chemistry; Richard, J.P., Ed.; Academic Press: New York, NY, USA, 2009; pp. 79–134.

27. Gómez-Barea, A.; Leckner, B.; Perales, A.V.; Nilsson, S.; Cano, D.F. Improving the performance of fluidized bed biomass/waste gasifiers for distributed electricity: A New Three-Stage gasification System. Appl. Therm. Eng. **2013**, 50, 1453–1462.

28. Han, J.; Kim, H. The reduction and control technology of tar during biomass gasification/pyrolysis: An Overview. Renew. Sustain. Energy Rev. **2008**, 12, 397–416.

29. Meng, X.; de Jong, W.; Fu, N.; Verkooijen, A.H.M. Biomass gasification in a 100 kW_{th} steam-oxygen blown circulating fluidized bed gasifier: Effects of operational conditions on product gas distribution and tar formation. Biomass Bioenergy **2011**, 35, 2910–2924.

30. Li, X.T.; Grace, J.R.; Lim, C.J.; Watkinson, A.P.; Chen, H.P.; Kim, J.R. Biomass gasification in a circulating fluidized bed. Biomass Bioenergy **2004**, 26, 171–193.

31. Sheth, P.N.; Babu, B.V. Experimental studies on producer gas generation from wood waste in a downdraft biomass gasifier. Bioresour. Technol. **2009**, 100, 3127–3133.

32. Dogru, M.; Howrath, C.R.; Akay, G.; Keskinler, B.; Malik, A.A. Gasification of hazelnut shells in a downdraft gasifier. Energy **2002**, 27, 415–427.

33. IEA (International Energy Agency). Task33: Small Scale Gasification: Gas Engine CHP for Biofuels, 2011. Swedish Energy Agency Report. Available online: http://www.ieatask33.org/app/webroot/files/file/publications/new/Small%20Small_scale_gasification_overview.pdf (accessed on 29 December 2015).

34. Seggiani, M.; Vitolo, S.; Pucclni, M.; Bellini, A. Cogasification of sewage sludge in an updraft gasifier. Fuel **2012**, 93, 486–491.

35. Plis, P.; Wilk, R.K. Theoretical and experimental investigation of biomass gasification process in a fixed bed gasifier. Energy **2011**, 36, 3838–3845.

36. Zwart, R.; van der Heijden, S.; Emmen, R.; Bentzen, J.D.; Ahrenfeldt, J.; Stoholm, P.; Krogh, J. Tar Removal from Low-Temperature Gasifiers, 2010. ECN Report. Available online: http://www.ecn.nl/docs/library/reort/2010/e10008.pdf (accessed on 29 December 2015).

7

CHAPTER

COST ESTIMATES OF COAL GASIFICATION FOR CHEMICALS AND MOTOR FUELS

Marek Sciazko[1] and Tomasz Chmielniak[1]

[1] *Institute for Chemical Processing of Coal, Zabrze, Poland*

1. INTRODUCTION

Solid fuels gasification technology has been understood and applied for a long time. The current directions in developing coal gasification technology are primarily related to power generation in combined systems involving steam and gas turbine implementation, which considerably increases fuel use efficiency. Compared to the first gasifying installations, the current solutions have a much higher conversion intensity and are more reliable. Integrated power generation-related gasification technology developments have created increased interest in chemical products, such as liquid motor fuels, methanol and hydrogen. At the present time, the basic reason for the increase of coal use as a raw material for chemical production is the dynamic industrial growth in countries with high economic potential that do not have their own natural gas and oil resources and have limited access to international sources of the above minerals. China is a good example of a country in this situation, and it constitutes the largest coal gasifying economy in the world. In China alone, more than 100 million tonnes of coal is gasified yearly. We expect that countries such as the USA and India will follow China in coal gasification-based production growth.

The crucial driver of gasification technology development is the necessity of a drastic reduction in CO_2 emission from anthropogenic sources, which is considered to be one of the main contributors to the greenhouse effect. Among fossil fuels, the most important CO_2 emitter is coal, which is characterised as having the highest concentration of carbon element compared to its caloric value. In the coal gasification process, carbon dioxide is removed from the processed gas by the absorption of acid components, which constitutes an inherent part of the technology. In case of chemical plant the acid gases, i.e. H_2S and CO_2 must be removed from the processed gas, regardless of the chemical facility's production profile because H_2S can damage the catalysts used during chemical synthesis, and the content of CO_2 is corrected to the expected composition of a synthesis gas. This removal step can alleviate the need for additional CO_2 separation so that the costs associated with dehydration and compression are the primary costs remaining. These two processes are critically important to system, as they ensure safe transport of the CO_2 to the storage (sequestration) area.

In the case of integrated gasification combined cycle (IGCC, power generation), the removal of sulphur compounds (H_2S, COS) is required to protect the gas turbine, and CO_2 removal is conducted only to reduce atmospheric emissions. However, because of the high concentrations of carbon dioxide and the high-pressure of the treated gas, the removal of CO_2 from syngas (i.e., pre-combustion removal) is less expensive than if the CO_2 were separated from the flue gases (post-combustion removal). Pre-combustion CO_2 removal results in better process and economic efficiency of IGCC systems (in case of CO_2 sequestration) compared to conventional power plants based on coal combustion.

The development state of coal gasification technology

A review of the global development state of gasification technologies has been performed based on a 2010 database developed by the U.S. Department of Energy (US Department of Energy & National Energy Technology Laboratory [US DOE & NETL], 2010a). The results of this analysis have been compared in three categories characterising the current status of technology development: plants that are operational, plants that are under construction (or start-up) and, plants that are in the development phase (this category includes plants in varying degrees of implementation, including plants at the stages of planning, conceptual work and designing). When analysing the data for the various systems, plants that use natural gas as a fuel have been omitted as these plants are not considered to be gasification systems but rather are plants for the partial oxidation of natural gas. The total power of the aforementioned systems (the thermal capacity of syngas output) amounts to 15,281 MW_{th}, of which 72 % (10,936 MW_{th}) is attributed to a plant using a Shell pressure reactor that is under construction in Qatar.

The published data show that there are 116 gasification plants equipped with 342 reactors with a total power of 50,104 MW_{th} are currently operating worldwide. The total power represents the chemical energy in the gas that is produced but does not include the systems for the partial oxidation of natural gas. Seventeen systems are under construction (28 gasification reactors, 16,289 MW_{th}; coal), and 37 plants (76 reactors, 40,432 MW_{th}) are at the planning stage with systems to be implemented in the years 2011 - 2016. Since the last review in 2007, the installed power increased by 7 %, resulting in the largest recorded increase for coal gasification at 18 %. For other fuels, there was a clear decrease in the amount of gas produced, particularly for biomass and petroleum coke (-68 % and -37 %, respectively) (Table 1). The implementation of all current and planned investment projects will contribute to more than a doubling of gas production (106,825 MW_{th}).

The largest percentage of gasification systems is operating in the Asia and Oceania region (39 % of total global gas production), primarily because of extremely dynamic technology developments in China (78 % of this region). In this region, which includes China, Australia, South Korea and Vietnam, the majority of systems are now under construction and planned for implementation in the next few years. Long-term plans exist for technology development in North America (primarily in the U.S.), the implementation of which would move this region into second place in the global production of gas from gasification (30.4 % of global gas production).

Coal, the basic feedstock for gasification, is used in gasification plants that are currently operating and accounts for 61.6 % of global gas production (Fig. 1). Petrochemical industry by-products rank second (35.8 %), and the remaining 2.6 % of gas production is attributed to petroleum coke and biomass. For plants that are under construction and planned for start-up by 2016, the role of coal as the basic fuel will be maintained, and the share of gas produced from this raw material will increase to 79 %.

The basic products of operational systems using gasification processes comprise chemicals such as ammonia, hydrogen and oxy-chemicals (46 % of world gas productions), products of Fischer-Tropsch synthesis (30 %), power (16 %) and gaseous fuels (8 %) (Fig. 2). Chemicals will also be the main products of the plants that are under construction (72 %). In the case of plants planned for implementation, the largest share will be power-generating plants (37.5 %), which is probably related to the attractiveness of power systems that are integrated with gasification, particularly in the context of the necessity for CO_2 emission reduction (Fig. 2).

Table 1. Comparison of the state of worldwide existing gasification technologies in the years 2007 and 2010 (US DOE & NETL, 2007, 2010a).

Feedstock		Operating 2010 (operating, construction, start-up)	Operating 2007 (operating, construction, start-up)	Difference	%
Coal	MW_{th}	36,315	30,825	5,490	18
	Gasifiers	201	212	-	-
	Plants	53	45	-	-
Petroleum	MW_{th}	17,938	18,454	-516	-3
	Gasifiers	138	145	-	-
	Plants	56	59	-	-
Petcoke	MW_{th}	911	1,441	-530	-37
	Gasifiers	5	8	-	-
	Plants	3	5	-	-
Biomass/ waste	MW_{th}	373	1,174	-801	-68
	Gasifiers	9	21	-	-
	Plants	9	13	-	-
Total	MW_{th}	55,537	51,894	3,643	7
	Gasifiers	353	386	-	-
	Plants	121	122	-	-

In the case of coal use, the most popular gasification plants are now fixed bed gasification technology, which is practically no longer developed but still accounts for 57 % of gas production due to the high manufacturing potential of the Sasol plant in South Africa.

Processes using entrained flow reactors are the most intensively developed technologies (operating plants, 43 % of gas production) as confirmed by the projects that are under construction and planned for start-up by 2016, which are nearly all related to this reactor design. Fig. 3 shows the structure of the operational plants and the coal gasifiers planned for start-up in terms of the technological solutions used.

Of the technologies used for coal gasification in entrained flow reactors (operating plants), the Shell (dry feeding) and GE/Texaco (slurry feeding) have the dominant share in gas production (77 %), followed by the ECUST (15.3 %) technology. The third place position of the use of ECUST (East China University of Science and Technology) technology in developing plants is noteworthy because of the rapid pace of the ECUST technology development. Beginning with a pilot plant (22 t/d of fuel) in 1996, the technology led to operational demonstration plants in the years 2001 – 2005 (750 and 1,150 t/d of coal) and 17 commercial gasifiers that were implemented by 2010 (capacity of up to 2,000 t/d of coal) (Liu, 2010).

The highest percentage of plants planned for start-up that are under construction and in the development phase will use the Shell gasification technology (26.7 %; 11,913.2 MW_{th}) followed by ECUST (20.8 %), Udhe PRENFLO (16.8 %), Siemens (14.7 %), ConocoPhillips E-Gas (11.3 %), GE Energy/Texaco (5.3 %) and MHI (3.7 %) (Fig. 3). The fluidised bed coal gasification reactor technologies GTI U-GAS and TRIG (KBR Transport Gasifier) will be developed apart from the entrained flow technologies.

2. TECHNOLOGICAL OPTION

The review of the global development state of gasification technologies shows that gasification systems will be used for syngas production in power generation systems (IGCC) and particularly in chemical synthesis to obtain liquid and gaseous fuels including methanol and hydrogen. The analysis of the above processes is the subject of this study. Four cases for coal gasification applications involving chemical synthesis and electricity generation have been analysed and discussed in detail. The options include:

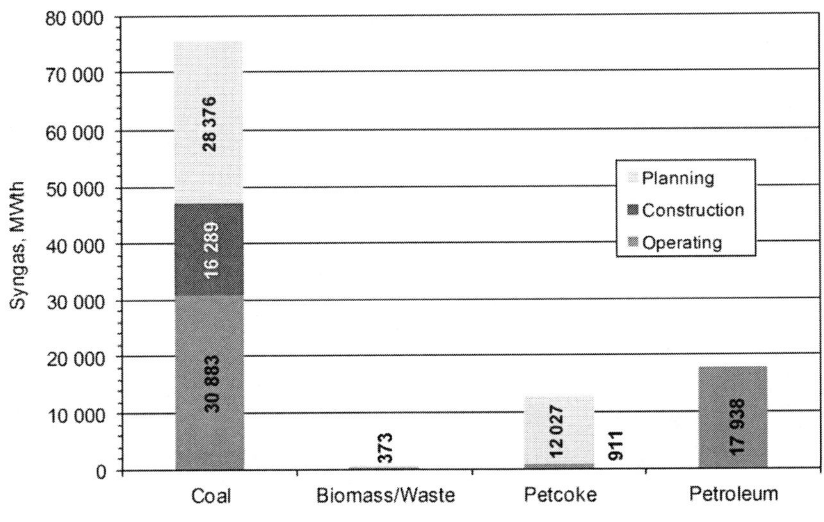

Figure 1. Total capacity of gasifiers versus fuel used (current and forecast by 2016).

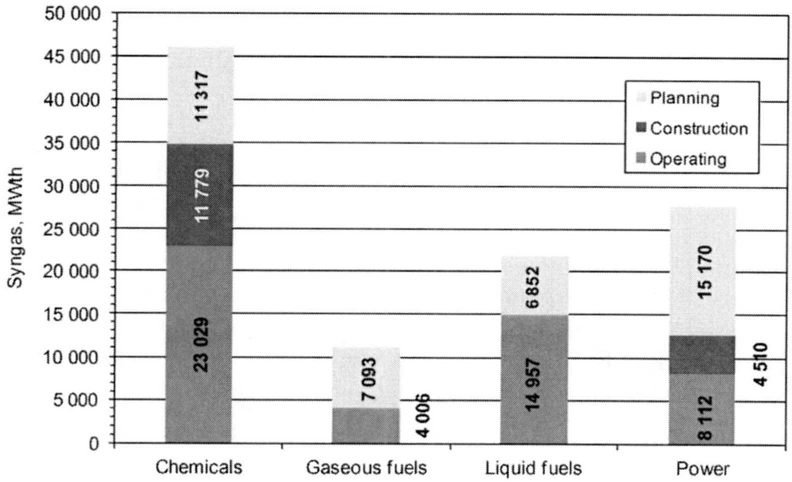

Figure 2. Total capacity of gasifiers versus product manufactured (current and forecast by 2016).

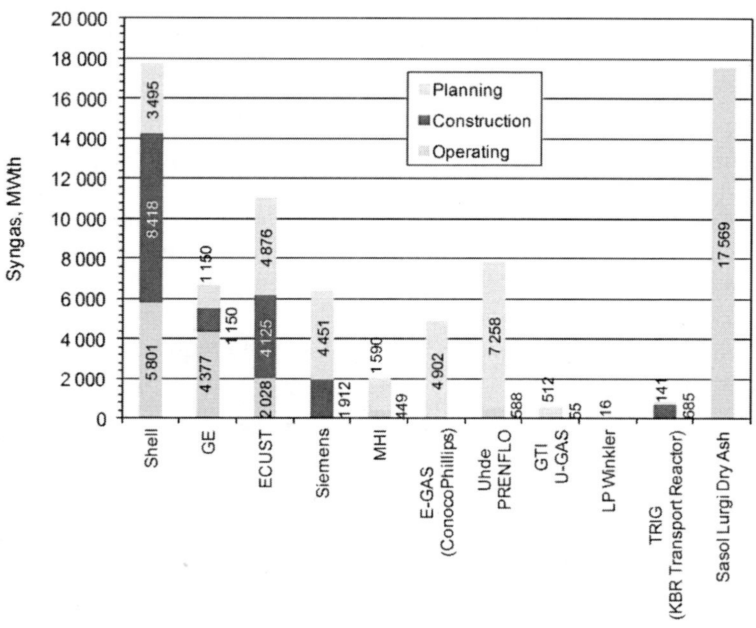

Figure 3. Total capacity of reactors using coal as the main fuel, breakdown by technological groups (current and forecast by 2016).

liquid fuel production, hydrogen generation, methanol production (options I-III) (Dreszer & Mikulska, 2009), and power and syngas production (Polygeneration Plant, option IV) (Chmielniak et al., 2008; Energoprojekt-Katowice S.A & Institute for Chemical Processing of Coal [EPK & IChPW], 2008).

Option I – A system of six gasifiers, which requires an annual coal consumption of approximately 5,600,000 t/y. The adopted scale of coal processing results from preliminary cost-effectiveness studies for liquid fuel production from coal, which have shown that the operation of a production plant starts to be profitable only at a production level exceeding 1 million tonnes of liquid fuels, which corresponds to the adopted scale of coal consumption. The plant products have been defined as technical propane-butane (LPG), diesel oil and a semi-product for the diesel oil that is not further processed into final commercial products.

Option II and III – one gasifier system. The adopted scale makes it possible to accomplish the following:

- cover the demand for hydrogen on the scale of a single standard chemical plant fertiliser production train (no network for high-volume hydrogen distribution was assumed) (option II).
- produce methanol from the gas originating from coal gasification on the scale of 500,000 t/y (option III).

Option IV – a system of two gasifiers operating in parallel technological trains to produce syngas (for methanol synthesis) and power (IGCC). Due to their identical capacity, gasifiers operating in an integrated system can provide mutual back-up functionality for each other, increasing the annual availability of syngas or electricity production units, depending on the adopted production programme. The scale of production allows to manufacture of approximately 500,000 t/y of methanol what ensures the profitability of its production.

Each of the analysed options consists of a syngas generation unit, i.e., a coal gasification system including units for converting and cleaning syngas.

Gasification technologies in the entrained flow reactors play an essential role in the production of syngas from coal and are offered by a number of providers. The final choice of gasification technology must therefore be made using a separate analysis based on detailed data from the technology providers, including investment and operational cost and the assessment of coal suitability for processing.

GE/Texaco technology has been selected for the analysis of the considered cases for the following factors:

- mature technology / solution used for the longest recorded period,
- one of the largest shares in the coal gasification sector (33 %, operational plants),
- absence of inert gases in syngas, which constitute a redundant ballast in chemical synthesis and result in an increase in equipment size needed due to the increased gas volume in the circuit.

The disadvantage of this technology is the lower energy efficiency of the gasification process compared to technologies using dry coal feeding. However, it has been assumed that to assess different fuel production systems based on coal gasification using conceptual studies, it will be less risky to assume process guidelines for GE/Texaco technology with a coal-water slurry feeding system.

3. DESCRIPTION OF CONSIDERED TECHNOLOGICAL SYSTEMS

3.1. Coal gasification – GE/Texaco technology

The coal-water slurry (62-68 % coal) and oxygen from the air separation system are fed through a system of valves to the feedstock injector in the top part of the reactor where gasification proceeds at a temperature of 1,260 – 1,480 °C. The hot processed gas with molten ash flows to the bottom part of the reactor, the radiant cooler, where it is cooled down to approximately 730 °C and then is taken off of the reactor to a convective cooler and a scrubber. After being cooled down to approximately 230 °C, the raw gas is directed to the gas conversion and/or cleaning systems. High-pressure (HP) steam is produced in the radiant and convective coolers. The molten ash flows down to the water bath in the bottom part of the reactor where, after solidification and cooling, it is taken off the system through a lockhoppers. Fly ash that is separated from the gas is also taken off together with slag (the ash separation from the gas occurs through a sudden change of its flow direction before leaving the reactor). After water separation, the slag is directed to a waste landfill. Separated fly ash with a carbon fraction of approximately 30 % is delivered to the coal-water slurry preparation system for recirculation to the reactor. The spraying water from the scrubber and the water from slag dewatering is returned to the scrubber after the removal of solid particles (fine slag/fly ash), and its excess is fed to a water treatment plant (US DOE & NETL, 2002).

In addition to the technology option described above, General Electric commercially offers two other configurations of gasification plants (US DOE & NETL, 2010b):

- a reactor with direct water cooling: in this system, the hot processed gas is cooled down to 260 °C through direct contact with water before leaving the reactor.
- a reactor with a radiant cooler: the processed gas leaving the gasification zone passes through a radiant cooler that produces high pressure steam where it is cooled down to approximately 800 °C and then passes through a water lock, which lowers its temperature to approximately 200 °C.

The gasification pressure was assumed to be 3, 5.6 and 7 MPa for the production systems of liquid fuels, hydrogen and methanol plants, respectively (the pressure was selected to match the process condition for F-T, hydrogen and methanol production units).

Oxygen for the gasification system is supplied from an air separation system based on cryogenic separation. Oxygen purity levels of 99.5 % for liquid fuel production and 95 % for all other cases were assumed.

3.2. System layout – Fuel and chemical production plants

Fig. 4 to 6 present the process diagrams of the considered plants based on coal gasification. The data on the technological configurations are summarised in Table 2.

3.2.1. Liquid fuel production plant

Gas from the gasification system is directed to the hydrolysis reactor where, in the presence of the catalyst, carbonyl sulphide (COS) is hydrolysed to hydrogen sulphide. Gas exiting the COS reactor is cooled to approximately 38 °C in several heat exchangers fed by boiler feed water (steam production) or cooling tower water. Entrained water (condensate) is separated and used for coal-water slurry production and for slag cooling in the gasifier. Cool gas is fed to the Selexol system, where hydrogen sulphide and carbon dioxide are removed. Hydrogen sulphide is directed to the Claus system for sulfur recovery. Clean gas is heated to approximately 313 °C, deep purified from the remaining hydrogen sulphide in the reactor, filled with zinc oxide and fed to Fischer-Tropsch synthesis reactors. Fischer-Tropsch synthesis is carried out in a slurry reactor at 250 °C under a pressure of 2 MPa in a presence of cobalt catalyst. Unreacted part of syngas is fed to the carbon dioxide separation system based on chemical absorption (MDEA) and then to the dehydration and compression system. After passing through the product separation system, the gas is then recirculated to autothermic reforming and sent back to the synthesis reactor. Separated carbon dioxide from the Selexol and amine units is compressed to 12 MPa and transported to a storage location.

3.2.2. Hydrogen production plant

Partially cleaned gas from the gasification island is directed to the Water Gas Shift (WGS) reactor where approximately 97 % of the CO is converted to CO_2 and hydrogen. Gas exiting the WGS reactor is cooled to approximately 38 °C and then fed to the Selexol unit. In the two-stage Selexol system, gas is divided into three streams: sour gas (primarily H_2S), carbon dioxide and hydrogen-rich processed gas. Sour gas from the first stage of the Selexol absorber is directed to the sulfur recovery unit (Claus, Scot). CO_2 is compressed to 12 MPa in preparation for transport and storage. Cleaned processed gas with approximately 90 % hydrogen content is fed to a PSA (Pressure Swing Adsorption) system, where hydrogen with >99 % purity is produced. The off gas from the PSA system is combusted in a steam boiler, and then steam from the boiler and from the gasification system is used for power generation in the steam turbine.

3.2.3. Methanol production plant

Partially cleaned gas from the gasifier is divided in two streams. One of them, which accounts for approximately 65 % of the total flow, is fed to the high temperature CO shift reactor, where, at temperatures between 400 °C and 410 °C, a carbon monoxide and steam reaction occurs, generating hydrogen and carbon dioxide and producing the required hydrogen concentration in syngas, which is directed to the methanol synthesis reactor. After being cooled to approximately 250 °C, the gas is then joined with the second stream and directed to the COS hydrolysis reactor. Next, hydrogen sulphide and carbon dioxide are removed in the Selexol system from the gas after it is cooled to 38 °C. The hydrogen sulphide that is removed from the gas is then transported to the Claus system for sulfur recovery. Carbon dioxide is separated with 78 % efficiency (the separation level is assumed to meet the stoichiometric ratio required for methanol synthesis ((H2–CO2)/(CO+CO2) = aprox. 2) and is then compressed to 12 MPa. The composition of the syngas leaving the synthesis system enables its direct use in methanol synthesis. Syngas that is purified in the Selexol process is then joined with circulating tail gas from the synthesis unit and, after being heated to approximately 210 °C, is conducted to the adiabatic, methanol synthesis reactor. The post-reaction mixture leaving the synthesis reactor is then cooled to 38 °C while heating the gas that is being directed to the synthesis reactor, and then it is separated into a liquid methanol fraction and off gas. The liquid fraction is decompressed and transported to a degasifying tank. The raw methanol is then directed to the rectification system, where methanol of high (>99 %) purity level is obtained. Part of the tail gas is compressed and redirected to the methanol synthesis system, and after being decompressed, the remaining gas is combusted in boiler burners where steam is overheated and directed to the steam turbine. The high pressure steam generated in the gasifying system also feeds the turbine.

3.3. Polygeneration plant

A schematic diagram of the Polygeneration Plant is presented in Fig. 6. The system enables simultaneous electricity, heat and syngas generation with sequestration of the carbon dioxide formed during the production process. Joining the combined power and heat generation with syngas production enables the high efficiency of fuel primary energy conversion, low emission indicators and high economic efficiency, also in the case of CO_2 sequestration. The presented solution was developed by Institute for Chemical Processing of Coal (IChPW) and Energoprojekt-Katowice SA (EPK) for TAURON SA (power producer, Poland) and Zakłady Azotowe Kedzierzyn SA (ZAK SA, chemical works, Poland) (Chmielniak et al., 2008 ; EPK & IChPW, 2008).

To demonstrate an alternative for clean coal technology, the concept of a polygeneration facility assumes possible complete elimination of atmospheric carbon dioxide emissions. Thus, the proper configuration of the IGCC system (energy island, Fig. 6) is necessary for the efficient removal of CO_2 (CO_2 removal in a chemical island is a technological need for the production of syngas). Regarding the IGCC plants that are currently under operation (without CO_2 removal), major changes include the introduction of CO shift reactors and CO_2 separation system. The CO conversion process allows to convert gasifier product (raw gas) to hydrogen-rich syngas and to

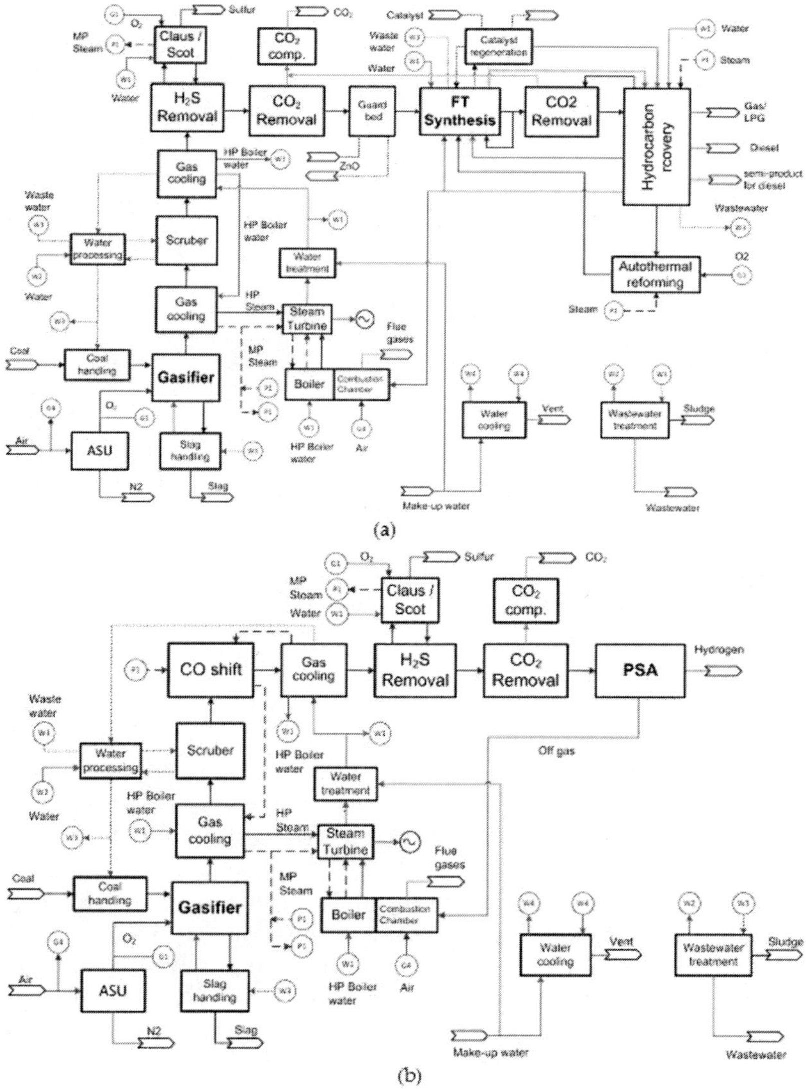

Figure 4. Process diagrams of A) a liquid fuel production plant (option I) and B) a hydrogen production plant (option II).

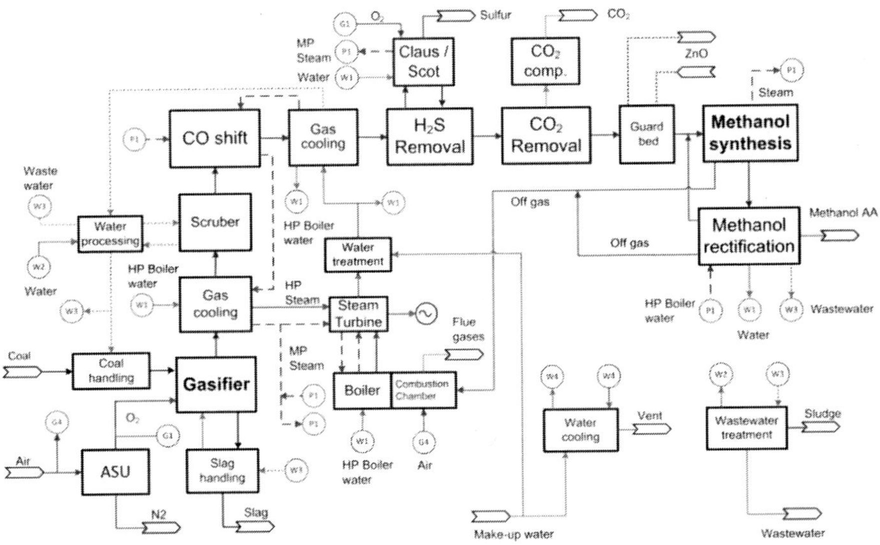

Figure 5. Process diagram of metanol production plant (option III).

concentrate most of the carbon contained in the gas in to a CO_2 stream. This allows for the removal of carbon from the syngas before the combustion process (a CO_2 stream is removed in the subsequent stages of syngas processing). Additionally, during the conversion process, the COS hydrolysis reaction takes place without requiring additional equipment (an IGCC facility without CO_2 removal requires systems for the hydrolysis of COS). CO_2 is removed from the syngas during an absorption process. Due to the high pressures under which the gasifier is typically operated, the most energy efficient method of gas separation is by physical absorption. A double stage physical absorption system is recommended for use in a gasification system when separation of CO_2 is required.

A Polygeneration Plant consists of three basic technological units:

- Chemical island: coal gasification system that is equipped with a gas conversion and purification system with a CO_2 separation unit and generates syngas for chemical production purposes and high pressure steam for power and heat generation. Technological configuration as in the case of the production of methanol (option III, see p. 3.2.3).

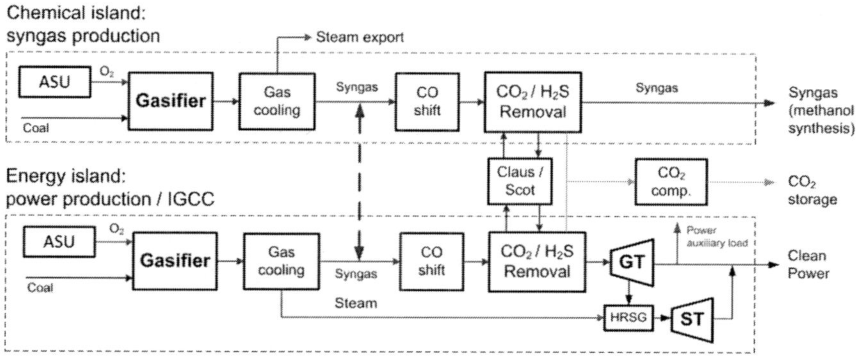

Figure 6. A scheme of Polygeneration Plant.

- Energy island: coal gasification system that is integrated with a combined cycle for power production (gas and steam turbines, HRSG - Heat Recovery Steam Generator) and is equipped with syngas conversion and purification systems, as well as with a pre-combustion CO_2 capture system. Configuration of gas treatment system as in the case of the production of hydrogen (option II, see p. 3.2.2) with the difference that after removal of CO_2 the gas is not enriched in hydrogen (no installation PSA) but is heated to about 240 °C and then mixed with nitrogen comes from the air separation unit[1] - in order to reduce gas lower heating value (LHV) to 4.7 MJ/m_n^3 (increase of power output of gas turbine as the result of mass flow increase and lowering of gas firing temperature for, i.a., control of NOx emission).
- CO_2 transport and storage system.

The design of the Polygeneration Plant assumes that the system is coupled with a classic CHP plant (not shown in Fig. 6) consisting of a circulating fluidised bed boiler and steam turbine power generator. A CHP plant uses high temperature steam produced in the chemical island of the Polygeneration Plant for additional power and heat production. The energy production in the form of heat and power covers the needs of local consumers, the town heat distribution network and industrial users. It is assumed that the presented conceptual facility will replace two actual operating heat and power plants. Due to their identical production capacity, the gasifiers working in the system may complete each other, increasing the yearly availability of syngas or power production units based on the assumed production programme.

Table 2. Data on the process configuration of fuel production plants and a Polygeneration Plant (Dreszer & Mikulska, 2009; EPK & IChPW, 2008).

Specification	option I	option II	option III	option IV
Product	Liquid fuels	Hydrogen	Methanol	Polygeneration Plant
ASU	Cryogenic separation			
Gasification island				
Reactor	Entrained flow, slurry feed; Technology: GEE/Texaco			
Gasification Pressure	3 MPa	5.6 MPa	7 MPa	5.6 MPa
Coal conversion	98 %	98 %	98 %	98 %
Gas cooling	Radiant and convective cooler			
Oxidiser	Oxygen 99.5 % (% vol.)	Oxygen 95 % (% vol.)	Oxygen 95 % (% vol.)	Oxygen 95 % (% vol.)
Fuel	Coal-water slurry; 63% dry solids concentration in the slurry			
Gas treatment				
CO shift	No	Yes, Sour gas shift two-stage CO conversion 97 %	Yes, Sour gas shift one-stage CO conversion ~68 %[a]	Yes, IGCC: see option II Methanol: see option III
Sulfur removal	COS hydrolysis Selexol I stage (99.7 %), ZnO sulfur polishing bed (< 1 ppb)	Selexol I stage (99.7 %)	Selexol I stage (99.7 %) COS hydrolysis (raw gas bypass)	Selexol I stage (99.7 %) COS[b] hydrolysis (raw gas bypass)
Sulfur recovery	Claus, Scot; elemental sulfur			
CO_2 separation	Selexol II stage MDEA	Selexol II stage	Selexol II stage	Selexol II stage
Liquid fuel	F-T synthesis slurry reactor	-	-	-
Hydrogen	-	PSA, 85 %	-	-
Methanol	-	-	Adiabatic, fixed bed reactor	Adiabatic, fixed bed reactor[c]
Power	Steam turbine excess heat, gas (hydrocarbon recovery unit) combustion,	Steam turbine excess heat, tail gas (PSA) combustion	Steam turbine excess heat, tail gas (methanol synthesis) combustion	Combined cycle IGCC, gas turbine, HRSG, steam turbine

[a] as the result of CO Shift and by pass of the raw gas; [b] methanol line; [c] Polygeneration Plant produce syngas with composition enabling its direct use in methanol synthesis.

3.4. Transport and storage of CO_2

Separated carbon dioxide is compressed in a multi-stage, intercooled compressor. During the compression, the CO_2 stream is dehydrated with triethylene glycol that is introduced into the compressed stream of CO_2. Dry

CO_2 is directed to an intermediate tank and then transported by a pipeline to underground storage units. The condensate from CO_2 drying is directed to a water purification system.

4. RESULTS OF PROCESS CALCULATIONS

Coal: For analysis, three hard coals produced in Poland[2] - were selected:

- "Ziemowit" and "Piast" coal mines: option I
- "Bogdanka" coal mines: option II
- "Janina" coal mines: options II, III and IV

For gasification, the chosen coals have acceptable water and ash contents and sufficient caloric value and ash fusion temperature. We should highlight, however, that the gasification of coals with lower quality parameters, such as high ash content, leads to gasification efficiency decrease and may cause technical problems in the slag feed system. The assessed properties of coal are presented in Table 3.

Table 3. Properties of selected coals for analysis of coal gasification for liquid and gaseous fuel production.

Lp.	Parameter	Coal mine			
		"Janina"	"Bogdanka"	"Piast"	"Ziemowit"
Proximate analysis					
1.	W^{ar}, %	19.1	11.3	13-16	14.8
2.	W^{ad}, %	8.6	5.5	4-6	7.3
3.	A^{ad}, %	19.8	21.0	20-25	20.1
4.	V^{ad} ,%	28.4	27.1	30.2	28.5
5.	C^{ad}_{fix}, %	43.2	46.4	43.2	44.0
6.	Q^{ar}_{i}, MJ/kg	18.16	21.28	18.0-20.0	19.83
Ultimate analysis, %					
1.	C^{ad}	54.00	59.45	55.26	56.01
2.	H^{ad}	4.04	3.47	3.56	3.50
3.	N^{ad}	0.94	1.26	0.82	0.69
	S^{ad}	2.00	1.07	0.91	0.93
4.	O^{ad}	10.62	8.20	14.32	11.40
Ash fusion temperatures, °C					
1.	Initial deformation temp. (IT)	920	900	910	910
2.	Softening temperature (ST)	1,260	1,220	1,250	1,310
3.	Hemispherical temp. (HT)	1,340	1,500	1,360	1,490
4.	Fluid temperature (FT)	1,360	1,500	1,360	1,500

^{ar} as received, ^{ad} air dried

Availability: a total yearly working time of 85 % has been assumed for all of the options, which is equal to 7,446 hours/year.

Gasifier: process calculations were made for gas generated in the gasifier using GE/Texaco technology. It was assumed that the gasification process would be carried out in a gasification reactor with 125 t/h of raw coal processing capacity. This value meets the processing capacity of operating and newly built entrained flow gasifiers, which are in the range of 100-130 t/h of coal. In typical gasification systems using GE/TEXACO technology, both radiant and convective coolers produce high pressure saturated steam. In the analysed cases, it was proposed that in the radiant cooler, the produced steam is overheated in a convective heat exchanger and then fed directly to a steam turbine for power generation.

Preparation of CO_2 for transport and storage: separated carbon dioxide is compressed to the pressure required for transport conditions, i.e., approximately 12 MPa, and then is transported to storage sites for underground storage.

Process calculation: for the considered technological options, mass balances have been determined on the basis of a calculation made in the ChemCAD v.6.0.2 process simulator for steady state conditions. For liquid fuel production by Fischer-Tropsch synthesis, process calculations were made using data from (US Department of Energy [US DOE], 1999).

4.1. Results of calculations

The summarised results of the process calculations are shown in Table 4.

After consuming 750 t/h (5.6 million t/y) of raw coal, a plant produces 146.2 t/h of Fisher-Tropsch synthesis products, including 14.6 t/h of LPG (liquefied petroleum gas), 25.3 t/h of diesel and 106.3 t/h of components for diesel production. In addition, sulfur (6.6 t/h) and carbon dioxide (883.7 t/h) are also produced in the system. The off gas from the F-T processes and the steam generated in the system are used to produce electricity (electric power: 350 MW_e). The electricity produced covers approximately 95 % of the system needs; to balance the power consumption, an additional 127 GWh (17 MW_e) of electric energy is needed.

Table 4. Results of the process calculations (option I -III).

Parameter	unit	option I "Ziemowit"/ "Piast" coal	option II "Janina" coal	 "Bogdanka" coal	option III "Janina" coal
Coal input	t/h	750	125	125	125
Thermal input	MWth	4,131	631	739	631
F-T liquid production	kg/h	146,200	-	-	-
Methanol	kg/h	-	-	-	62,138
Hydrogen	kg/h	-	10,941	12,197	-
Gross power output	kWe	349,920	73,470	80,040	71,965
Auxiliary load	kWe	366,957	69,204	79,864	72,778
Net power output	kWe	-17,037	4,266	176	-813
Production efficiency	%	N.A.	57.7 hydrogen LHV	54.8 hydrogen LHV	54.6 methanol LHV
CO_2 sequestration (total)	kg/h	883,660	188,448	220,039	210,462
geological	kg/h	883,660	188,448	220,039	125,022
chemical	kg/h	N.A.	-	-	85,440
CO_2 capture[a]	kg/h	62[b]	86	86	96
CO_2 emission[c]	kg/h	40,800 (56,866)	25,800 (21,777)	29,620 (29,454)	4,143 (4,910)

[a] including geological and chemical sequestration, [b] chemical sequestration not included, [c] including the necessary purchase of electricity (943 kg CO_2/MWh) (Finkenrath, 2011)

In the case of hydrogen production, which depends on coal, the plant produces 10.9 and 12.2 t/h of hydrogen from "Janina" and "Bogdanka" coal, respectively. The application of lower quality coal decreased the hydrogen production by approximately 11 %. The gross electricity production also decreased, but due to the growing auxiliary needs in the case of "Bogdanka" coal, which has a greater oxygen demand, a facility using lower quality fuel produces more net energy. In both cases, the electricity production covers the needs of the system. The system also produces sulfur (2.2 and 1.1 t /h) and carbon dioxide (188 and 220 t/h). The efficiency of hydrogen production is 58 % and 55 % (based on LHV) for "Janina" and "Bogdanka" coal, respectively.

A methanol production plant produces 62 t/h of methanol with a high grade purity level. The efficiency of methanol production is approximately 55 % (based on LHV). The energy generated in the system nearly covers the system needs (approximately 99 %). The sulfur production amount is 2.2 t/h. For all of the analysed options, methanol production is characterised by the lowest CO_2 emissions to the atmosphere and, consequently, the highest efficiency CO_2 removal (96 %).

This is because "chemical sequestration" takes place in the methanol production process and part of the CO_2 formed during coal gasification and the conversion of synthesis gas is "stored" in the final product, i.e., methanol.

Case IV involving the Polygeneration Plant is described and analysed in a later section of the paper.

5. INVESTMENT EXPENSES

To calculate the investment expenses, an exponential investment assessment method was used based on the following function:

$$C_1 = C_0 \left(\frac{S_1}{S_0} \right)^f \tag{1}$$

where: C_1 is the calculated investment for the system component, C_0 is the reference investment cost, S_1 is the scale of the system component, S_0 is the base scale parameter and f is the scaling exponent.

The base scales and scaling exponents for the components of the production facilities based on coal gasification are shown in Table 5.

Capital expenditures specified for the base year were calculated for the current year using the method of indices according to equation (2):

$$C_2 = C_1 \left(\frac{I_2}{I_1} \right) \tag{2}$$

where: C_2 is the current investment. C_1 is the base investment. I_2 is the current index value and I_1 is the base index value.

The indices used in this study were from the M&S (Marshall & Swift Equipment Cost Index) and CEPCI indices (Chemical Engineering Plant Cost Index) as published in Chemical Engineering. Having assessed the main equipment investments (machines, instruments, devices), the factor analysis has been used by adding relevant coefficients to the coordinates positions and obtaining fixed assets investment estimation results. For total fixed assets investment estimation, the following equation has been used:

$$C_n = E + \sum_{i=1}^{m} f_i E \tag{3}$$

where: C_n is the fixed assets investment, E is the equipment purchasing costs, and f_i are the coefficients for instruments and devices, fittings, foundations assembly cost, etc.

Table 5. Base scales and scaling exponents for coal conversion system equipment investments.

Plant component	Scaling parameter	Base scale	Exponent
Coal handling	Coal feed	100 t/h	0.67
Gasifier	Coal thermal input	697 MW_{th}	0.67
Oxygen plant – ASU	O2 flow	76.6 t/h	0.50
O_2 compression	Compression power	10 MW_e	0.67
N_2 compression	Compression power	10 MW_e	0.67
Selexol –H_2S removal	Sulfur feed	3.4 t/h	0.67
Selexol –CO_2 removal	CO_2 removed	327 Mg/h	0.67
CO_2 drying and compression	Compression power	13 MW_e	0.67
CO Shift (WGS)	Thermal input	1,377 MW_{th}	0.67
Claus. SCOT	Sulfur feed	3.4 t/h	0.67
Boiler	Heat transfer surface	225, 000 m^2	0.67
Steam turbine	Turbine output	136 MW_e	0.67
Gas turbine	Turbine output	266 MW_e	–
FT synthesis reactor	Thermal output	100 MW_{th}	1.00
FT product upgrading	FT product production	286 m^3/h	0.7
MeOH synthesis reactor– w/o recirculation	Syngas flow	2.89 kmol/s	0.65
MeOH synthesis reactor – w/ recirculation	Syngas flow	10.81 kmol/s	0.65
MeOH separation and purification	Methanol production	4.66 kg/s	0.29
PSA – hydrogen separation	Hydrogen production	0.294 kmol/s	0.74
CO_2 removal	CO_2 flow	3, 280 mol/h	0.60

The investment costs were calculated assuming expenditures presented as "overnight costs" on the basis of the second quarter of the year 2006 and taking into account an investment cost growth of approximately 60 % by mid-2008. To determine the escalation of capital costs a 30 % increase in the cost of engineering services (60 % share in cost increase) and a 100 % increase in steel price[3] - (40 % share in cost increase) were assumed.

The costs of instruments and devices include the initial equipment plus chemical substances and catalysts. Unpredictable expenses include process costs and project risk.

To calculate investment costs for CO_2 transport and storage 40 km (option I) and 100 km (options II and III) pipelines were assumed.

The investment estimation was conducted with the same accuracy as the pre-feasibility study. i.e., ± 30 %. The investment estimation results are presented in Tables 6 and 7.

Table 6. The investment estimation results for the technological part of the considered plants.

Investment component	Thousands $ (1 $ =2.2531 PLN; 2008)		
	option I	option II	option III
Instruments and devices supply[a]	1,766,211	390,751	400,737
Instruments and devices assembly[b]	671,160	148,507	152,279
Instrumentation and control equipment	105,973	23,434	24,056
Electric installation	162,491	35,950	36,883
Construction works	264,931	58,630	60,095
Land development	105,973	23,434	24,056
Total direct investments	**3,076,740**	**680,707**	**698,105**
Design and supervision	370,904	82,065	84,151
Total direct and indirect investments	**3 447,644**	**762,771**	**782,256**
Unpredictable expenses	635,836	140,695	144,246
Total investment in **Fixed capital**	**4 083,480**	**903,466**	**926,501**
Start-up	68,953	15,268	15,623
Total investments	**4,152,433**	**918,734**	**942,124**
Total investments, Thousands $/TPD (Investments 10^3 $/coal input in tonne per day)	**230.7**	**306.2**	**314.0**

[a] – includes auxiliary equipment, [b] – includes foundations and piping

Table 7. Total CO_2 transport and storage related investments.

Description	Thousands $ (1 $ =2.2531 PLN; 2008)		
	option I	option II	option III
CO_2 pipeline construction	146,082	113,419	76,932
CO_2 storing facility	47,601	17,309	11,784
Total	193,683	130,729	88,716

6. FINANCIAL AND ECONOMIC ANALYSIS

The base year for finacial and economic analysis is assumed to be 2008 (Q4). The analyses have been prepared using fixed prices, without consideration for inflation prognoses or other changes that may constitute factors influencing future prices of the elements involved in the production process. Any prognoses for the coal, gaseous and oil based fuel processing sector bears considerable risk, which convinced us to use actual prices (base year) and keep the relationships between individual assisting factor prices in our analysis. All of the prices used in the calculations are net with VAT excluded. In the calculation, the unit prices were estimated according to the prudence rule for both sales income and for enterprise working cost, which creates a safety margin in terms of possible price fluctuations and other unexpected expenses. At the time of analysis was performed 1 $ =2.2531 PLN and 1 € = 3.438 PLN. The limit value of the internal rate of return assumed at 6.4 and 8.2 % respectively for the models FCFF (Free Cash Flow to Equity) and FCFA (Free Cash Flow to Firm). The analysis was performed using the UNIDO method (COMFAR III Expert software).

Regarding the foreseen changes in compulsory CO_2 emission allowances starting in 2013, the efficiency calculation is based on three development scenarios:

- basic, assumes project functioning in the present conditions with no regulations on CO_2 (no necessity to buy rights) – hereinafter referred to as scenario 1.
- reference, where a plant owner buys 100 % of the CO_2 emission rights at a price of 39 €/t – hereinafter referred to as scenario 2.
- prospective, assumes the necessity of building CO_2 transport and storage facilities. In this scenario, we include the costs of purchasing and assembling systems for carbon dioxide sequestration, which enable the majority of emitted carbon dioxide to be stored in designated geological structures. For the remaining CO_2 emitted to the atmosphere, there is a requirement to purchase 100 % of the emission rights at a price of 39 €/t – hereinafter referred to as scenario 3.

Assumptions for the calculation are summarised in Table 8 and Table 9 show the adopted total operational costs for the chemicals, the transport and storage of CO_2 and environmental protection costs (waste disposal, emission fees: NO_x, SO_2, dust, CO_2). The results of the economic analysis are presented in Fig. 7.

The liquid fuel production does not reach the required return rate of the invested capital in the predicted scenarios. The reasons for this situation are the large initial investments for building the plant and production start-up. In case of scenarios which assume the necessity to purchase CO_2 emissions, and especially in the scenario 2 weak financial result is due to the large amounts of CO_2 formed in relation to the manufactured product which is about 6 t/t.

Hydrogen generation enables invested capital return in both analysed cases ("Janina" coal and "Bogdanka" coal); however, considering the possibility of CO_2 emission rights fee implementation, it will be necessary to build additional carbon dioxide transport and storing facilities. Whenever a project lacks these structures, there is no profitability (results – scenario 2).

The methanol production option produced the best results among all of the options analysed for scenarios 2 and 3. This is related to the lower CO_2 amount that is emitted (option II) or designed for sequestration (option III) compared to the hydrogen generation options. It is associated with the "chemical sequestration" i.e. the use of CO_2 for methanol synthesis.

A lack of economic effectiveness in scenario 2 for options I and II and, at a lower rate, for option III with respect to scenario 3 confirms the desirability of CO_2 sequestration (capture. transport and storage), particularly from the perspective of the probability of 100 % emission rights duty after 2012.

The results of the calculations of DPBT (Dynamic Pay Back Time) for the FCFF models allow us to make the following conclusions:

- liquid fuel production does not allow a return on investment expenditures in the assumed lifetime of the installation (30 years).
- for the hydrogen generation project, the discounted pay back times are the following: "Janina" coal: scenario 1 – 9 years from the operation start-up, scenario 3 – 13 years from the operation start-up; "Bogdanka" coal: scenario 1 – 8 years from the operation start-up, scenario 3 – 12 years from the operation start-up.
- methanol generation enables the achievement of financial results that guarantee invested capital return within 9 years from the operation start-up in scenario 1 and 10 years from the start-up in scenario 3.

Project profitability and liquidity assessment

In scenarios 1 and 3, the projects generate positive financial results, which constitute the basis for project stability and for getting the surplus necessary for invested capital return. Scenario 1 assures slightly higher profitability; nevertheless, we may potentially face CO_2 emission rights purchasing after 2012. For option I, the financial performance is insufficient to ensure a return on the invested capital.

Table 8. Unitary costs and prices.

Specification	Unit	Cost/ price	Comment
Unitary prices			
Liquefied petroleum gas (LPG)	\$/t	1,556	The basis for the technical propane unit price calculation was its market price, less the excise tax (2008 Polish market).
Diesel	\$/t	936	Fuel oil wholesale price (2008 Polish market).
Component for diesel production	\$/t	749	Price was determined by the fuel oil wholesale price, decreased by 20 % for the value added for its final processing.
Sulfur	\$/t	266	Sulfur prices grew considerably from 2007-2008 from 50 to 500 \$, which made us choose a safe price level considering possible speculative fluctuations. Additionally, price decreases caused by an oversupply in the market are usually small in this product segment.
Hydrogen	\$/t	3,106	Costs of hydrogen production from natural gas (NG) were calculated according to the equation presented in (Stiegel & Ramezan, 2006). The NG suppliers' price parameters have been used in this equation, using the prices for large buyers. As the equation structure primarily considers investment amortisation values, which drastically grew during 2007 and 2008, the results have been increased by 30 % for investment growth compensation. Chemical business specialists were consulted on the calculation methodology and estimated total production cost.
Power	\$/MWh	89	The power sales price, has been accepted as competitive in comparison with prices offered by the CHP plants to the industry, (2008 Polish market).
Methanol	\$/t	596	Average price on the European market for the 2007-2008 period.
Unitary costs			
Coal	\$/GJ	3.99	Market price (2008 Polish market)
Power	\$/MWh	111	See above
CO_2 emission cost	€/t	39	Related to data published directly by the European Commission (SEC(2008) 85/3)
Water	\$/t	0.11	-
Solid gasification product	\$/t	-	For the prudence rule, the solid product is given away for free, which eliminates the costs of its treatment and disposal.

Table 9. Operational costs related to chemical consumption, CO_2 transport and storage and environmental protection.

Specification	Thousands $		
	option I	option IIa (IIb)	option III
Chemical substances	2,663	444 (444)	444
CO_2 pipeline operation cost (scenario 3)	4, 674	3, 629	2,462
CO_2 storage operation costs (scenario 3)	2,799	1,156	858
Emission fees (scenarios 1 and 2)	1,062	239 (275)	166
Emission fees (scenario 3)	421	115 (130)	74
Waste disposal	2,219	444 (444)	444

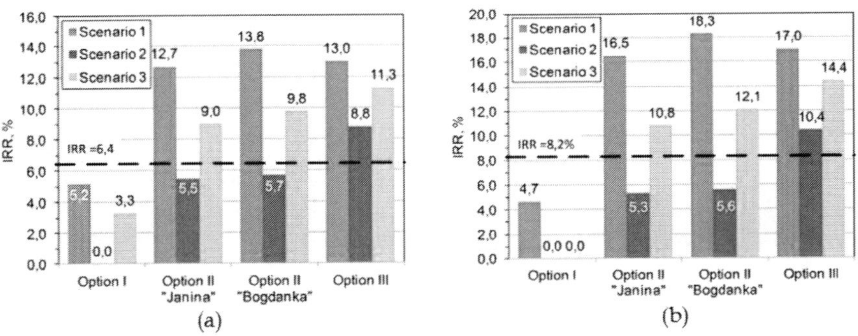

Figure 7. Internal return rate according to FCFF (A) and FCFA (B)

Risk assessment – Project sensibility

The project sensibility has been examined for all options in scenarios 1 and 3 (Fig. 8).

The following parameters have been subject to analysis:

- coal purchase prices: ± 10 % and their 20 % increase.
- investments: ± 10 % and 20 % and 30 % growth.

- basic product sales price in all of the options: ± 10 % and 20 %.
- CO_2 emission rights: 10 % and 20 % growth.

The results of the calculations enable us to formulate the following conclusions:

- coal prices changed in a given area do not implicate large deviations from the calculated efficiency indicators. A basic fuel price increase of 20 % does not cause any loss of liquidity in options II and III using both scenarios. For option I, a 15 % coal price drop in scenario 1 and a 40 % drop in scenario 3 is necessary to obtain a minimum level of profitability,
- an investment level growth of 30 % causes a loss of efficiency in option II using scenario 3. Achieving efficiency for option I is related to a necessity to reduce investments by 25 % in scenario 3 and by approximately 10 % in scenario 1,
- hydrogen sales prices drop by 20% will cause loss of efficiency in scenario 3. For the production of methanol, the lower limit for price level is 23% below the price which was assumed for the calculations. Achieving efficiency measures for option I is related to a necessity to raise sales prices by 15% and 27% respectively for scenario 1 and 3.
- thanks to a CO_2 transport and storage system, the project is not excessively price sensitive in terms of emission rights purchasing in scenario 3. Even with 20 % growth, the project efficiency is preserved. Option III is characterised by the smallest fluctuation and lowest carbon dioxide emission indicator.

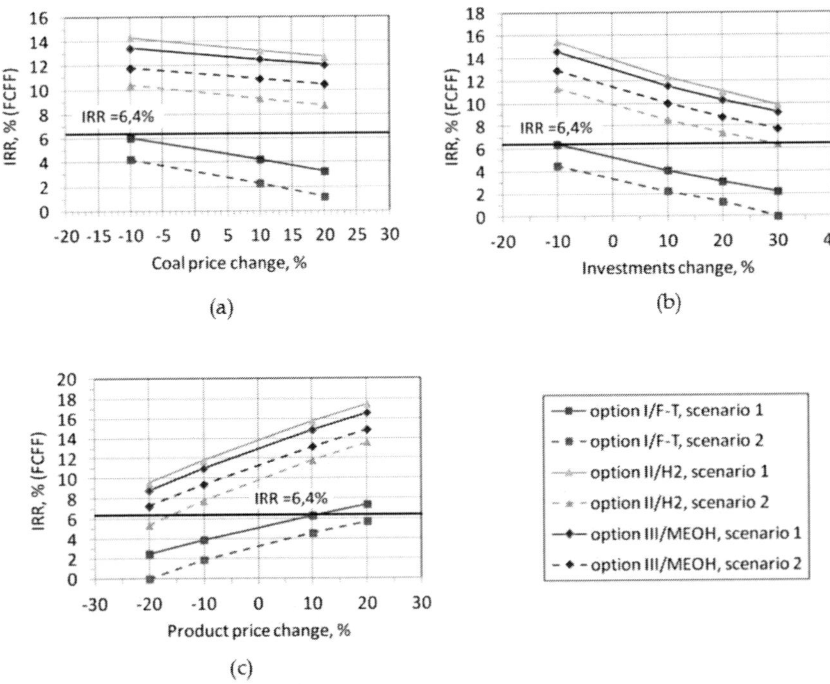

Figure 8. Sensitivity analysis A) coal price, B) investments, C) product price.

Additionally, for the defined scenarios using the considered options, the basic product minimum prices have been determined to assure profitability limit achievement. i.e., IRR equal 6.4 % (Table 10.). The prices of the analysed gasification products have been referred to the oil and natural gas prices in the following manner:

- option I – a motor oil semi-finished component is the basic product of the system. The unit price of the semi-finished motor oil component that was used in this analysis was the motor oil wholesale price decreased by 20 % (see Table 8). We have assumed that the motor oil semi-finished component will be equivalent to crude oil.
- option II – Hydrogen is the basic product of the system. The basis for the hydrogen generation cost calculation was the price of natural gas. The costs of hydrogen production from natural gas (NG) were

calculated according to the equation presented in (Stiegel & Ramezan, 2006).

- option III – Methanol is the basic product of the system has been compared to the equivalent natural gas prices on the basis of available projects and consultations with Polish Chemical Industry Chamber experts. We should, however, highlight that methanol prices in the market are subject to considerable fluctuations, which are not always caused by natural gas prices changes. The calculations above may be burdened with methodological error that is difficult to define.

The results of calculations show minimum oil and natural gas prices, which assure the profitability of products included in individual options and according to assumed scenarios. The methanol production project has the best relationship in this matter as hydrogen production marketability is more dependent on natural gas prices. For coal-based liquid fuel production (motor oil semi-finished component) to be attractive with the different scenarios considered, oil prices must exceed 87 $/bbl.

Table 10. Minimum selling prices of manufactured products assuming minimum profitability (IRR = 6.4 %).

Option	unit	Scenario 1	Scenario 2	Scenario 3
Liquid fuel production	$/t	832	1,338	948
	$/bbl oil equivalent	87	140	99
Hydrogen production	$/t	2,173 (2,090)	3,220 (3,192)	2,699 (2,617)
	$/1000 m³ NG equivalent	364 (350)	553 (549)	459 (444)
Methanol production	$/t	418	539	455
	$/1000 m³ NG equivalent	373	481	406

7. POLYGENERATION PLANT

Polygeneration systems mediate the simultaneous production of chemicals and electricity from syngas. The purpose of these systems is to make maximum use of the chemical energy of coal by maximising the total energy efficiency of the transformation of primary fuel into useful products while minimising the capital expenditure and operating costs. Syngas may be used independently to produce chemicals and electricity, most advantageously in IGCC (integrated gasification combined cycle) systems.

Polygeneration usually include electricity production that is integrated with the generation of hydrogen, methanol or the products of Fischer-Tropsch synthesis.The principal advantages of a polygeneration system include:

- increased economic flexibility (two or more products);
- lower production costs due to more efficient use of syngas and of the technological heat produced in the course of the production process.

The integration of the processes of power and chemical production in a polygeneration system allows the achievement of high rates of fuel conversion, low emission rates and high economic efficiency, as in the case of CO_2 sequestration.

Table 11 presents the basic process data of the considered system. Consuming about 257 t/h of coal, the production of syngas amount to 85.1 t/h. This is enough to obtain 63.4 t/h of methanol. Net power and heat (in form of HP steam) output is 142 MW_e and 130 MW_{th} respectively. Geological sequestration of CO_2 will be 311 t/h. The amount of CO_2 stored in the chemical end product (methanol) will be 87 t/h.

Tables 12 and 13 show the investment costs and the minimal energy and synthesis gas prices to ensure the viability of a project (NPV> = 0 and IRR> = 7 %. with an amortisation period of 20 years). The calculation results are presented separately for the main technological units and for the whole facility of the Polygeneration Plant for the base case (CO_2 emissions within the scope of given emission rights) and in the case of CO_2 sequestration (separation; transport and storage of CO_2; fees for the remaining CO_2 emissions 39 €/t).

Fig. 9 shows the results of the calculations of produced synthesis gas prices against electricity prices (NPV> = 0 and IRR> = 7 %. with an amortisation period of 20 years) and the area of the economic attractiveness of the project.

The calculation results clearly indicate the attractiveness of the polygeneration process. The combination of electricity generation and synthesis gas production for the presented technological configuration (use of gasification technologies for energy production and syngas) causes a significant reduction in the minimum price of energy in comparison to the IGCC system (production of electricity) to 49 and 21 $/MWh without and with CO_2 sequestration, respectively (Tables 12 and 13).

Table 11. Results of process calculations for option IV.

Parameter	unit	Option IV Poygeneration Plant "Janina" coal
Coal input	t/h	257
Thermal input	MW_{th}	1,296
Syngas production	kg/h	85,079
equivalnt methanol production	kg/h	63,400
Gross power output	kW_e	282,700
Auxiliary load	kW_e	140,591
Net power output	kW_e	142,109
Thermal output[a]	kW_{th}	130,000
Production efficiency (mixed)	%	57.6[b]
Syngas production efficiency	%	73.2
Power production efficiency (IGCC)	%	31.4
CO_2 sequestration	kg/h	397,811
geological	kg/h	310,636
chemical	kg/h	87,175
CO_2 capture [c]	kg/h	88
CO_2 emission	kg/h	38,802

[a] high pressure steam from chemical module (see Fig. 6), [b] including syngas (chemical enthalpy), heat (HP steam) and power production, [c] including geological and chemical sequestration

Table 12. Investments and price limits for manufactured products (power and syngas); Polygeneration Plant without CCS.

Specification		Unit	IGCC (power island)	Syngas production unit (chemical island)	Poligeneration Plant
Investments		mln $	**1,105**	**670**	**1, 776**
		10^3 $/TPD[a]	**358.3**	**217.3**	**287.9**
Price limits:	Power	$/MWh	**131**	111[b]	**82**
	Syngas	$/1000 m_n^3		**144**	202[c]

[a] Investments 10^3 $/coal input in tonne per day, [b] the approved purchase price of electricity reflects the price level of december 2008, [c] Adopted the maximum price of synthesis gas (Q4 2008), considered to be commercially attractive (the price of the synthesis gas produced from natural gas).

Table 13. Investments and price limits for manufactured products (power and syngas) Case: CO_2 sequestration.

Specification		Unit	IGCC (power island)	Syngas production unit (chemical island)	Poligeneration Plant
Investments:		mln $ 10^3 $/TPD[a]	1,256 368	804 221.5	2,060 294.8
Including:					
CO_2 Transport and Storage		mln $	121	121	242
Price limits:	Power	$/MWh	191	111[b]	170
	Syngas	$/1000 m_n^3		167	202[c]

[a] Investments 10^3 $/coal input in tonne per day - technological part only without CO_2 Transport and Storage, [b] and [c] see table 13

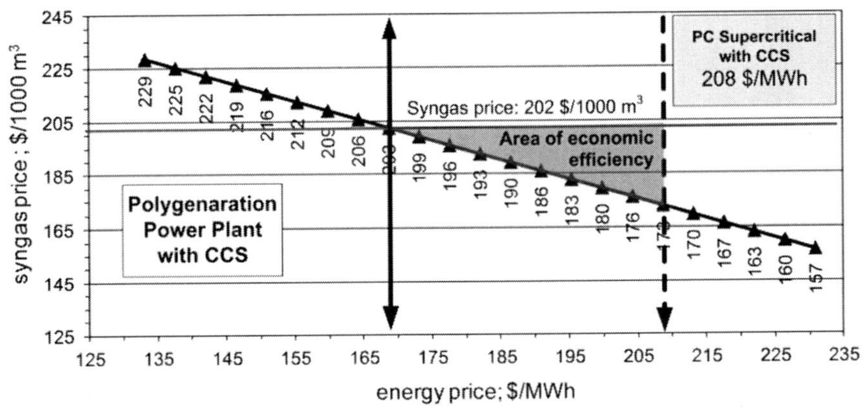

Figure 9. Estimated cost of synthesis gas in relation to the price of electricity and the area of economic efficiency of the Polygeneration Plant.

The CO_2 sequestration benefits of the proposed solution are also visible when comparing the Polygeneration Plant with a Supercritical Power Plant based on coal combustion. A comparison of the energy price limits for both cases at the same production level shows that with polygeneration we obtain lower energy prices by 38 $/MWh (energy price forecast for the supercritical coal unit with CCS amounts to 208 $/MWh). This underlines the attractiveness of the presented solution and the need to develop the proposed concept under appropriate technological conditions with the existence of a recipient for the produced synthesis gas as an alternative to traditional solutions.

8. CONCLUSION

The analysis concerned the installations for gaseous and liquid fuel production based on coal gasification using commercially available technologies of coal gasification, gas cleaning and conversion and chemical synthesis.

Systems for liquid fuels, hydrogen and methanol production were analysed in detail assuming three scenarios: basic (with no necessity to buy rights for CO_2 emission), reference (purchase 100 % of CO_2 emission rights at a price of 39 €/t), and prospective (assuming construction of CO_2 transport and storage facilities).

The analysis of the examined cases shows that with the adopted assumptions, the most favourable option is definitely the production of methanol, which shows economic effectiveness in all of the scenarios and, in the case of scenarios 2 and 3, gives the best results among the options analysed. The reason for this superiority among other options is related to low CO_2 emission, associated with the "chemical sequestration"i.e. the use of CO_2 for methanol synthesis.

The economic attractiveness of the production of hydrogen is significantly more dependent on natural gas prices. Hydrogen production is economically feasible only in scenarios 1 (base) and 3 (prospective). Developments in this direction and, consequently, the hydrogen economy seem to be limited due to a lack of cost-effective storage technology and transport infrastructure. At present, hydrogen from coal can effectively be used in chemical plants for the production of ammonia and fertiliser by substitution of the hydrogen produced from natural gas.

The coal to liquid fuels process based on Fischer-Tropsch synthesis is attractive only when exceed 87, 140 and 100 $/bbl for scenarios 1, 2 and 3, respectively.

Among the analysed technological options, the production of liquid fuels from coal using FT synthesis is definitely the least attractive and, on the basis of the obtained results, is not recommended as a potential direction for the application of coal gasification technology.

However, the idea of the production of liquid fuels from coal is still attractive, and the production of liquid fuels from coal using methanol seems to be a reasonable option. Methanol is used directly as motor fuel or is added to liquid motor fuels to improve their operational performance (methyl tertiary butyl ether, MTBE). Moreover, technologies for the production of motor fuels from methanol (MTG - methanol to gasoline and MTO/MOGD – methanol to olefines/Mobile olefines to gasoline and destilate) are being intensively developed and are commercially available at the industrial scale.

A lack of economic effectiveness in scenario 2 for options I and II and, at lower rates, for option III with respect to scenario 3, confirm the desirability of CO_2 sequestration (capture, transport and storage), particularly from the perspective of the necessity to purchase CO_2 emission rights after the year 2012.

The analysis of the Polygeneration Plant clearly shows the attractiveness of the solutions and the need to develop the proposed concept in appropriate technological conditions with the existence of a recipient for the synthesis gas produced as an alternative to traditional solutions. The realisation of this production process would give the possibility of significant reductions in the price of electricity generated, even in the case of CO_2 sequestration, compared to traditional technologies, including IGCC, while maintaining cost-effective production of synthesis gas for chemical applications. Also important from the economic point of view is installation flexibility in terms of the final product. i.e., the ability to design a production profile according to market demand for the manufactured products.

NOTES

1) If the amount of available nitrogen is not sufficient, gas is diluted through the humidification and the third option is steam injection (US DOE & NETL, 2010b)
2) The dominant share - 67% of coal production in the EU27 (Lorenz, 2008).
3) Steel Business Briefing Ltd, september 2008

REFERENCES

1. T. Chmielniak, M. Ściążko, M. Uliniarz, 2008Poligeneration power plant with CO2 capture. Archives of Energetics, 38245540006-6684X.

2. K. Dreszer, B. . Mikulska, Ed., 2009Feasibility study of the installation for the production of gaseous and liquid fuels from coal, Instytut Chemicznej Przeróbki Węgla & Energoprojekt-Katowice S.A., 978-8-39134-347-0Zabrze, Poland (in polish).

3. Energoprojekt-Katowice S.A (EPK) & Institute for Chemical Processing of Coal (IChPW).2008The concept of polygeneration plant for power, heat and syngas production to meets the needs of PKE SA Blachownia Power Plant and chemical works Kędzierzyn SA; Project 2007K, Contractor Południowy Koncern Energetyczny S.A. (TAURON Wytwarzanie S.A ; power producer) and Zakłady Azotowe Kędzierzyn S.A. (chemical works), (In polish).

4. M. Finkenrath, 2011Cost and Performance of Carbon Dioxide Capture from Power Generation. International Energy Agency IEA 2011, 02.2012, Available from http://www.iea.org/papers/2011/costperf_ccs_powergen.pdf

5. H. Liu, 2010OMB Gasification- Industrial Application Updates of Slurry Feeding & Developments of Dry Feeding, Gasification Technologies Conference, San Francisco 2010, 02.2012, Available from http://www.gasification.org/library/overview.aspx

6. U. Lorenz, 2008Main world steam coal exporters to the European market- some aspect of supply and prices. Energy Policy Journal, 111255272000-1429In polish).

7. G. J. Stiegel, M. Ramezan, M. , 2006Hydrogen from coal gasification: An economical pathway to a sustainable energy future, International Journal of Coal Geology, 651731900166-5162

8. US Department of Energy1999Baseline Design/Economics For Advenced Fischer-Tropsch Technology (1991÷1999), Raport (1-14Project DE-AC22PC90027, Available from www.fischertropsch .org/DOE/DOE_reportstoc.htm.

9. US Department of Energy & National Energy Technology Laboratory2002Tampa Electric Polk Power Station Integrated Gasification Combined Cycle Project, Final Technical Report, Project no. DE-FC-2191MC27363 US DOE/NETL, 03.2012,

Available from http://www.netl.doe.gov/technologies/coalpowercctc/cctdp/bibliography/demonstration/pdfs/tampa/TampaFinal.pdf

10. US Department of Energy & National Energy Technology Laboratory2007Gasification World Database 2007 Current Industry Status, Robust Growth Forecast, (2007). Available from http://www.netl.doe.gov.

11. US Department of Energy & National Energy Technology Laboratory2010aWorldwide Gasification Database (Excel file); Available from http://www.netl.doe.gov/

12. US Department of Energy & National Energy Technology Laboratory2010bCost and Performance Baseline for Fossil Energy Plants, 1Bituminous Coal and Natural Gas to Electricity, Revision 2, November 2010, DOE/NETL-2010/1397, Available from http://www.netl.doe.gov/

8

CHAPTER

MATHEMATICAL MODELLING OF BIOMASS GASIFICATION IN A CIRCULATING FLUIDIZED BED CFB REACTOR

Roberto Capata, Mario Di Veroli

Department of Mechanical and Aerospace Engineering, University of Roma "La Sapienza", Rome, Italy

ABSTRACT

The scope of the present paper is to investigate the suitability of a mathematical model for Circulating Fluidized Bed (CFB) coal combustion (developed by the International Energy Agency), to predict and simulate the performance of the 100 kW_{th} CFB for air-blown biomass gasification. The development of a mathematical model allows to simulate the operative conditions during biomass gasification, control the quality of the synthesis gas and improve the gasifier design. The geometrical, mechanical, hydro dynamical and thermo chemical features were introduced in the model by properly setting the input file and, some changes have been made in the code to assure the final convergence. A sensitivity analysis has been performed to study the variation in the input parameters of the program, and it has been finally verified by comparing the results with the empirical data collected during coal and wood combustion tests. The program, in the same case, could not successfully run; probably depending on wood char density value. For these reason the influence of char density will be investigated. The model predicts the development of tar and other hydrocarbons, valuating the agreement between the measured and calculated efficiency. A further development, to consider solid biomass, with a certain volatile percentages (20% - 40%), as a fuel has been previewed and analyzed. Finally some investigations have been carried out to provide some useful indications for future developments of the code, in the biomass gasification modelling.

Keywords: Biomass; Gasification; Fluidized Bed Reactor; Syngas; Mathematical Model

1. INTRODUCTION

Processes based on the thermo chemical conversion of biomass are nowadays rising importance, since they show an higher energy efficiency than the direct combustion. The biomass gasification appears as one of the most promising. It consists in the conversion, in a low O_2-content atmosphere, of the original biomass feedstock into a combustible gas mixture, the synthesis gas, which can be used in several applications, like gas turbines. A complete knowledge of the biomass gasification process is necessary to satisfy the increasing demand on the quality of the produced gas. The development of a mathematical model provides an indispensable tool to control the operational parameters and improve the gasifier design.

A mathematical model for Circulating Fluidized Bed (CFB) coal combustion, developed by the International Energy Agency (IEA), is, hereby, used to predict and simulate the performance of a 100 kW_{th} CFB gasifier. It derives from the initial considerations that the gasification process can be considered, on the chemical point of view, as "backward" step to the combustion, since it occurs in partial oxidation conditions. Moreover both processes can take place in a CFB reactor, presenting similar hydro dynamical and mechanical features. It means that the sub models used to describe such parameters, as the fluidization pattern of solid and gas flows, the particles size distribution, the thermal analysis, the chemical process of combustion and the post-process reactions, can be also useful to reproduce the complex of phenomena occurring during biomass gasification in a CFB reactor.

2. BIOMASS GASIFICATION IN A CIRCULATING FLUIDIZED BED REACTOR

Gasification is defined as "the thermo chemical process which converts solid or liquid fuels into a combustible gas mixture of low or medium heating value (4 - 10 MJ/Nm^3;), by partial oxidation of the feedstock, in the presence of a gasifying agent (air, O_2, CO_2 and/or steam)". The heat released by the partial oxidation is used to maintain the operating temperature and to develop the endothermic reactions in the gasifier. Stable gasification conditions are usually achieved for process temperature between 700°C and 1100°C and stochiometric air ratio (λ) between 0.2 and 0.4 [1]. Before introducing and describing the "core" of the gasifier we have to analyze the pyrolysis or devolatilization of biomass heating the biomass to a temperature between 400°C and 600°C (pyrolysis or devolatilization). Pyrolysis is defined as "the thermal decomposition of solid fuel in an oxygen-free atmosphere and it results in the production of solid char (10% - 25% wt. of the original biomass) and volatile matters, mainly composed of water, CO, CO_2, H_2, CH_4, C_2H_6 and tar". The presence of tar in the synthesis gas can produce several problems in the plant components: for this reason a particular attention has to be put into the control of tar formation and to the strategies for its removal. The final composition and efficiency of the synthesis gas is defined by the combined effects of pyrolysis, partial combustion and the complex equilibrium conditions given by competing reactions. Furthermore it is influenced by a number of parameters, such as the fuel composition, the stochiometric air ratio, the gasifying medium, the operating pressure and temperature and by the process technology.

Considering the available technologies for gasification, the CFB was considered to be suitable for medium-large scale gasification. Due to the unique hydro dynamical conditions in a fluidized bed furnace, CFB reactors present some interesting features which make more attracttive than other solid fuel fired boilers (excellent gas-solid and solid-solid mixing, homogeneous distribution of temperature, high conversion efficiency, in-situ sulphur removal and low NO_x emissions) [2].

3. DESCRIPTION OF THE IEA MODEL

The original model has been developed by the IEA to simulate the coal combustion process in industrial size CFB boilers; it has been tested on several plants designed for heat and steam generation, having thermal power output within 13 and 300 MW_{th}. It has been developed and designed to describe the overall process occurring in the reactor, including the gas evolution and solid flows, the determination of the particle size distribution, the coal conversion, the post-process reactions and the heat transfer process. It can be considered as an "engineering" model, resulting by a combination of mathematical models, available in literature, and empirical feedback [1].

Such approach is, sometimes, called "one and halfdimensional": the Riser is discretized in vertical direction by simple cells, while a distinction between core and annulus cells is assumed in radial direction. In the dense bed region, the core cell, is divided in emulsion and bubbling phase cells, with different balancing equations. The cell structure of the CFB and the modelled gas and solid flows are reported in Figure 1. Each cell is assumed to be homogeneous in concentration and temperature.

It is organized in a series of following blocks, each one including the set of routines and subroutines necessary to reproduce a particular physic sub model; the different blocks are linked together by balancing soubroutines, until equilibrium is reached. All the demanded data to run the program have to be introduced in an input file; some of the required parameters have to be considered as input parameters, which should be properly adjusted in accordance with the experimental measures referred to the particular modelled case. The model is capable to predict the composition of the outlet gas mixture from the reactor [1].

4. SCALE-DOWN OF IEA MODEL TO THE LABORATORY SYSTEMS

The performances of the IEA model have been tested on a laboratory-scale CFB (see Figure 2) unit designed for biomass gasification. It is, operated at atmospheric pressure, with process temperature within 700°C and 900°C and maximum thermal output of 100 kW_{th}. It basically consists of a Riser (height: 5.5 m; circular cross section,O: 83 mm) connected to a gas-solid cyclone separator: the solid particles are separated and recirculated by a downcomer into the furnace, while the gas flow is cleaned by a ceramic filter before being collected.

The P&I diagram is shown in Figure 3.

The test bench generally operates under the following operational conditions:

- Only primary air supplied

- No flue gas recirculation
- No limestone addition
- No bottom bed discharge

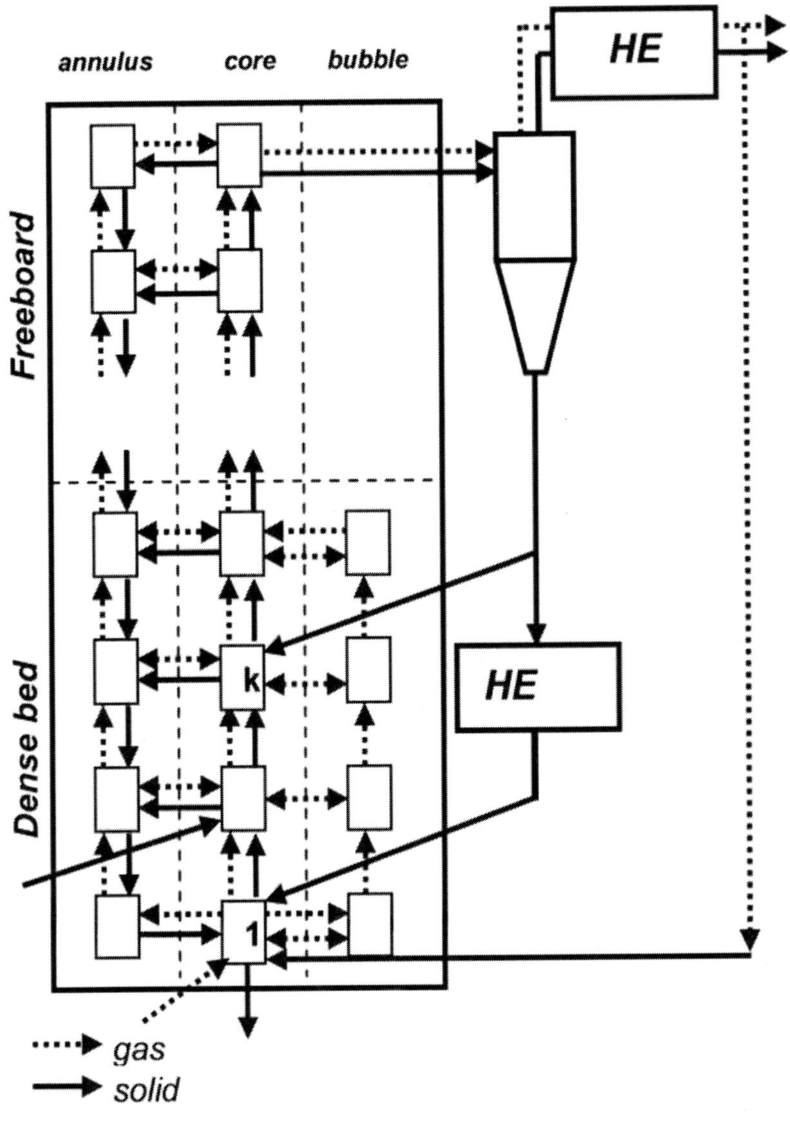

Figure 1. IEA model CFBC cell structure.

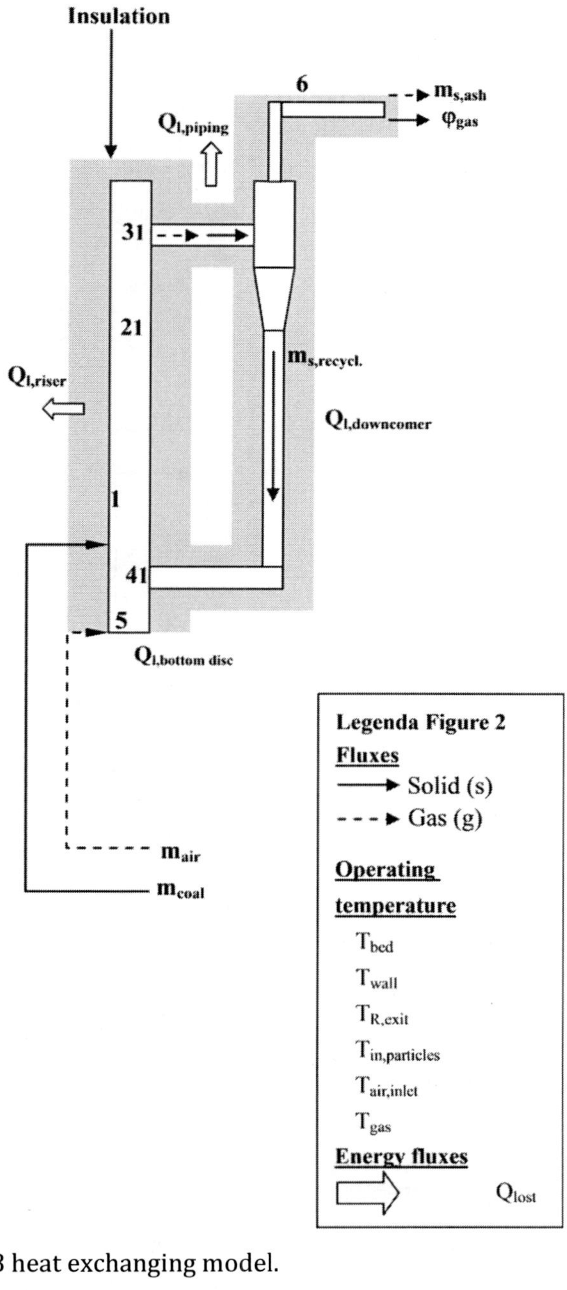

Figure 2. CFB heat exchanging model.

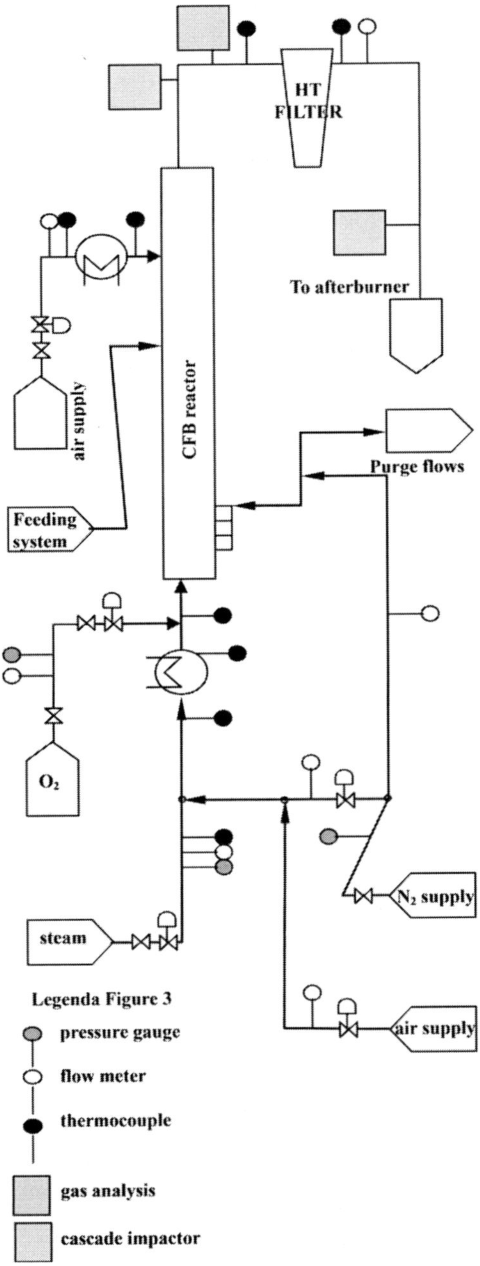

Figure 3. Gasifier P&ID model, with measure points and devices [3].

Unlike the boilers used for the validation of the IEA model, in this case the heat generated by oxidation of fuel has to be used into the furnace to sustain the gasification reactions; for this reason the gasifier is completely thermally insulated and it is not equipped with any other devices for heat exchange.

The design and operating characteristics were introduced in this work by proper input parameters.

5. MODIFICATION OF THE MODEL FOR THE BIOMASS GASIFICATION

The consistency of the approach proposed to scale down the model has been initially verified by using the IEA model to simulate coal combustion. Several experiments have been performed in the past by using the CFB test bench [2] and comparing the calculated and empirical results.

The wood characterization (composition, size particle distribution, physical properties) was introduced in the input file and two empirical cases of wood combustion were simulated, to verify the capacity of the model to simulate solid fuels other than coal.

Moreover, to adapt the IEA model to simulate the wood gasification process in a small size unit, some modification have to be necessary. It has been observed that the chemical and fluidization conditions occurring during gasification can be easily reproduced by reducing λ, fule/ air ratio. However, to achieve a reliable prediction of the process, it is essential to consider the influence of the pyrolysis gas composition and the endothermic gasification reactions (not modelled or simulated by the program) on the chemical equilibrium and on the synthesis gas composition.

6. SIMULATION RESULTS

6.1. Coal and Wood Combustion Simulations

The IEA model successfully runs and reaches the final convergence. The final gas composition (limited to the main components: CO, CO_2 and O_2) and the temperature profile along the Riser show a consistence with the available measurements. As shown in Figure 4, the calculated concentrations of CO_2 and O_2 deviate respecttively of 7.5% and 11.1% from the experimental ones, while a larger difference has been verified between the calculated and measured CO concentration (57.5%). These results are strongly influenced by the parameters related to the combustion kinetics, namely the char reaction constant ($k_{reac,0}$) and CO homogeneous oxidation constant ($k_{0,CO}$), and by the attrition constant (k_{attr}), which in the simulations could be just guessed, due to the lack of empirical data. A better approximation could be achieved by adjusting these parameters to the particular case study. These results are summerized in Table 1.

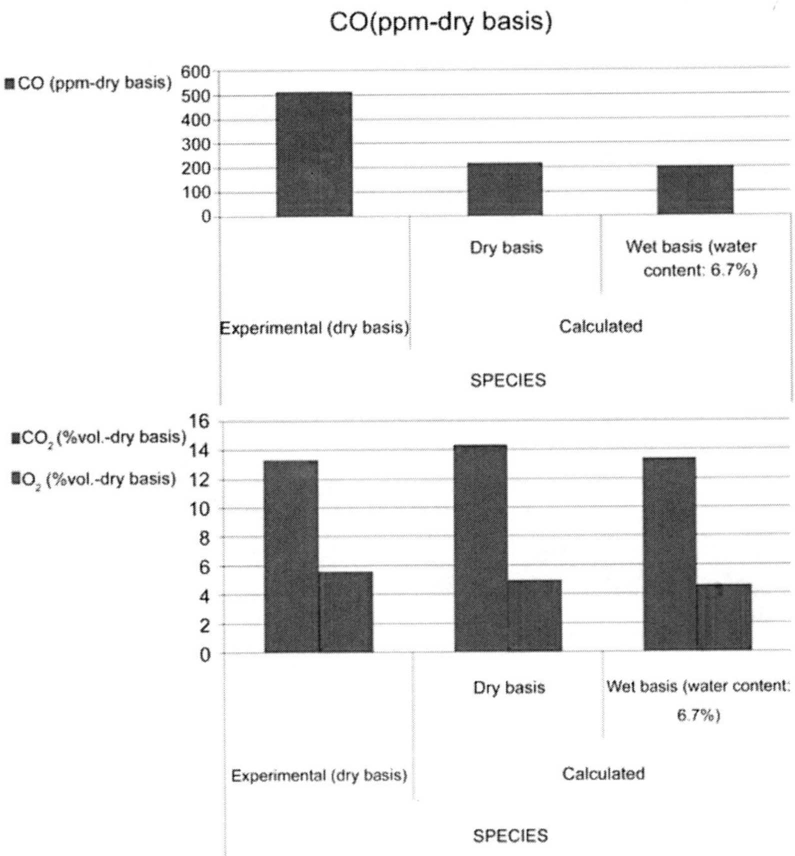

Figure 4. Comparison of experimental and calculated gas composition.

In Figure 5, the calculated temperature distribution well approaches the measured one, and a difference within 1.0% and 1.5% is noticed. The sum of the heat lost through Riser wall, and downcomer pipes, was 6.2 kW_{th}, which is lower than the total energy losses observed during the experiments, usually between 10 - 20 kW_{th}. However, such estimation is based on the indicative values of temperature (the wall temperature and re-entering particles temperature) assumed in this work. A more reliable validation of the heat transfer scheme could be achieved in the future by introducing the measured values of such quantities.

In case of wood combustion, the program does not, sometimes, reach the final convergence. The introduction of the wood physical properties (in particular the wood char density) in the input file leads, in the most of the cases, the program to crash and the final convergence can be reached only in correspondence of certain particular values (55 - 110 kg/m^3).

It does not allow to draw general conclusions about the capacity of the IEA model to simulate solid fuel, like wood, having char density lower than coal and for this reason the obtained results are not reported in this paper.

Table 1. Comparison of experimental and calculated gas concentrations.

	Experimental (dry basis)	Calculated	
		Dry basis	Wet basis (water content: 6.7%)
CO_2 (% vol.)	13.3	14.3	13.4
CO (ppm)	513	218	204
O_2 (% vol.)	5.53	4.9	4.5

6.2. Wood Gasification Simulations

The effects of a λ reduction on the model performance has been tested and the devolatilization model analyzed in detail. At this stage of the work the gasification reactions have not been implemented into the code yet, thus the model can not modelling and simulate the effects of such reactions on the synthesis gas composition. The input file is modified to reproduce, as best as possible, the process conditions observed during gasification experiments by using the test bench [3] and some tests have been run. Since the gasification reactions are not modelled, the results of the simulations could not be compared with the empirical ones. However useful indications could be obtained concerning the capacity of the IEA model to simulate low values of λ, enlighten the effects that a reduction of such parameter produces on the temperature distribution along the Riser and on the final gas composition. The results are shown in Figures 6 and 7.

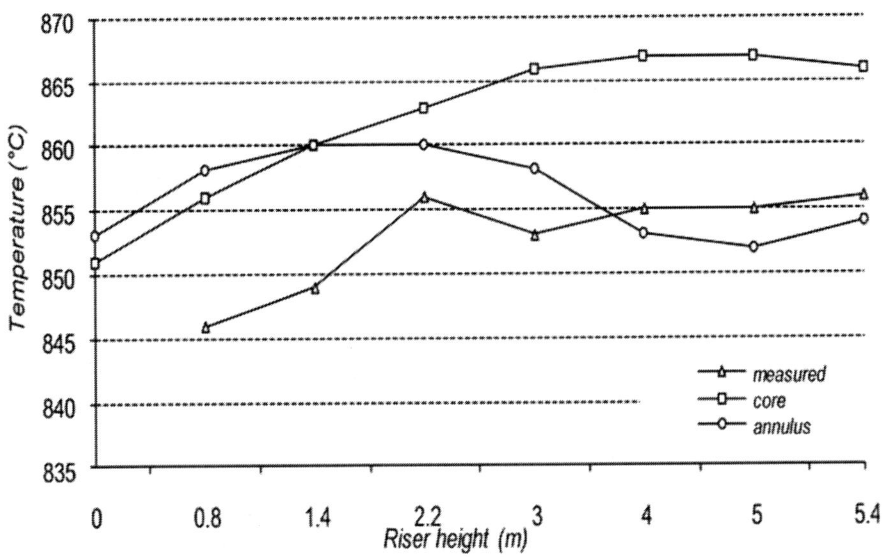

Figure 5. Comparison of measured and calculated temperature profiles.

The reduction of λ decreases the process temperature and, due to the low O_2-content atmosphere, only partial combustion can take place into the Riser. It means that decreasing λ produces a reduction of the CO_2 concentration and a simultaneous increase of CO and H_2, which can not be completely oxidized into CO_2 and water. The peaks observed in the temperature profiles are due to the fact that the amount of heat released by partial combustion in the lower section of the Riser is not sufficient to maintain higher level of temperature.

By the way these results represent only the expected consequences of the partial oxidation of fuel, thus they are still far to be considered as representative of the whole gasification process.

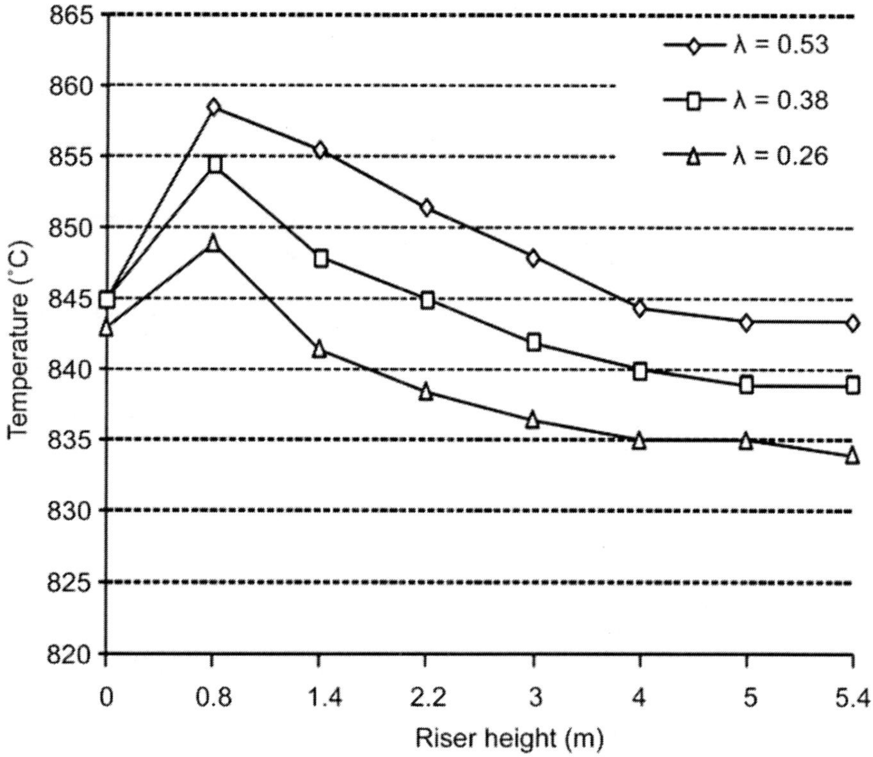

Figure 6. Effects of λ on the process temperature profile.

6.3. Pyrolysis Submodel

The procedure implemented in the IEA code to model the coal pyrolysis is not adequate, in case of thermal decomposition of biomass, since it does not consider the possible formation of tar and higher hydrocarbons, like methane and ethane, during the process. Dealing with coal combustion, such species can be neglected because they are assumed to be completely burned and not released in the flue gas; in case of gasification, they play a significant role in the evolution of the gas mixture and they have to be considered.

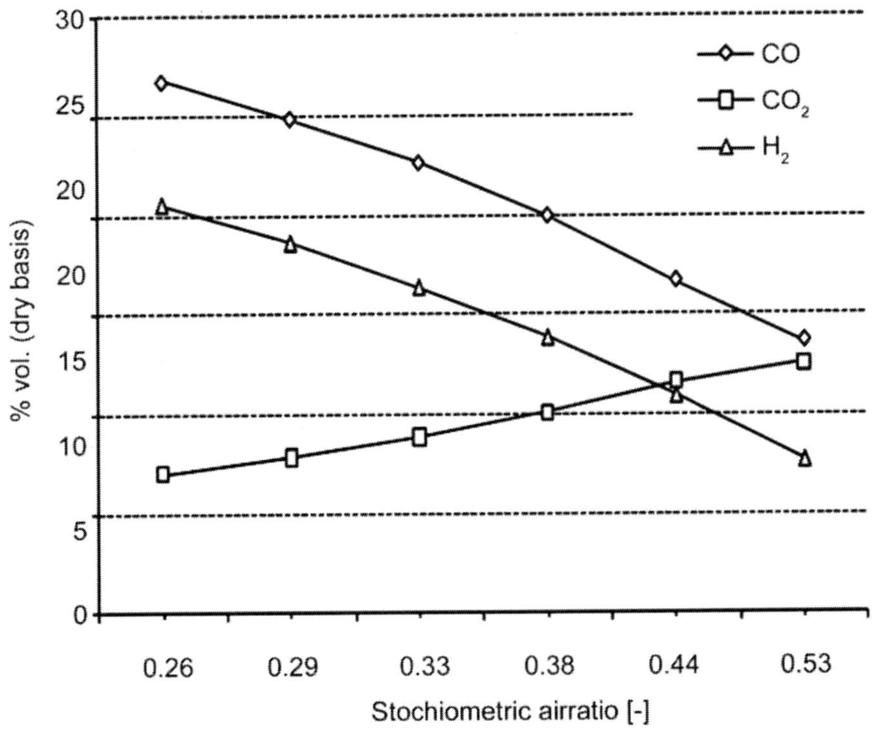

Figure 7. Effects of λ on the final gas composition.

An alternative pattern is included in the IEA model, based on a mathematical model for thermal decomposition of coal proposed by D. Merrick [4], which allows to predict the production of char and volatile matters (on dry and ash free basis). The volatile matter is assumed to be formed by nine components, CH_4, C_2H_6, CO, CO_2, H_2, H_2O, NH_3, H_2S and tars. This approach results a more convenient basis for the future modelling of the gasification process. It mainly consists of a set of ten linear equations, where the mass fractions of the formed gaseous species and char are expressed as unknown variables in the vector m_j:

$$\sum_{j=0}^{9} A_{ij} m_j = B_i \left(i = 1, \cdots, 10\right) \tag{1}$$

In the equations, the first six ones represent, respectively, the elemental mass balances of C, H, O, N, S and the coal mass balance. The latter four are defined by empirical correlations derived in accordance with the following statistical findings, verified by Merrick for coals having volatile content ranging between 16.3% and 37%:

- 32.7 % of H in coal evolved in CH_4
- 4.4% of H in coal evolved in C_2H_6
- 18.5% of O in coal evolved in CO
- 11 % of O in coal evolved in CO_2

In this work the Merrick model was activated, partially modified to link it to the main code and then implemented. The linear equations system, as finally defined, is reported in its matrix form:

$$\begin{pmatrix} C_{char} & 0.75 & 0.8 & 0.42 & 0.27 & C_{tar} & 0 & 0 & 0 & 0 \\ H_{char} & 0.25 & 0.2 & 0 & 0 & H_{tar} & 1 & 0.11 & 0.17 & 0.058 \\ O_{char} & 0 & 0 & 0.57 & 0.72 & O_{tar} & 0 & 0.89 & 0 & 0 \\ N_{char} & 0 & 0 & 0 & 0 & N_{tar} & 0 & 0 & 0.82 & 0 \\ S_{char} & 0 & 0 & 0 & 0 & S_{tar} & 0 & 0 & 0 & 0.94 \\ 1 & 0 & 0 & 0 & 0 & 0 & 0 & 0 & 0 & 0 \\ 0 & 1 & 0 & 0 & 0 & 0 & 0 & 0 & 0 & 0 \\ 0 & 0 & 1 & 0 & 0 & 0 & 0 & 0 & 0 & 0 \\ 0 & 0 & 0 & 1 & 0 & 0 & 0 & 0 & 0 & 0 \\ 0 & 0 & 0 & 0 & 1 & 0 & 0 & 0 & 0 & 0 \end{pmatrix} \cdot \begin{pmatrix} Char \\ CH_4 \\ C_2H_6 \\ CO \\ CO_2 \\ Tar \\ H_2 \\ H_2O \\ NH_3 \\ H_2S \end{pmatrix} = \begin{pmatrix} C \\ H \\ O \\ N \\ S \\ 1\text{-}Vol \\ 1.3H \\ 0.2H \\ 0.3O \\ 0.1O \end{pmatrix}$$

The results can be seen in Table 2.

Due to the lack of empirical results, the reliability of such approach could be just verified by the simulation of two experimental cases of laboratory-scale fluidized bed coal pyrolysis taken from literature [5,6]. The coal used in the experiments presents a volatile content respectively equal to 47.1% and 34.6% on dry and ash free basis. In both cases the model successfully reaches convergence, and a certain amount of tar and higher hydrocarbons have been calculated as components of the volatile mixture, as shown in the Table 3 below.

The calculated molar concentrations had to be processed to present the results in suitable form to be compared with the empirical data (see Table 4). Even though the graphs in Figures 8 and 9 show a reasonable agreement

Table 2. Char and tar composition.

	C	H	O	N	S
Char (wt.%)	0.981	0.0	0.0	0.019	0.0
Tar (wt.%)	0.85	0.082	0.049	0.009	0.01

Table 3. Pyrolysis gas composition in terms of molar fraction.

	Test 1	Test 2
CO	0.031	0.016
CO_2	0.009	0.005
H_2O	0.108	0.051
H_2	0.047	0.167
CH_4	0.087	0.103
C_2H_6	0.008	0.009
NH_3	0.006	0.008
H_2S	0.006	0.008
Tar	0.623	0.594
O_2 in volatiles	0.075	0.039

between the calculated results and the experimental ones, however deviations have been reported between the calculated efficiency of pyrolysis gas (including tar) and the volatile content.. Moreover, it was noticed that only a small fraction (11% in both cases) of the O_2 content of coal is used to form volatiles, which is far from the expected behaviour.

Table 4. Mass balances.

	Test 1		Test 2	
(%wt. d.a.f.)	experimental	calculated	experimental	calculated
Fixed carbon	0.529	0.529	0.654	0.65.4
Volatiles	0.471	0.591*	0.346	0.399*
Tot.	1	1.12	1	1.053
Deviation (%)	12		5.3	

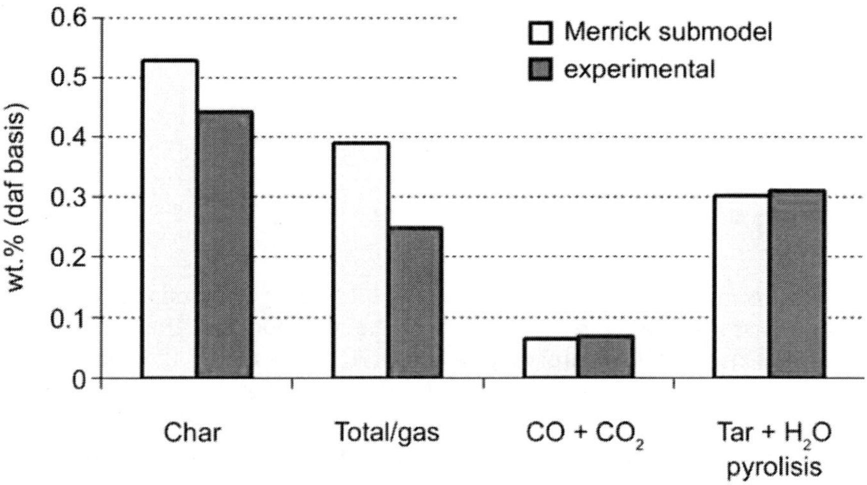

Figure 8. Validation of Merrick submodel: test 1 [5].

The reason of such degree of inaccuracy is to ascribe to those equations of the linear system defined by empirical correlations, tested for the of coal pyrolysis with volatile content varying in the particular range previously mentioned, but not necessarily valid in case of fuel with higher volatile content like those used in the simulated cases [7,8].

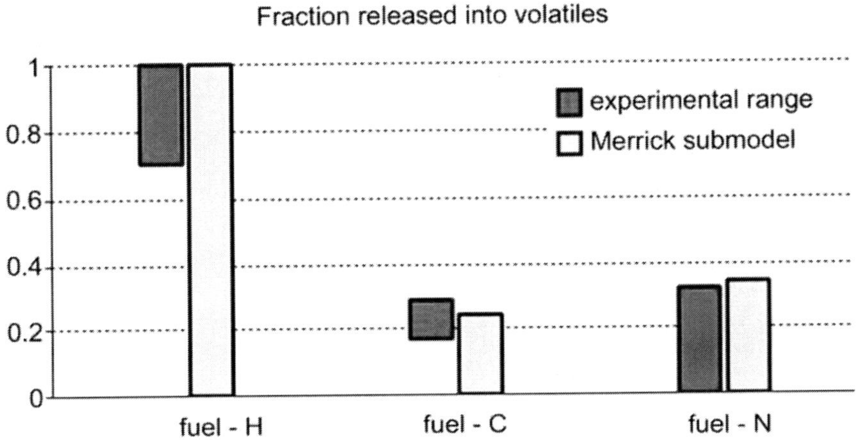

Figure 9. Validation of merrick submodel: test 2 [6].

7. CONCLUSIONS

The IEA model, initially designed to simulate the performance of industrial size installations, results to be an helpful tool in the analysis and optimization of laboratory scale units, like the tester CFB gasifier. The scalingdown approach adopted in this work can be considered to be adequate to model the particular design and operational conditions of the reactor. In case of coal combustion simulations, the program successfully achieves the final convergence.

The results confirm a good compromise between calculation values and empirical results. This condition will be optimized by choosing better input data On the other hand, the introduction of the biomass physical properties generates computational errors, and the solution of this condition represents the main problem to deal with. The results obtained at this stage of the work can only provide a rough estimation of the wood combustion process, however may be useful as starting point for future developments and studies.

The pyrolysis model based on the approach proposed by Merrick and tested in this work can be suggested as promising basis for a more accurate modelling of the biomass thermal decomposition. It is capable to predict the formation of tar and hydrocarbons as components of the volatile matter, but still requires further improvements. To improve such procedure, the empirical correlations used to build the linear equations system should be re-defined and validated in accordance with the gas composition observed during the experiments of biomass pyrolysis, in the case of solid fuels with a volatile content typically around 80%, surely higher than coal [9].

The proposed model can simulate and manage low values of λ and well approaches the expected consequences in case of incomplete combustion, but it is not suitable to reproduce the whole gasification process yet, since the gasification reactions and their reaction kinetics are not modelled in the code. Considerable efforts will be necessary to introduce in the original model a reliable model to consider the effects of these reactions on the synthesis gas composition; by the way, the analysis carried out in this work seems to encourage further investigations, with a promising future application of the IEA model, as helpful tool, in the optimization of the CFB biomass gasification process [10,11].

REFERENCES

1. J. Hannes, "Mathematical Modelling of Circulating Fluidized Bed Combustion," Kartographie und Druck Peter List, Achen, 1996.

2. M. Siedlecki, "Alkali Measurements: Start-Up of the CFB Combustor and Development of Gas Sample Procedure," M.Sc. Thesis, Delft University of Technology, Delft, 2003.

3. M. Siedlecki, K. Van der Nat, E. Simeone and W. De Jong, "The First Results of Gas and Solids Characterization Obtained during Steam-Oxygen Gasification of Biomass in a 100kWth CFB Gasifier," World Renewable Energy Congress IX, Florence, 19-25 August 2006.

4. D. Merrick, "Mathematical Models of the Thermal Decomposition of Coal. 1. The Evolution of Volatile Matters," Fuel, Vol. 62, No. 5, 1983, pp. 534-539.

5. J. F. Stubington and Sumaryono, "Release of Volatiles from Large Coal Particles in a Hot Fluidized Bed Fuel," Vol. 63, No. 7, 1984, pp. 1013-1019.

6. A. N. Hayhurst and A. D. Lawrence, "The Devolatilization of Coal and a Comparison of Chars Produced in Oxidizing and Inert Atmospheres in Fluidized Beds," Combustion and flame, Vol. 100, No. 4, 1995, pp. 591-604.

7. L. Devi, K. J. Ptasinski and F. J. J. G.Jansen. "A Review of the Primary Measures for Tar Elimination in Biomass Gasification Process," Biomass and Bioenergy, Vol. 24, No. 2, 2003, pp. 125-140.

8. P. Basu, "Combustion and gasification in fluidized beds," Taylor and Francis, Oxford, 2006.

9. C. Franco, F. Pinto, I. Gulyurtlu and I. Cabrita, "The Study of Reactions Influencing the Biomass Steam Gasification Process," Fuel, Vol. 82, No. 7, 2003, pp. 835-842.

10. X. T. Li, J. R. Grace, C. J. Lim, A. P. Watkinson, H. P. Chen and J. R. Kim, "Biomass Gasification in a Circulating Fluidized Bed," Biomass and Bioenergy, Vol. 26, No. 2, 2004, pp. 171-193.

11. K. W. Ragland, D. J. Aerts and A. J. Baker, "Properties of Wood for Combustion Analysis," Bioresource Technology, Vol. 37, No. 4, 1991, pp. 446-452.

9
CHAPTER
FLUIDIZED-BED GASIFICATION OF PLASTIC WASTE, WOOD, AND THEIR BLENDS WITH COAL

Lucio Zaccariello * and Maria Laura Mastellone

Department of Environmental, Biological and Pharmaceutical Sciences and Technologies, Second University of Naples, Via Antonio Vivaldi 43, Caserta 81100, Italy

ABSTRACT

The effect of fuel composition on gasification process performance was investigated by performing mass and energy balances on a pre-pilot scale bubbling fluidized bed reactor fed with mixtures of plastic waste, wood, and coal. The fuels containing plastic waste produced less H_2, CO, and CO_2 and more light hydrocarbons than the fuels including biomass. The lower heating value (LHV) progressively increased from 5.1 to 7.9 MJ/Nm^3 when the plastic waste fraction was moved from 0% to 100%. Higher carbonaceous fines production was associated with the fuel containing a large fraction of coal (60%), producing 87.5 g/kg$_{Fuel}$ compared to only 1.0 g/kg$_{Fuel}$ obtained during the gasification test with just plastic waste. Conversely, plastic waste gasification produced the highest tar yield, 161.9 g/kg$_{Fuel}$, while woody biomass generated only 13.4 g/kg$_{Fuel}$. Wood gasification showed a carbon conversion efficiency (CCE) of 0.93, while the tests with two fuels containing coal showed lowest CCE values (0.78 and 0.70, respectively). Plastic waste and wood gasification presented similar cold gas efficiency (CGE) values (0.75 and 0.76, respectively), while that obtained during the co-gasification tests varied from 0.53 to 0.73.

Keywords: fluidized bed gasifier; co-gasification; plastic waste; wood; coal; mass and energy balances

1. INTRODUCTION

There is a growing interest to apply thermo-chemical process to different kind of wastes, considering the environmental impact of these materials and economic aspects of power plants. Gasification is a viable technology for the thermo-chemical conversion of biomass and wastes due to its greater environmental sustainability and to the production of valuable products from different fuels. Among all gasification technologies, fluidization is often chosen as a reference for its great operating flexibility: the good mixing properties that ensure uniform process conditions, while also simultaneously feeding different fuels; the possibility to utilize various fluidizing agents [1,2,3,4]; to operate with or without a specific bed catalyst [5,6,7,8,9,10]; to add reagents along the reactor height [11,12,13,14], and to feed fuels in different positions of the reactor [15,16,17]. On the other hand, during the gasification process, the unavoidable formation of contaminants such as tar, carbonaceous particles, and inorganics leads to an increase in operating costs and efficiency loss. Since the formation of these by-products is strictly correlated to the fuel structure and composition, the possible synergy between the products and the intermediates produced during the gasification of different materials could lead to improving the process performance, reducing carbon losses, and increasing producer gas energy content. Gasification plants can be conducted by co-feeding different fuels in order to produce better results in term of producer gas quality and energy saving respect to those obtained by utilizing a single material.

Co-gasification is a relatively new process where the industrial know-how is far from exhaustive, even if the scientific literature is full of interesting research studies on the co-feeding effect of different fuels into fluidized bed gasifiers. Authors do not always agree about the effects of co-feeding on the gasification process performance [18,19,20,21,22]. These differences are probably also due to the difficult comparison of the results obtained from different gasification technologies, different fuels, and different operating conditions. Generally, synergistic effects are attributed to the interaction between the produced volatiles or between the volatiles and the ashes contained in the char of the gasifying fuels [23,24]. Some authors reported that the composition of the feedstock affects the producer gas quality by means of non-additive models with a synergistic interaction among different materials during co-gasification test. The changes in the gas composition were non-linear and, consequently, it could not be predicted on the basis of gasification of the individual materials [18,19,20]. On the contrary, several studies did not observe any interaction [21,22].

Wilk and Hofbauer [18] performed co-gasification tests in a pilot dual fluidized bed reactor by using different mixtures of wood and different types of plastic material as feedstock. The results indicated that the product gas composition was strongly influenced by fuel mixtures. During the gasification tests more CO and CO_2 were produced from co-gasification than would be expected from linear interpolation of mono-gasification of wood and plastic. On the other hand, light hydrocarbons and tar in the product gas were considerably lower than presumed. Saw and Pang [19] conducted co-gasification experiment of blended lignite and wood in a pilot-scale dual fluidized bed steam gasifier. The experimental results showed that the producer gas yield and gas compositions were non-linearly correlated to the lignite to wood ratio, which indicated a synergy effect of the blending. The authors attributed synergistic effect on the tar production to the catalytic elements (Ca and Fe) contained in the blended chars. Fermoso et al. [20], utilizing a lab-scale high-pressure gasification reactor, studied the effect of several operating variables and that of blending bituminous coal with petcoke and biomass on gasification process performances. The results showed a positive deviation from the linear additive rule in the case of H_2 and CO production for the blends of coal with petcoke.

Conversely, Kumabe and co-workers [21] gasified woody biomass and coal with air and steam in a downdraft fixed bed reactor without observing an apparent synergy in terms of carbon distribution of products. The authors reported changes in the producer gas composition related to the increasing of biomass fraction in the mixture: a decrease in H_2 together with an increase in CO_2 content was observed. In addition, CO and hydrocarbons concentrations appeared independent by the fuel compositions. Aigner et al. [22] observed a linear relationship between the producer gas composition and wood ratio when a mixture of coal and wood was gasified. In particular, when H/O ratio in the fuel decreased, the H_2/CO and H_2/CO_2 ratios in producer gas decreased as well.

The aim of this study was to evaluate the effect of fuel composition on the gasification process by performing mass and energy balances on a pre-pilot bubbling fluidized bed reactor. The experimental runs were carried out by feeding five alternative fuels including mono-gasification tests of plastic waste and woody biomass, and co-gasification tests of plastic waste, wood, and coal mixtures.

2. EXPERIMENTAL APPARATUS AND PROCEDURE

2.1. Experimental Apparatus

The experimental work was carried out utilizing a pre-pilot scale bubbling fluidized bed gasifier (BFBG) with a maximum feeding capacity of 5 kg/h, depending on the type of fuel (Figure 1).

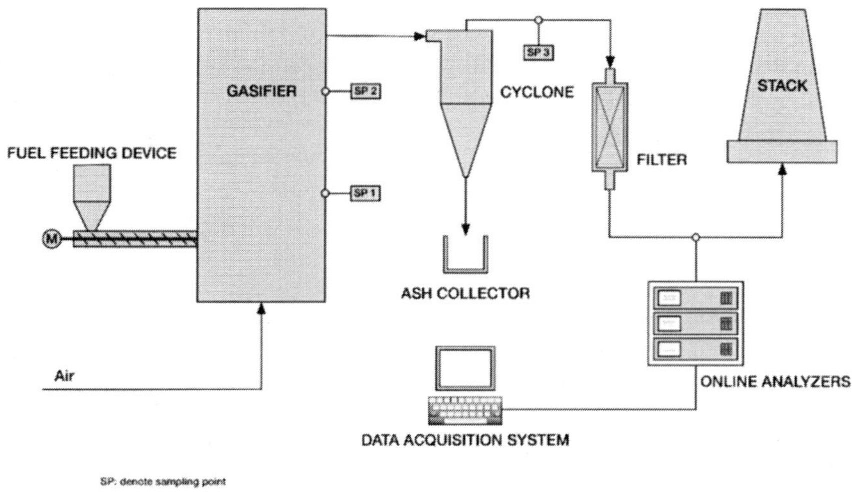

Figure 1. Schematic illustration of the pre-pilot bubbling fluidized bed gasifier.

The BFBG is a 10 cm internal diameter cylindrical column, made of AISI 316L and electrically heated by five shell furnaces. Each furnace is controlled by a data acquisition system connected to five thermocouples, located in the reactor internal wall, which allow for independently setting the temperature of each reactor section (plenum, bed, and freeboard).

The air utilized as the fluidizing agent was injected at the bed bottom through a distributor plate composed of three nozzles. These have a truncate pyramidal shape and were specifically designed in order to ensure a homogeneous distribution of the fluidizing gas in the bed cross-section. The total height is 2.5 m from the distributor plate to the product gas outlet. The feedstock was over-bed fed, continuously, by means of a screw-feeder device. A nitrogen flow of 0.32 kg/h was used to help the fuel feeding and to avoid the back flow of the hot gas. At the syngas outlet a high-efficiency cyclone allows the collection of elutriated fines. Downstream of this are two alternative symmetric gas conditioning lines, each one is composed by a bubbler filled with water and by a cartridge filter, which provide for tar, residual fly ash, and acid and basic gases removal. The gas coming from the producer gas treatment section is then sent to a stack.

2.2. Operating Conditions

The pre-pilot scale BFBG was fed with the five different fuels by keeping fixed the type and size range of the bed material (silica sand, 0.2–0.4 mm), the gasifying agent (air), the fluidized bed velocity (0.4 m/s) and the equivalence ratio (0.25), in order to obtain information about the role of fuel composition on the gasification process performance. Table 1 lists the values chosen for the complete set of experiments.

Table 1. Operating conditions of experimental runs.

Fuel	Bed Material	U_g (m/s)	W_{Air} (kg/h)	W_{Fuel} (kg/h)	A/F (kg_{Air}/kg_{Fuel})	ER
RP	Silica sand	0.42	3.42	1.08	3.17	0.24
WRP	Silica sand	0.42	3.42	1.14	2.99	0.25
WRPC	Silica sand	0.42	3.43	1.56	2.20	0.25
WC	Silica sand	0.41	3.44	2.09	1.65	0.25
WD	Silica sand	0.42	3.42	2.46	1.39	0.25

With the purpose of obtain reliable data to perform accurate material and energy balances, sampling procedures of producer gas, elutriated fines, and tar were activated when the values of gas composition, temperature, and pressure were at steady state conditions and last for not less than 1 h.

2.3. Analytical Equipment

The main product gas compounds (CO_2, CO, H_2, CH_4, C_2H_2, C_2H_4, C_2H_6, C_3H_6, C_3H_8, and N_2) were measured by using an Agilent 3000 micro gas chromatograph (micro-GC, Santa Clara, CA, USA) located downstream of the tar sampling line. In addition, an ABB AO2020 (for total hydrocarbons) (Zurich, Switzerland) and two HORIBA VA-3000 (for CO, CO_2, and O_2) (Kyoto, Japan) on-line analyzers were used as a check of accuracy of the micro-GC measurements. This double system allows a high reliability of measured gas composition.

The producer gas was further sampled by means of tedlar bags for off-line analyses in two other points along the reactor height (0.9 and 1.8 m). The flow rate of producer gas was determined by the tie component method applied to the value of nitrogen content in the dry gas, as obtained by on-line GC measurements.

Elutriated fines, consisting of char and fragmented bed particles, collected by the cyclone and particulate filter, were analyzed in a LECO TruSpec Elemental Analyzer (St. Joseph, MI, USA) in order to evaluate the content of carbon, hydrogen, nitrogen, and sulfur. For the sampling of condensable species, a system consisted of four in-series cooling coils plugged into an ice bath; a suction pump and a flow meter were installed and operated with a gas flow rate of 2 NL/min for 1 h to obtain tar and water phase. The condensate is then washed from the coils using dichloromethane as solvent and collected in dedicated glass bottles. After a few minutes the condensate stratifies, forming two distinct solutions: water and hydrocarbons and dichloromethane. Water was separated from tar by means of a graduated syringe in order to evaluate its content in the producer gas. After water separation, the condensed hydrocarbons were analyzed off-line in a Perkin-Elmer Clarus 500 gas chromatograph (Waltham, MA, USA) coupled with a mass spectrometer (GC-MS). Elutriated fines and tar flow rate were obtained by dividing the total mass collected to the sampling time.

Hydrogen chloride, hydrogen sulfide, and ammonia were collected by bubbling the product gas through a pair of gas bubblers; each of them containing a solution of 50 mL of NaOH (0.5 M) and 50 mL of HCl (0.5 M). Subsequently, these solutions were analyzed by means of a Dionex DX-120 ion chromatograph (Sunnyvale, CA, USA).

Data obtained from on-line and off-line gas measurements and from chemical analyses of solid and condensed samples were processed to develop mass balance on atomic species and the related energy balance for each gasification tests.

2.4. Feedstock

Gasification tests were carried out using five different materials including plastic waste, wood and coal. Two fuels were selected for the mono-gasification tests: recycled plastic (RP), a mixture of several plastic wastes obtained from the separate collection of post-consumer packaging materials and natural wood (WD), generally utilized to prepare fuel for domestic heating. Co-gasification tests were conducted utilizing three mixtures as fuels: the blend named WRP, composed of plastic waste and virgin wood; the fuel indicated as WRPC, obtained by blending recycled polyethylene, virgin wood, and brown coal and, finally, the fuel specified as WC, a mixture of virgin wood and brown coal. An overview of the employed fuels is given in Table 2.

Table 2. Composition and main physical properties of the tested fuels.

Items	RP	WRP	WRPC	WC	WD
Plastic waste, $\%_{wb}$	100	80	30	—	—
Wood, $\%_{wb}$	—	20	20	40	100
Coal, $\%_{wb}$	—	—	50	60	—
Size (diameter and length), mm	Irregular	6, 20	6, 20	6, 20	6, 20
Bulk density, kg/m^3	590	580	615	620	570

The RP fuel has an irregular-spheroidal shape with a particle diameter of about 15–20 mm, while the other ones have a cylindrical shape, with a diameter of 6 mm and a length of about 15 mm. Plastic waste was selected as fuel because it saves the use of natural resources, for its large availability, and for its high calorific value. Woody biomass was chosen for the experimental tests since it offers credits resulting from the utilization of a renewable zero-emission energy resource. Lastly, coal utilization provides the benefits of wide fuel availability. Design and operation of thermochemical conversion systems need fuel composition as well as its chemical energy. In this context two types of compositions are used: proximate and ultimate analyses. The proximate analysis provides moisture, volatile matter, fixed carbon, and ash content, while the ultimate analysis gives the fuel composition in terms of its basic elements such as carbon, hydrogen, nitrogen, sulfur, and oxygen. In this work the proximate analysis was performed as follows: the gross fuel sample was heated in air to 105 °C for 12 h to obtain moisture content, to 950 °C in inert ambient (nitrogen) for 5 h to obtain volatile matter, and to 750 °C in air for 2 h to obtain ash amount. Finally, the carbon-rich residue (fixed carbon) that remains after drying and devolatilization was calculated by subtracting the percentage of moisture, volatile matter, and ash from 100%. The ultimate analysis was carried out processing the fuel sample in the LECO TruSpec CHN/S Analyzer. Results of proximate and ultimate analyses of the tested fuels are listed in Table 3.

Table 3. Proximate and elemental analyses of the tested fuels.

Items	RP	WRP	WRPC	WC	WD
Proximate analysis, $\%_{wb, ar}$					
Volatile matter	94.50	92.64	68.92	59.53	86.74
Fixed carbon	2.89	3.42	19.14	20.82	3.12
Moisture	0.67	2.11	4.11	9.42	9.93
Ash	1.94	1.83	7.83	10.23	0.21
Ultimate analysis, $\%_{wb, ar}$					
C	79.54	73.20	62.28	51.93	45.31
H	13.06	11.15	8.11	5.44	5.59
N	0.18	0.30	0.19	0.22	0.26
S	0.08	0.10	0.13	0.16	<0.01
O (by difference)	4.53	11.31	17.35	22.60	38.70
Heating value, MJ/kg_{Fuel}					
HHV^{a}_{db}	42.69	37.45	29.35	22.00	18.40
LHV^{b}_{ar}	36.95	32.57	25.73	19.41	15.73

wb = weight basis; db = dry basis; ar = as received; [a] Evaluated by means of relationship proposed by Channiwala and Parikh [25]; [b] Evaluated by the HHV on dry basis by taking into account the latent heats of vaporization of the fuel moisture and the water obtained as product of hydrogen combustion.

3. RESULTS AND DISCUSSION

3.1. Producer Gas

The chemical characteristics of the fuels selected for the experimental tests differ considerably. Plastic waste contains a larger fraction of carbon and hydrogen than woody biomass that, conversely, contains more oxygen. The chemical energy of the fuels is greatly affected by the oxygen and water contents: the higher their values, the lower the chemical energy content. This agrees with the results reported in Table 3, which shows a decreasing of the lower heating value (LHV) when oxygen and moisture amounts in the feedstock increase. Plastic waste gives more volatile matter while the mixtures containing coal show the lower volatile content and a larger amount of fixed carbon. These marked differences in the fuels composition may present significant variations in the gasification products.

Gas measurements results from the micro-GC, that provides reading every 200 s, were used for all gasification tests calculations. The data acquisition system provides measurement data of temperature and pressure obtained from the gasifier and cleaning section in intervals of 1 s. These values were recorded and averaged over the entire sampling period. Figure 2 and Figure

3 show gas composition, temperature, and pressure data averaged every ten minutes during the gasification test performed by feeding woody biomass (WD).

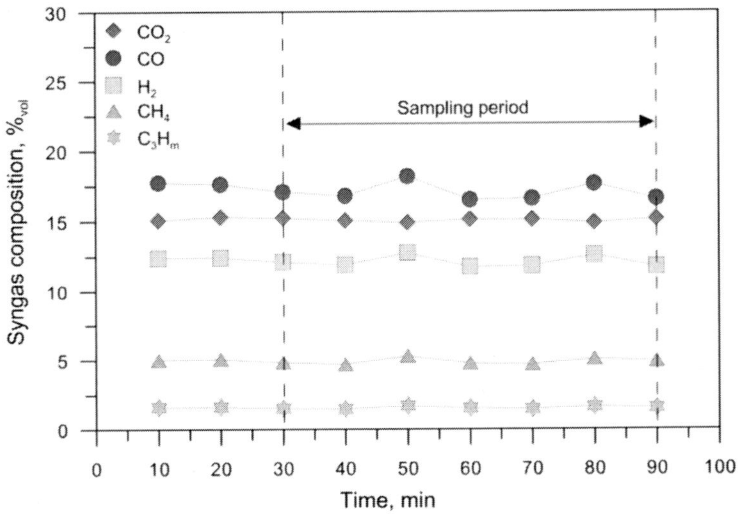

Figure 2. Producer gas composition profile obtained during wood gasification test.

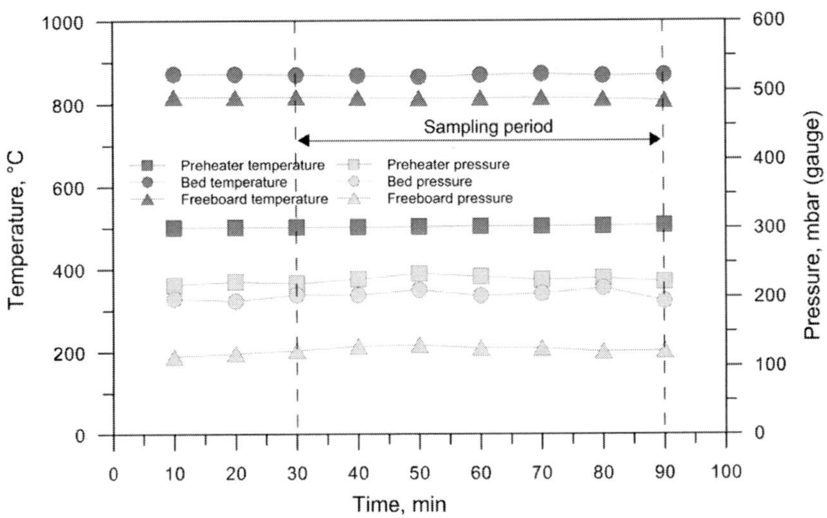

Figure 3. Temperature and pressure profiles obtained during wood gasification test.

Fuel composition strongly affects the characteristics of the producer gas obtained during the gasification tests. The results reported in Figure 4 show that the fuels containing plastic waste yield less hydrogen, carbon monoxide, and carbon dioxide and more methane and light hydrocarbons (sum of ethane, ethylene, acetylene, propane and propylene, C_3H_m) than the fuels including woody biomass. The considerable increase of carbon monoxide (from 4.9% to 17.1%) and carbon dioxide (from 11.2% to 15.1%) content in the product gas, as the wood fraction increases from 0% to 100%, can be attributed to the remarkable amount of oxygen and/or C–O groups in the biomass. On the other hand, the increase of plastic waste fraction, mainly made of polyethylene and polypropylene (consisting of long monomeric chains of C_2H_4 and C_3H_6, respectively), promotes the production of methane and light hydrocarbons (that increase from 4.9% to 9.1% and from 1.6% to 5.1%, respectively), which lead to a reduction in hydrogen content (from 12.2% to 9.2%) in the producer gas.

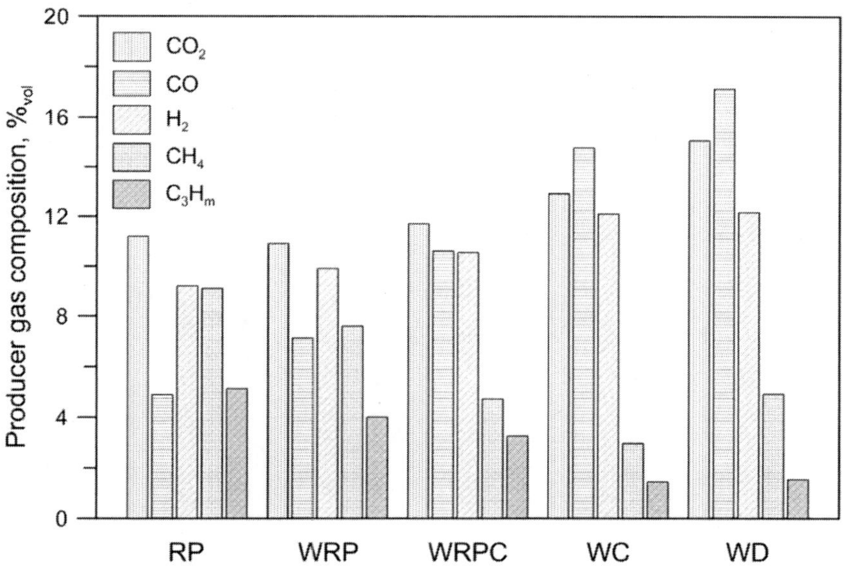

Figure 4. Concentration of the main components of the dry product gas.

The above presented results are in agreement with the data obtained in other gasification studies using plastics and woody biomass as fuels. Pinto et al. [26] carried out experimental runs by means of a steam bubbling fluidized bed gasifier fed with mixtures of plastic and pine wood. They observed a decrease in the production of carbon monoxide as the fraction of plastic in the mixture increases. Wilk and Hofbauer [18] obtained similar results in a dual fluidized bed gasifier. In addition, the authors also reported an increase of methane and light hydrocarbon concentrations as the fuel plastic fraction increases from 0% to 100%.

The gasification tests carried out by utilizing mixtures containing coal further highlight the predominant role of plastic waste and woody biomass on the producer gas composition. They confirm that the production of hydrogen, carbon monoxide, and carbon dioxide can be attributed to wood degradation, whereas methane and light hydrocarbon generation can be assigned to plastic waste decomposition. Gasification tests carried out with WRPC, which contains 50% of coal, 30% of plastic, and 20% of wood, yields lower concentrations of hydrogen and carbon monoxide, and higher fractions of methane and light hydrocarbons than WC, which contains 60% of coal and a larger fraction of wood (40%). These results indicate that coal addition to the blends does not affect the trends shown by the main producer gas constituents obtained during the tests without coal. A possible explanation is that both lower coal reactivity and reduced volatile matter content, than that of plastic and wood, lead to a higher production of heavy tar and char [21,22]. As a consequence, coal addition slightly contributes to the change in the gas composition and to the increasing gas yield. This hypothesis is confirmed by the results displayed in Figure 5, which reports that the gasification tests with the blends containing coal (WRPC and WC) show a larger production of these by-products.

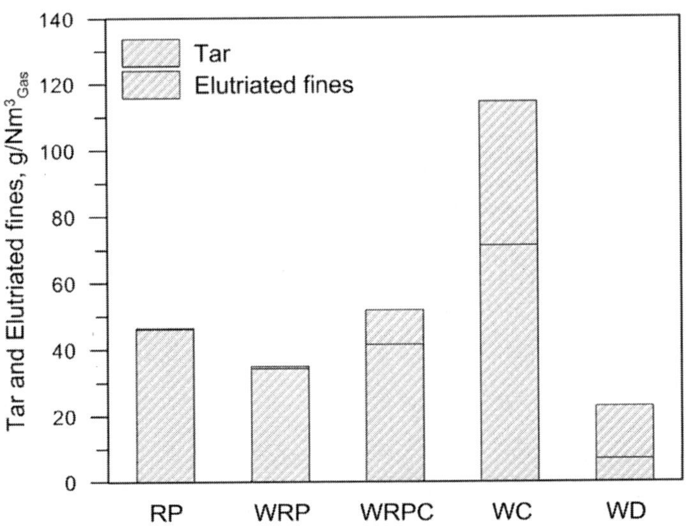

Figure 5. Elutriated fines and tar produced during the gasification tests.

Table 4. Main results obtained during the experimental runs.

Fuel	T °C	CO_2 %vol	CO %vol	H_2 %vol	CH_4 %vol	C_3H_m %vol	Q_{PG} Nm^3/h	C_{EF} g/Nm^3	C_{Tar} g/Nm^3	LHV_{PG} MJ/Nm^3	SE MJ/kg_{Fuel}	HRE	CCE	CGE
RP	877	11.20	4.90	9.20	9.10	5.14	3.79	0.29	46.06	7.90	27.77	0.91 ± 0.03	0.84 ± 0.11	0.75 ± 0.04
WRP	872	10.90	7.12	9.91	7.60	4.01	3.86	0.67	34.24	7.06	23.84	0.90 ± 0.19	0.83 ± 0.22	0.73 ± 0.15
WRPC	868	11.71	10.62	10.55	4.73	3.25	4.20	10.31	41.38	6.09	16.40	0.79 ± 0.23	0.78 ± 0.17	0.64 ± 0.18
WC	862	12.91	14.77	12.09	2.96	1.45	4.22	43.26	71.20	5.09	10.29	0.70 ± 0.22	0.70 ± 0.14	0.53 ± 0.16
WD	870	15.07	17.13	12.16	4.93	1.55	4.80	15.94	6.86	6.15	12.03	0.79 ± 0.10	0.93 ± 0.10	0.76 ± 0.10

PG = producer gas; EF = elutriated fines; ± = ±standard deviation.

As previously reported, the features of the feedstock affect the producer gas composition and, as a consequence, its LHV. The results reported in Table 4 show that the LHV trend is consistent to that obtained for methane and light hydrocarbons due to their high calorific value. Since plastic gasification generated a large amount of hydrocarbons, the LHV progressively increases up to 7.9 MJ/Nm^3 (gasification test with RP fuel) when more plastic waste is used in the fuel mixtures. The mono-gasification test with wood (WD) shows intermediate LHV value (6.2 MJ/Nm^3) that is halfway between those obtained in the tests with the fuels containing plastic (RP and WRP) and those that include coal (WRPC and WC).

3.2. Mass Balance

As already mentioned, data obtained from on-line gas measurements and from chemical analyses of the elutriated fines and tar, produced during steady operation of the gasification plant, were processed to develop mass and energy balances for each gasification test. In particular, in this section, the acquired data were elaborated to perform overall mass and elemental balances. Mass inputs involve fuel, dry air, and nitrogen purge. The industrial area that houses the gasifier is equipped with an air dryer device; therefore, dry air was utilized as the gasifying agent during the experimental runs. Mass outputs include dry gas, reaction water, elutriated fines, and tar.

For accurate mass and energy balance calculations, it is also necessary to determine the composition of the elutriated fines and tar. The composition of elutriated fines sampled during the gasification tests was determined utilizing a LECO TruSpec Elemental Analyzer, while the tar elemental composition was evaluated from the molecular tar composition as obtained from the GC-MS analyzer (Table 5).

Table 5. Elutriated fines composition produced during the gasification tests.

Fuels	Ultimate Analysis, $\%_{wb}$									
	C		H		N		O		S	
	EF	Tar	EF	Tar	EF	Tar	EF	Tar	EF	Tar
RP	42.80	93.25	1.80	6.17	<0.01	<0.01	<0.01	<0.01	<0.01	<0.01
WRP	48.10	93.34	2.00	6.00	<0.01	<0.01	<0.01	<0.01	<0.01	<0.01
WRPC	39.70	94.13	0.90	5.85	0.50	<0.01	<0.01	0.01	0.16	<0.01
WC	49.70	93.49	0.80	6.08	0.71	<0.01	<0.01	0.02	0.59	<0.01
WD	65.45	93.82	1.10	6.15	0.39	<0.01	<0.01	0.02	<0.01	<0.01

Table 6 reports mass and elemental balances performed for the five gasification tests. The results indicate that the overall mass balances show values ranging between 96.1% and 99.6%. The non-closure of the mass balances is due to the sum of the systematic errors that affect the measurement instrumentations and those due to the effectiveness of the sampling procedures. Assuming that these errors were regularly repeated during all the experimental runs and that the overall maximum error is quite low (less than 4%), it is possible to carry out the following observations.

Plastic addition to the fuel mixture leads to an increase in gas specific yield (GSY). The results reported in Table 6 show that the GSY increases from about 2360 g/kg_{Fuel}, with the fuels without plastic (WC and WD), to values of 3185.4, 3959.8, and 4126.8 g/kg_{Fuel}, as the plastic content increases from 30% to 100% (WRPC, WRP and RP, respectively). Similar results were obtained in a previous study [27] and by other researchers. Ahmed et al. [28] conducted co-gasification experiments feeding polyethylene and woodchip mixtures in a semi-batch reactor, using steam as the gasifying agent. The authors observed a progressive increase in gas yield as the amount of plastic rises up to 80%.

The experimental runs conducted by feeding the fuels containing large amount of plastic (RP and WRP) yield small specific quantities of elutriated fines (1.0 and 2.3 g/kg_{Fuel}, respectively), while the fuels containing coal (WRPC and WC) generate huge amounts of fines (27.8, and 87.5 g/kg_{Fuel}, respectively). This could be due to the large quantity of fixed carbon contained in these fuels that produce more char, which is then carried out from the BFBG in the form of carbonaceous fines. This phenomenon in the test with WRPC appears partially suppressed. This evidence could be explained considering the structure of the WRPC pellets: plastic, wood and coal were crumbled and intimately blended before being pelletized. When the WRPC pellets drop down in the reactor, the plastic contained in this fuel melts and binds the particles of coal and wood, protracting the permanence of these aggregates in the most reactive zone of the reactor (bed and splashing zone), favoring their conversion into gaseous products.

Table 6. Mass balance performed during the gasification tests.

Item	Input, kg/h					Item	Output, kg/h				
	Mass	C	H	O	N		Mass	C	H	O	N
Fuel, RP	1.08	0.86	0.14	0.05	<0.01	Dry gas	4.45	0.72	0.13	0.74	2.86
Air	3.42	–	–	0.80	2.62	Water	0.01	–	<0.01	0.01	–
Nitrogen purge	0.32	–	–	–	0.32	Elutriated fines	<0.01	<0.01	<0.01	–	<0.01
–	–	–	–	–	–	Tar	0.17	0.16	0.01	0.00	<0.01
Total	*4.82*	*0.86*	*0.14*	*0.85*	*2.95*	*Total*	*4.64*	*0.86*	*0.14*	*0.75*	*2.86*
Mass balance, %			**96.28**			–	–	–	–	–	–
Fuel, WRP	1.14	0.84	0.13	0.13	<0.01	Dry gas	4.52	0.70	0.12	0.80	2.92
Air	3.42	–	–	0.80	2.62	Water	0.03	–	<0.01	0.03	–
Nitrogen purge	0.32	–	–	–	0.32	Elutriated fines	<0.01	<0.01	<0.01	–	<0.01
–	–	–	–	–	–	Tar	0.13	0.12	0.01	0.00	0.00
Total	*4.88*	*0.84*	*0.13*	*0.93*	*2.95*	*Total*	*4.69*	*0.82*	*0.13*	*0.83*	*2.92*
Mass balance, %			**96.08**			–	–	–	–	–	–
Fuel, WRPC	1.56	0.97	0.13	0.27	<0.01	Dry gas	4.98	0.75	0.10	1.02	3.10
Air	3.43	–	–	0.80	2.63	Water	0.10	–	0.01	0.09	–
Nitrogen purge	0.32	–	–	–	0.32	Elutriated fines	0.04	0.02	<0.01	–	<0.01
–	–	–	–	–	–	Tar	0.17	0.16	0.01	<0.01	0.00
Total	*5.31*	*0.97*	*0.13*	*1.07*	*2.95*	*Total*	*5.30*	*0.94*	*0.12*	*1.11*	*3.10*
Mass balance, %			**99.75**			–	–	–	–	–	–
Fuel, WC	2.09	1.08	0.11	0.47	<0.01	Dry gas	5.00	0.76	0.08	1.22	2.94
Air	3.44	–	–	0.80	2.64	Water	0.13	–	0.01	0.11	–
Nitrogen purge	0.32	–	–	–	0.32	Elutriated fines	0.18	0.09	<0.01	–	<0.01
–	–	–	–	–	–	Tar	0.30	0.28	0.02	<0.01	<0.01
Total	*5.85*	*1.08*	*0.11*	*1.27*	*2.96*	*Total*	*5.62*	*1.13*	*0.11*	*1.33*	*2.94*
Mass balance, %			**96.06**			–	–	–	–	–	–
Fuel, WD	2.46	1.11	0.14	0.95	0.01	Dry gas	5.71	1.03	0.11	1.62	2.95
Air	3.42	–	–	0.80	2.62	Water	0.21	–	0.02	0.19	–
Nitrogen purge	0.32	–	–	–	0.32	Elutriated fines	0.08	0.05	<0.01	–	<0.01
–	–	–	–	–	–	Tar	0.03	0.03	<0.01	<0.01	<0.01
Total	*6.20*	*1.11*	*0.14*	*1.75*	*2.95*	*Total*	*6.03*	*1.11*	*0.13*	*1.81*	*2.95*
Mass balance, %			**97.33**			–	–	–	–	–	–

The mono-gasification test of plastic waste (RP) produces the highest specific amount of tar, 161.9 g/kg_{Fuel}, while woody biomass (WD) gasification generates only 13.4 g/kg_{Fuel}. The co-gasification tests yield similar (high) amounts of tar, i.e., 115.7, 111.4 and 144.1 g/kg_{Fuel} in the runs with WRP, WRPC, and WC, respectively.

Water content in the producer gas, as previously reported, was determined by applying the condensation method. In the runs with RP and WC fuels this procedure failed; thus, water content was estimated by applying the atomic balance method on the hydrogen element. Table 6 shows that woody biomass promotes the generation of reaction water. In particular, water content increases from 13.4 to 85.5 g/kg_{Fuel} as the wood fraction in the fuel rises from 0% to 100%. Probably, water generation from the fuels containing a higher biomass fraction is due to both physical and chemical mechanisms: by evaporation of the larger fuel moisture content (Table 3) and by oxidation of hydroxyl and carboxyl groups widely present in lignocellulosic materials.

Data obtained from fuels characterization and from on-line GC measurements were utilized to evaluate the hydrogen recovery efficiency (HRE) and the carbon conversion efficiency (CCE). In this work HRE and CCE were calculated according to Equations (1) and (2):

$$HRE = \frac{m_{H,H_2} + m_{H,CH_4} + m_{H,C_3H_m}}{m_{H,Fuel}} \tag{1}$$

$$CCE = \frac{m_{C,CO_2} + m_{C,CO} + m_{C,CH_4} + m_{C,C_3H_m}}{m_{C,Fuel}} \tag{2}$$

where $m_{H,i}$ and $m_{C,i}$ are hydrogen and carbon mass flow rates of the constituent i, respectively.

The fuels containing high plastic fractions transfer more hydrogen in the producer gas than the blends without plastic. In particular, Table 4 shows that the HRE values obtained in the runs with RP, WRPC and WRP are 0.91, 0.90 and 0.79; while those found in the runs with WC and WD are 0.70 and 0.79, respectively. On the other hand, the gasification test conducted by feeding wood (WD) exhibits higher a CCE value (0.93). Furthermore, coal seems to have a negative effect on CCE. In fact, the gasification tests with the fuels WRPC and WC show the lowest CCE values (0.78 and 0.70, respectively). The low conversion efficiency in terms of HRE and CCE recorded in tests carried out with the blend WC is a direct consequence of carbon and hydrogen losses determined by the considerable production of elutriated fines and tar (Figure 5 and Table 6).

3.3. Energy Balance

Data obtained from material balances were utilized to perform energy balances for the gasification tests carried out by feeding all the five tested fuels. Energy input to the gasifier considered the chemical enthalpy (CE) of the fuel, while energy outputs took into account the CEs of the producer gas, elutriated fines, and tar. The CE was calculated as indicated in the Equation (3):

$$CE_i = m_i \times LHV_i \tag{3}$$

where m_i and LHV_i are volume or mass flow rate and lover heating value of the constituent i, respectively.

The LHVs of the fuels, elutriated fines, and tar were estimated from the HHV by taking into account the latent heats of vaporization of the fuel moisture and the water obtained as product of hydrogen combustion. The HHV was evaluated using the correlation, with a wide range of applicability, proposed by Channiwala and Parikh [25]:

$$HHV = 349.1C + 1178.3H + 100.5S - 103.4O - 15.1N - 21.1Ash \tag{4}$$

where C, H, S, O, N, and Ash are percentages of carbon, hydrogen, sulfur, oxygen, nitrogen, and ash as determined by ultimate analysis on a dry basis.

Table 7 reports the chemical enthalpy balances associated to input and output streams. As it shows, energy balance closure ranges from 78.4% to 88.5%.

Table 7. Energy balance performed during the gasification tests.

Item	Input, MJ/kg$_{Fuel}$					Item	Output, MJ/kg$_{Fuel}$				
	RP	WRP	WRPC	WC	WD		RP	WRP	WRPC	WC	WD
Fuel	36.95	32.57	25.73	19.41	15.73	Dry syngas	27.77	23.84	16.40	10.29	12.03
Air	–	–	–	–	–	Water	–	–	–	–	–
Nitrogen purge	–	–	–	–	–	Elutriated fines	0.02	0.04	0.38	1.49	0.72
–	–	–	–	–	–	Tar	4.92	3.52	3.39	4.38	0.41
Total	*36.95*	*32.57*	*25.73*	*19.41*	*15.73*	*Total*	*32.71*	*27.40*	*20.17*	*16.15*	*13.15*
Energy balance, %	**88.51**	**84.11**	**78.40**	**83.25**	**83.60**	–	–	–	–	–	–

During the gasification of plastic a large amount of light hydrocarbons are generated, resulting in a producer gas with high calorific value. As a consequence, the synthetic gas produced from the experimental runs carried out feeding the fuels containing plastic waste (RP, WRP and WRPC) exhibit higher specific energy (SE) values (27.8, 23.8 and 16.4 MJ/kg$_{Fuel}$, respectively), than those obtained with the fuels without plastic (WC and WD; 10.3 and 12.0 MJ/kg$_{Fuel}$, respectively). These results are in agreement with those obtained during co-gasification studies utilizing fluidized bed reactors [18,26,29]. These authors observed that the addition of plastic waste to the fuel mixture generate a gas with higher hydrocarbon content, which results in a producer gas with high LHV and, thus, larger SE values.

The energy loss due to tar formation increases from 0.4 MJ/kg_{Fuel}, for the test with woody biomass (WD), to 4.9 MJ/kg_{Fuel}, for the test carried out by feeding plastic waste (RP). On the contrary, woody biomass gasification presents a greater energy loss, determined by a large production of carbonaceous fines (0.7 MJ/kg_{Fuel}) than that observed during the gasification of plastic waste (<0.1 MJ/kg_{Fuel}).

A remarkable amount of energy loss during the gasification tests is due to the energy needed to transform the fuel into gasification products; here, this energy is indicated as gasification loss (GL) and evaluated according to Equation (5):

$$GL = 1 - \frac{CE_{PG} + CE_{EF} + CE_{Tar}}{CE_{Fuel}} \tag{5}$$

where CE_{PG} is the producer gas chemical enthalpy, CE_{EF} is the elutriated fines chemical enthalpy, CE_{Tar} is the tar chemical enthalpy, and CE_{Fuel} is the fuel chemical enthalpy.

Mono-gasification tests with RP and WD display GL values of 11.5% and 16.4%, respectively. These results indicate that plastic degradation requires lower amount of energy to be converted in the gasification products. This could be a consequence of the peculiar thermal behavior of plastic as a result of its different structure and composition. This experimental evidence is supported by the results obtained from the co-gasification tests with WRP and WRPC, which show increasing values of GL as the plastic fraction decreases from 80% to 30% (15.9% and 21.6%, respectively).

The chemical energy of the fuel transferred to the product gas, also known as cold gas efficiency (CGE), was evaluated as indicated in Equation (6):

$$CGE = \frac{Q_{PG} \times LHV_{PG}}{W_{Fuel} \times LHV_{Fuel}} \tag{6}$$

where Q_{PG} is the volumetric flow rate of the producer gas, W_{Fuel} is the mass flow rate of the fuel, LHV_{PG} is the lover heating value of the producer gas and LHV_{Fuel} is the lover heating value of the fuel.

In accordance with the results presented above, the product gas generated during the mono-gasification tests of RP and WD shows higher values of CGE (75.2% and 76.5%, respectively), whereas the gas obtained from the co-gasification tests of WRP, WRPC, and WC, characterized by a greater production of by-products and by high GL values, yields lower CGEs (73.2%, 63.7% and 53.0%, respectively) as can be deduced from Table 7 (and reported in Table 4).

4. CONCLUSIONS

The gasification process performance was evaluated by performing mass and energy balances on the BFB reactor fed with plastic waste, woody biomass, and their blends with coal. The results indicated that plastic waste induces the production of light hydrocarbons, which led to a reduction in hydrogen content in the producer gas. Additionally, woody biomass promotes the generation of CO_2, CO, and H_2 and depresses the production of light hydrocarbons. Coal addition to the blends does not seem to affect the trends shown by the main producer gas constituents obtained during the tests without coal. The experimental runs conducted by feeding the fuels containing large amount of plastic (RP and WRP) yield small specific quantities of elutriated fines and the highest amount of tar, similar to those generated with the fuels containing coal. On the contrary, the mono-gasification test of woody biomass (WD) generates the lowest amount of tar. The fuels containing high plastic fractions transfer more hydrogen in the producer gas than the blends without plastic, while the gasification test conducted by feeding wood (WD) exhibits higher CCE value.

The mono-gasification tests (RP and WD) show higher CGE values, whereas the co-gasification tests (WRP, WRPC and WC), which are characterized by a greater energy loss due to high GL values and by a large production of char and tar, inevitably transfer less fuel energy in the producer gas.

The results reported in this work indicate that the air gasification process of the tested fuels is technically feasible even if the co-gasification tests show the worst performance. These results suggest that the co-gasification process requires further research efforts with the purpose of reducing the production of char and tar in order to obtain higher conversion efficiencies in terms of HRE, CCE, and CGE.

ACKNOWLEDGMENTS

The authors are indebted to Giuseppe Barbato who performed the off-line analyses of the reported experimental runs.

AUTHOR CONTRIBUTIONS

Both authors contributed to publishing this paper. Lucio Zaccariello has contributed significantly to the conception, design, execution, analysis and interpretation of the data; participate in drafting, reviewing, and revising the manuscript for intellectual content; and approve the manuscript to be published. Maria Laura Mastellone, head of the project, has contributed to the analysis, interpretation and evaluation of the data; and approve the manuscript to be published.

CONFLICTS OF INTEREST

The authors declare no conflict of interest.

REFERENCES

1. Mastellone, M.L.; Zaccariello, L.; Santoro, D.; Arena, U. The O_2-enriched air gasification of coal, plastics and wood in a fluidized bed reactor. Waste Manag. **2012**, 32, 733–742.

2. Li, K.; Zhang, R.; Bi, J. Experimental study on syngas production by co-gasification of coal and biomass in a fluidized bed. Int. J. Hydrog. Energy **2010**, 35, 2722–2726.

3. Pinto, F.; Franco, C.; André, R.N.; Tavares, C.; Dias, M.; Gulyurtlu, I.; Cabrita, I. Effect of experimental conditions on co-gasification of coal, biomass and plastics wastes with air/steam mixtures in a fluidized bed system. Fuel **2003**, 82, 1967–1976.

4. Minkova, V.; Marinov, S.P.; Zanzi, R.; Björnbom, E.; Budinova, T.; Stefanova, M.; Lakov, L. Thermochemical treatment of biomass in a flow of steam or in a mixture of steam and carbon dioxide. Fuel Process. Technol. **2000**, 62, 45–52.

5. Mastellone, M.L.; Zaccariello, L. Metals flow analysis applied to the hydrogen production by catalytic gasification of plastics. Int. J. Hydrog. Energy **2013**, 38, 3621–3629.

6. Arena, U.; Zaccariello, L.; Mastellone, M.L. Gasification of natural and waste biomass in a pilot scale fluidized bed reactor. Combust. Sci. Technol. **2010**, 182, 625–639.

7. Devi, L.; Ptasinski, K.J.; Janssen, F.J.J.G. Pretreated olivine as tar removal catalyst for biomass gasifiers: investigation using naphthalene as model biomass tar. Fuel Process. Technol. **2005**, 86, 707–730.

8. Corella, J.; Toledo, J.M.; Padilla, R. Olivine or dolomite as in-bed additive in biomass gasification with air in a fluidized bed: Which is better? Energy Fuels **2004**, 18, 713–720.

9. Pfeifer, C.; Rauch, R.; Hofbauer, H. In-bed catalytic tar reduction in a dual fluidized bed biomass steam gasifier. Ind. Eng. Chem. Res. **2004**, 43, 1634–1640.

10. Rapagnà, S.; Jand, N.; Kiennemann, A.; Foscolo, P.U. Steam-gasification of biomass in a fluidised-bed of olivine particles. Biomass Bioenergy **2000**, 19, 187–197.

11. Mastellone, M.L.; Zaccariello, L. Gasification of polyethylene in a bubbling fluidized bed operated with the air staging. Fuel **2013**, 106, 226–233.

12. Aznar, M.P.; Caballero, M.A.; Sancho, J.A.; Francés, E. Plastic waste elimination by co-gasification with coal and biomass in fluidized bed with air in pilot plant. Fuel Process. Technol. **2006**, 87, 409–420.

13. Pan, Y.G.; Roca, X.; Velo, E.; Puigjaner, L. Removal of tar by secondary air in fluidised bed gasification of residual biomass and coal. Fuel **1999**, 78, 1703–1709.

14. Narváez, I.; Orío, A.; Aznar, M.P.; Corella, J. Biomass gasification with air in an atmospheric bubbling fluidized bed. Effect of six operational variables on the quality of produced raw gas. Ind. Eng. Chem. Res. **1996**, 35, 2110–2120.

15. Wilk, V.; Schmid, J.C.; Hofbauer, H. Influence of fuel feeding positions on gasification in dual fluidized bed gasifiers. Biomass Bioenergy **2013**, 54, 46–58.

16. Kern, S.; Pfeifer, C.; Hofbauer, H. Gasification of wood in a dual fluidized bed gasifier: Influence of fuel feeding on process performance. Chem. Eng. Sci. **2013**, 90, 284–298.

17. Corella, J.; Herguido, J.; Alday, F. Pyrolysis and steam gasification of biomass in fluidized beds. Influence of the type and location of the biomass feeding point on the product distribution. In Research in Thermochemical Biomass Conversion; Bridgwater, A.V., Kuester, J.L., Eds.; Springer: Dordrecht, The Netherlands, 1988; pp. 384–398.

18. Wilk, V.; Hofbauer, H. Co-gasification of plastics and biomass in a dual fluidized-bed steam gasifier: Possible interactions of fuels. Energy Fuels **2013**, 27, 3261–3273.

19. Saw, W.L.; Pang, S. Co-gasification of blended lignite and wood pellets in a 100 kW dual fluidised bed steam gasifier: The influence of lignite ratio on producer gas composition and tar content. Fuel **2013**, 112, 117–124.

20. Fermoso, J.; Arias, B.; Plaza, M.G.; Pevida, C.; Rubiera, F.; Pis, J.J.; García-Peña, F.; Casero, P. High-pressure co-gasification of coal with biomass and petroleum coke. Fuel Process. Technol. **2009**, 90, 926–932. [

21. Kumabe, K.; Hanaoka, T.; Fujimoto, S.; Minowa, T.; Sakanishi, K. Co-gasification of woody biomass and coal with air and steam. Fuel **2007**, 86, 684–689.

22. Aigner, I.; Pfeifer, C.; Hofbauer, H. Co-gasification of coal and wood in a dual fluidized bed gasifier. Fuel **2011**, 90, 2404–2412.

23. Howaniec, N.; Smolinski, A.; Stanczyk, K.; Pichlak, M. Steam co-gasification of coal and biomass derived chars with synergy effect as an innovative way of hydrogen-rich gas production. Int. J. Hydrog. Energy **2011**, 36, 14455–14463.

24. Sjöström, K.; Chen, G.; Yu, Q.; Brage, C.; Rosén, C. Promoted reactivity of char in co-gasification of biomass and coal: Synergies in the thermochemical process. Fuel **1999**, 78, 1189–1194.

25. Channiwala, S.A.; Parikh, P.P. A unified correlation for estimating HHV for solid, liquid and gaseous fuel. Fuel **2002**, 81, 1051–1063.

26. Pinto, F.; Franco, C.; André, R.N.; Miranda, M.; Gulyurtlu, I.; Cabrita, I. Co-gasification study of biomass mixed with plastic wastes. Fuel **2002**, 81, 291–297.

27. Mastellone, M.L.; Zaccariello, L.; Arena, U. Co-gasification of coal, plastic waste and wood in a bubbling fluidized bed reactor. Fuel **2010**, 89, 2991–3000.

28. Ahmed, I.I.; Nipattummakul, N.; Gupta, A.K. Characteristics of syngas from co-gasification of polyethylene and woodchips. Appl. Energy **2011**, 88, 165–174.

29. Pinto, F.; André, R.N.; Franco, C.; Lopes, H.; Gulyurtlu, I.; Cabrita, I. Co-gasification of coal and wastes in a pilot-scale installation 1: Effect of catalysts in syngas treatment to achieve tar abatement. Fuel **2009**, 88, 2392–2402.

10

CHAPTER

GASIFICATION STUDIES ON ARGENTINE SOLID FUELS

G.G. Fouga[1, 2], G. De Micco[1, 2], H.E. Nassini [1] and A.E. Bohé[1, 2]

[1] Comisión Nacional de Energía Atómica (CNEA), Republic of Argentina

[2] Consejo Nacional de Investigaciones Científicas y Técnicas (CONICET), Republic of Argentina

1. INTRODUCTION

As the global population growth and energy demand are steadily raising and the industry is forced to reduce the greenhouse gas emissions due to the global warming, there is an increasing pressure to improve the overall efficiency of the energy production systems. In this challenging framework, a renewed interest on coal gasification technologies has recently emerged worldwide, since they offer the potential of clean and efficient energy. One attractive characteristic of coal gasification technology is the possibility of co-production of electricity, hydrogen, liquid fuels and high-value chemicals that contributes to the improvement of power generation efficiency compared with conventional pulverised coal fired plants as well as the reduction of emissions of greenhouse gases and particulates to the atmosphere (Minchener, 2005). Gasification has also the additional advantage of accommodating a wide range of feed stocks, including low-cost fuels like petroleum coke, biomass, and municipal wastes (Higman & Van der Burgt, 2003).

As it will be explained in Section 2, Argentina is presently investigating the application of the concept of co-production for the integral exploitation of its coal reserves. Co-production of power, fuels and chemicals offers an innovative, economically advantageous mean of achieving the long-term energy goals of our country since it involves the integration in a single energy complex of three major building blocks: (1) gasification of coal to produce synthesis gas; (2) conversion of a portion of the synthesis gas to high-value products, such as high-purity hydrogen and liquid fuels; and (3) combustion of the remaining synthesis gas and unreacted gas from the conversion processes to produce electric power in a combined-cycle system. In the co-production concept, an energy complex produces not only power, but also fuels and/or chemicals. This concept greatly increases the flexibility of the complex and offers economic advantages compared with separate plants, one producing only power and the other only fuels or chemicals.

Following this objective, an extensive research and development program is being implemented in our country on solid fuel gasification technologies, beginning with both theoretical and experimental studies for understanding the mechanisms of the gasification reactions, in order to determine the optimum parameter conditions for the synthesis gas production and the further cleanup steps for the harmful contaminants removal. For providing indirect heating to the gasification reactors, replacing the partial combustion of the feed material that is needed to drive the endothermic gasification reactions, the alternative of using a nuclear high temperature gas reactor is being also evaluated (Nassini et al., 2011).

It is well-known that the chemical composition, the heating value and, then, the future use of the synthesis gas produced by solid fuel gasification is variable with the gasification technology employed, depending on a lot of factors such as solid fuel composition and rank; pre-processing and feeding procedures; gasification agents; operational conditions in the gasification reactor, i.e. temperature, pressure, heating rate, and residence time; and plant configuration characteristics like the flow geometry, ash removal method and gas cleaning system. There is a large number of gasification processes implemented at commercial level and the choice of a given gasification technology is difficult because it depends on diverse factors such as solid fuel availability, type and cost; size constraints; and production rate of energy. Even, in principle, all types of solid fuels can be gasified, the properties of the material to be processed are the least flexible factor to be considered in the analysis and, then, the gasification technology should be primarily matched to the properties of the solid fuels available for gasification (Collot, 2006).

According to that, a theoretical and experimental study is being now performed at laboratory scale, addressed to characterize the behaviour of Argentine solid fuels under typical gasification conditions and to identify the most suitable gasification process for the production of hydrogen and liquid fuels, respectively. The research program that is described below was designed to simulate in laboratory, as close as possible, the operational conditions of large-scale gasification plants and to provide the necessary information about fundamental mechanisms and kinetics of the gasification reactions for a further scaling up of experimental facilities.

2. ARGENTINE ENERGY SITUATION AND SCIENTIFIC BACKGROUND

The current energy matrix of Argentina is largely based on fossil fuels, i.e. petroleum oil and natural gas, but the preservation of non-renewable resources and the minimization of pollution are goals which today determine decisively further development of fossil fuel-fired power stations. In this sense, a so-called Hydrogen Law was dictated by the Argentine Congress in 2006 declaring of national interest the development of technologies needed for the progressive introduction of hydrogen as a clean energy carrier that can be used to meet the increasing residential, transportation and industrial demands. According to that, the national government is promoting all scientific activities related with the production, purification, safe storage and applications of hydrogen, as well as the development of more efficient energy production systems (Bohe & Nassini, 2011).

In order to bring together the requirements of a sustainable economic growth with the environmental protection, our country is then encouraging strategies for the rational and integral utilization of domestic coal reserves and this tendency is expected to increase with time, as natural gas and petroleum resources are becoming exhausted. The main domestic coal reserve accounting more than 7% of conventional energy resources of Argentina is a high-volatile sub bituminous coal that is extracted from the Río Turbio minefield, located in Santa Cruz province, in the south of the country (Carrizo, 2002). Another materials containing carbon and amenable to be gasified are asphaltites arising from minefields located in Mendoza (Beloff, 1972) and Neuquen (Savelev et al, 2008). Asphaltites would be an excellent raw material for the production of synthesis gas through solid fuel gasification due to their low content of ashes and high percentage of elemental carbon (Fouga et al., 2011).

It is well-understood that solid fuel gasification is a two-step process. In the first step, pyrolysis, volatile components of feed material are rapidly released at temperatures between 300 and 500 ºC, leaving residual char and mineral matter as by-products. The second step, char conversion, involves the gasification of residual char and it is much slower than devolatilization step, becoming then the rate-limiting step of the overall process. Even gasification reactions have been extensively studied during years worldwide, a better understanding of the fundamental reaction mechanisms and kinetics is still required for optimizing the design and operation of large-scale gasifiers in order to maximize the efficiency and economics of the overall gasification process.

Earlier studies demonstrated that the reactivity of chars to gasifying agents is very dependent on their formation conditions, particularly temperature, pressure, heating rate, time at peak temperature, and the gaseous environment. When volatile matter is generated, the physical structure of char changes significantly and swelling of fuel particles may occur. The complexity of char structure lies in the facts that the structure of a char itself is highly heterogeneous inside an individual particle and between different particles and the chemistry of a char is strongly dependent on the raw material properties. Then, a good understanding of the swelling of particles and the formation of the char pore structure during the devolatilization step, as well as the further evolution of the released volatile matter is essential to the development of advanced gasification technologies (Yu et al., 2007).

On the other hand, even coal is generally classified by its rank with fixed carbon content and calorific value as the major indicators, coal rank related parameters do not always provide adequate predictors for gasification reactivity since coals of similar rank may undergo quite different extents of reaction when they are gasified at a particular condition. Additionally to coal rank, reaction conditions and sample preparation procedures, several other factors are thought to influence the coal gasification reactivity such as the mineral matter content of coals which is known to influence the gasification reactivity because of the presence of reportedly catalytically active components (Domazetis et al., 2005).

3. EXPERIMENTAL APPROACH

When introduced into a high-temperature atmosphere in a gasification reactor, solid fuel particles are heated at high heating rates (above 10^3 ºC/sec) and they undergo devolatilization and gasification simultaneously under more or less the same condition. In spite of this evidence, most of the char reactivity data reported in literature was obtained under gasification conditions that were different from the devolatilization conditions under which the chars were prepared. According to that, a high spread in char reactivity measurements is found, even for chars prepared from the same parent coal but under different pyrolysis conditions (Peng et al., 1995).

As earlier studies have demonstrated that the reactivity of chars to gasifying agents is very dependent on their formation conditions, to get meaningful data about kinetics of gasification reactions it is essential, at least, to produce chars in laboratory that replicate, as close as possible, the real conditions of char formation in large-scale gasifiers, i.e. high heating rates and intense gas convection around individual char particles.

The experimental approach followed to achieve both objectives in the char preparation is the so-called "two-stage" experiments in which the gasification reactivities are determined on char samples prepared in a previous pyrolysis step where parent coal particles are heated in an inert atmosphere at high heating rates and short residence times at high temperatures (Megaritis et al., 1998). A drop tube furnace was designed and built up for producing chars in laboratory at temperatures up to 1100 ºC and heating rates in the order of 10^3 ºC/sec, while the CO_2 and steam gasification reactivities of these ex-situ chars were measured in a thermo-gravimetric system adapted to work with corrosive gases and in tubular reactors coupled with gas chromatography. The experimental setups used for pyrolysis and gasification experiments are described in more detail in the following section.

4. EXPERIMENTAL PROCEDURES AND METHODS

4.1. Characterization of solid fuels for gasification experiments

The first step of the experimental program consisted of a detailed physical and chemical characterization of the Río Turbio coal and several asphaltites called Emanuel, Susanita, Fortuna 4 and Toribia, and the main results are summarized in Table 1. It can be seen that Toribia and Fortuna 4 asphaltites have the highest volatile content (above 50 wt%) while Emanuel asphaltite has the highest fixed carbon content. Furthermore, the Río Turbio coal has the greatest ash content and porosity. BET areas were measured by N_2 adsorption/desorption according to Barrett–Joyner–Halenda (BJH) method (Barrett, 1951), and using Digisorb 2600 equipment (Micrometrics Ins. Corporation). The analysis of elemental composition indicates the presence of nickel and vanadium in three of the asphaltites, and the recovery of theses valuable metals could be of economical interest. Calcium and sodium, silicon and iron are present in most of the samples. The XRD measurements indicate that those elements are forming the following majority phases: quartz, calcium sulfate, hematite, and aluminum silicates.

Table 1. Physical and chemical characterization of the Río Turbio coal and asphaltites.

Determination	Solid Fuels				
	Coal	Asphaltites			
	Río Turbio	Emanuel	Susanita	Fortuna 4	Toribia
Moisture	3.5 (wt %)	11.47 (wt %)	10.93 (wt %)	0.26 (wt %)	0.58 (wt %)
Volatile matter[a]	36.4 (wt %)	26.18 (wt %)	33.18 (wt %)	58.97 (wt %)	56.06 (wt %)
Fixed carbon[a]	51.2 (wt %)	68.67 (wt %)	55.50 (wt %)	40.57 (wt %)	43.25 (wt %)
Ash[a]	12.3 (wt %)	5.13 (wt %)	11.32 (wt %)	0.46 (wt %)	0.69 (wt %)
Density	1.107(g·cm³)	0.679 (g·cm³)	0.642 (g·cm³)	0.412 (g·cm³)	0.427 (g·cm³)
C_T	59.8	64.3	63.6	78.0	75.3
N_T	2.78	3.27	3.24	2.92	3.14
S_T	0.86	2.36	0.7	4.5	4.40
Determination	**Char**				
BET area	96 ($m^2 \cdot g^{-1}$)	3.5 ($m^2 \cdot g^{-1}$)	3.17 ($m^2 g^{-1}$)	0.44 ($m^2 \cdot g^{-1}$)	na
Pore volume	0.064319	0.01495 ($cm^3 g^{-1}$)	na	na	na
Total porosity, ε_0	12 %	1.6 %	na	na	na
Ash content	20 %	6.95 %	16.94 %	1.12 %	1.57 %
Determination	**Ash**				
Elements present in Ash[b]	Na, Mg, Al, Si, K, S, Ca, Ti, Fe.	S, Ca, V, Fe, Si, Al, Ba, Ni, K, Sr, Mo, P, Cu.	Mg, Al, Si, S, Ca, V, Fe, Ni, Zn.	Na, Mg, Al, Si, K, Ca, V, Fe, Ni, Cu.	Na, Al, Si, K, Ca, V, Fe, Ni.
Main phases in Ash[c]	Fe_2O_3, SiO_2	SiO_2; Fe_2O_3; $CaSO_4$; $Ca_3V_2O_8$; $CaSiO_3$; $(Na,Ca)Al(Si,Al)_3O_8$	$CaSO_4$; SiO_2; $Ca_2Al_2SiO_7$.	SiO_2; NaV_6O_{15}	SiO_2; CaV_2O_6; $Al_6Si_2O_{13}$

[a] Moisture free
[b] Energy dispersive spectroscopy (EDS) and Energy dispersive X-Ray fluorescence spectroscopy (ED-XRF).
[c] X-Ray diffraction (XRD).
[na] not available

4.2. Drop tube furnace for pyrolysis experiments

The drop tube furnace (DTF) that is shown in Figure 1 was used for preparing chars at high heating rates and short residence times at high temperatures from the Río Turbio coal and asphaltites. The reactor has a three-zone electric furnace able to operate up to 1100 ºC, which surrounds two concentric quartz tubes of 41 and 26 mm inner diameter, 1.30 and 1.20 m long, respectively. Primary nitrogen gas is injected at the bottom of the outer tube and is preheated while flowing upwards. When at the top of the outer tube, the gas is forced onto the inner tube through a flow rectifier and the gas flows downwards and leaves the reactor through a water-cooled collection probe. The solid fuel particles are entrained by a non-preheated secondary nitrogen gas jet to a water-cooled injection probe placed on top of the inner tube. The heating rate is estimated to be higher than 10^3 ºC/sec and the residence time of particles in the reactor less than 0.3 sec. The chars leave the reactor through the collection probe, and an extra nitrogen flow is added to the exhausted gases in order to quench the reaction and improve the collection efficiency in the cyclone.

The major operating parameters in the reactor were: (1) temperature of pyrolysis, ranging between 700 to 1100 ºC; (2) mass flow of solid fuel particles, through variations in the secondary nitrogen gas flow; and (3) particle residence time at high temperature, derived from the heated tube length which can be varied since the three axial zones have independent electric power supply.

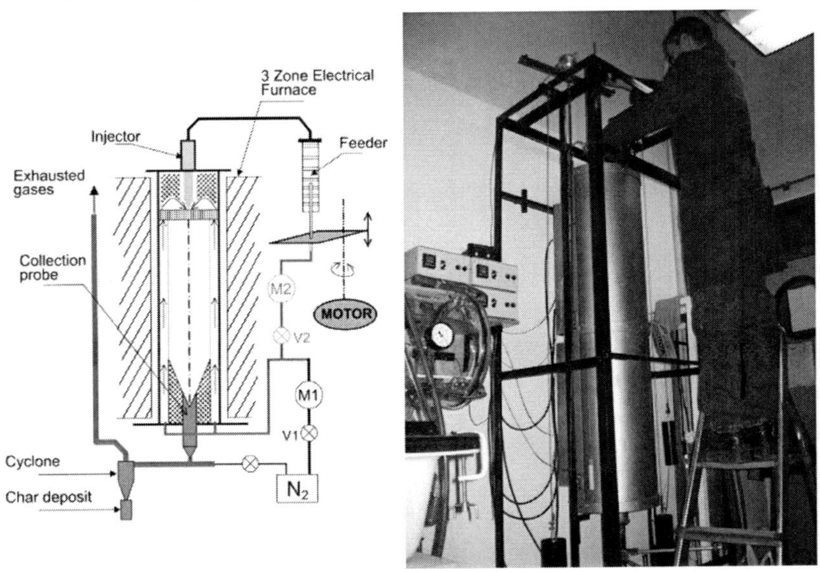

Figure 1. Drop tube furnace for pyrolysis experiments: (a) schematic view; (b) photograph taken during assembly.

4.3. Thermo-gravimetric system for gasification experiments

Gasification experiments using carbon dioxide and steam as gasifying agents were carried out in a thermo-gravimetric analyzer (TGA) that is schematically shown in Figure 2. This experimental setup consists of an electro-balance (Model 2000, Cahn Instruments, Inc.), a gas line, and a data acquisition system, having a sensitivity of ± 5 μg while operating at 950 ºC under a flow of 8 L/h. In a typical TGA run, the weight of the char sample is measured as a function of time and temperature as it is subjected to a controlled temperature program. TGA tests are usually carried out in two ways: (i) isothermal, where the sample is heated at a constant temperature, and (ii) non-isothermal with linear heating, where the sample is heated at a constant temperature rate.

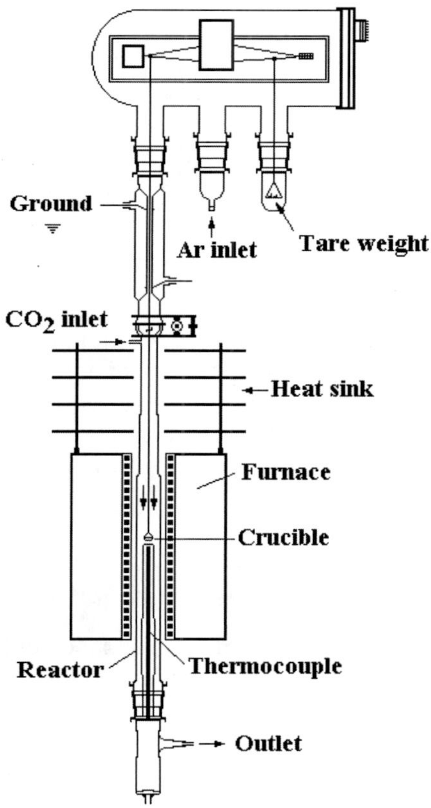

Figure 2. Thermo-gravimetric system for gasification experiments with carbon dioxide and steam.

The gasification rate under several experimental conditions of temperature, partial pressure of gasifying agent and sample mass, can be obtained from the temporal evolution of relative mass loss of char, as follows:

$$\alpha = \frac{m_0 - m}{m_0 - m_{ash}} \tag{1}$$

where m_0 is the initial mass of char, m is the mass of char at time t, and m_{ash} is the mass at the end of the gasification reaction when there is no more fixed carbon and corresponds to the ash content. According to equation (1) α takes values in the range between 0 and 1 and, hence, the gasification rate, R, can be expressed as:

$$R = \frac{d\alpha}{dt} = -\frac{1}{(m_0 - m_{ash})}\frac{dm}{dt} \tag{2}$$

In mathematical form, R is expressed as a function of temperature (T), partial pressure of gas (p_{gas}) and reaction degree (α), as follows:

$$R = \frac{d\alpha}{dt} = K(T)F(p_{gas})G(\alpha) \tag{3}$$

where K(T) refers to an Arrhenius type equation, $F(p_{gas})$ expresses the dependence of R with the partial pressure of gasifying agent, and G(α) is a function that describes the geometric evolution of the reacting solid. This procedure allows to exclude mass sample effects and represents an appropriate approach for the analysis of gas-solid heterogeneous reactions (De Micco et al., 2010).

4.4. Tubular reactors coupled with gas chromatography

Gasification experiments using carbon dioxide and steam as gasifying agents were also carried out in tubular reactors coupled with gas chromatography. The experimental setup for gasification experiments with carbon dioxide is shown in Figure 3 and consists of a horizontal quartz tube surrounded by an electrical furnace, a gas control panel, and a gas chromatograph (SRI 8610 C) with a packed column Alltech CTR I and helium as carrier gas. Solid char samples of 10 mg were placed on a flat quartz crucible forming a loose packed bed and inside the tubular reactor where an argon flow of 3.5 L/h was maintained. For the isothermal experiments, char samples were heated at the working temperature for about 1 hour after which carbon dioxide was introduced into the reactor. At the same time, the exhausted gases were injected in a gas chromatograph every 5 minutes. To inject the gases into the chromatograph, the exhausted gas stream was connected to a 1 ml loop and, according to the gaseous flow used, the time required to fill the loop was 0.86 seconds. The Reynolds number corresponding to the experimental conditions indicates that the gaseous flow inside the reactor is laminar (De Micco et al., 2012).

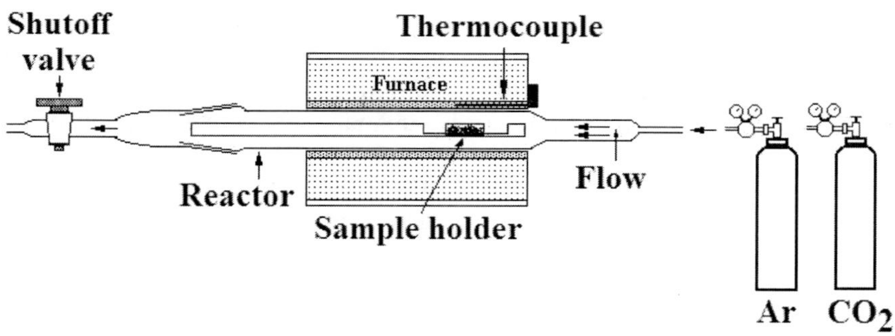

Figure 3. Tubular reactor coupled with gas chromatography for CO_2 gasification experiments.

The gasification rate is determined by monitoring the evolution of the concentration of reaction product, i.e. carbon monoxide (CO(g)), as a function of time. To follow the gasification kinetics, the peak areas corresponding to CO(g) concentration from the chromatograms registered every 5 minutes during the reactions are used. These areas are proportional to the amount of CO(g) moles formed during the time interval required to fill the loop. Since this time interval of 0.86 s is very small compared to the total time needed to achieve the complete reaction (more than 3000 sec), and assuming that no significant axial mixing occurs under laminar flow conditions, it can be considered that the peak areas are proportional to the instantaneous gasification rate. Plots of *CO-Area vs. time* were constructed for each gasification reaction and these experimental data were fitted with appropriated curves.

The number of moles formed at the time t can be calculated by integrating the curves from $t = 0$ to t, and the degree of reaction at time t can be obtained from the ratio of the previous result and the value of integrating the whole *CO-Area* vs. *time* for the complete gasification reaction, according to the following equation:

$$X(t) = \frac{n_{CO}(t)}{n_{CO}(t_f)} \quad (4)$$

where $X(t)$ is the degree of reaction at time t, $n_{CO}(t)$ is the number of moles of CO(g) formed from the beginning of reaction until time t, and $n_{CO}(t_f)$ is the total number of moles of CO(g) formed during the whole reaction.

The experimental setup for gasification experiments with steam is shown in Figure 4 and consists of a horizontal quartz tube surrounded by an electrical furnace, a gas control panel, a steam generator, a set of thermal and chemical traps for retaining the water molecules from the gaseous stream, and a gas chromatograph for analysing the gasification products. The kinetics of gasification reactions can be followed either by measuring the concentration of reaction products with the chromatograph or by gravimetric measurements in which case the gasification reaction is stopped at different reaction degrees.

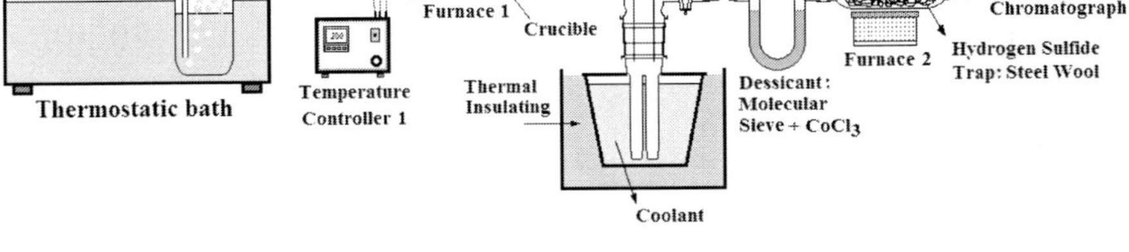

Figure 4. Experimental setup for gasification experiments with steam.

5. MODELLING OF GASIFICATION REACTIONS

A good understanding of solid fuel reactivity and reaction kinetics with carbon dioxide and steam is required for careful optimization of gasification processes. For this reason, numerous studies are being performed worldwide in order to determine the kinetic parameters and reaction mechanisms of the gasification reactions, for each type of parent coal and char. In general, the gasification reaction is a heterogeneous gas-solid reaction where a porous solid is consumed leading to the formation of gaseous products such as carbon dioxide, carbon monoxide and hydrogen, and ash as a solid residue.

In order to obtain the kinetic parameters of gasification reactions, it is useful to measure the reaction rate under chemical control regime. To do that, it is necessary to find the experimental conditions under which mass-transfer resistance is absent. This is accomplished by changing the experimental parameters that influence the rate of the mass transfer processes occurring during the reaction.

Three main kinds of mass transport processes involving different physical phenomenon can be distinguished in this type of reaction: (a) transport of the gaseous reactant by bulk motion (mass convection); (b) transport of the gaseous reactant and products through the gaseous boundary layer (mass gaseous diffusion); and (c) transport of gaseous reactant and products within the solid pores (ordinary or Knudsen gaseous diffusion). The experimental parameters that can be systematically modified in order to make the mass transfer rate faster than the chemical reaction rate are the gaseous flow rate and the initial amount of solid reactant. Once the conditions are achieved to measure the reaction rate under chemical control in the selected range of temperatures, it is possible to apply different reaction models for describing the solid evolution during the reaction, and for obtaining the mathematical expression for the reaction rate.

In general, porosity, surface area and particle size of the solid fuel may vary during the reaction. There are many models that consider the effect of these changes to a different extent and, depending on the hypotheses the models can face various degrees of complexity. Furthermore, due to the porous nature of coal and char, it is not always possible to achieve complete chemical control of the reaction because the diffusion within the pores limits the overall rate of reaction. When this happens, both processes, chemical reaction and pore diffusion, exert an influence on the progress of reaction. Consequently, it is necessary to take into account mass transfer effects in the reaction rate expression.

Many models were developed and published for modeling the coal gasification reactions since the 1950s up to now. One of first approaches was done by Pettersen (Pettersen, 1957) who presented a method for a linear kinetic expression in the concentration and where appreciable concentration gradients were established in the pore system. He assumed uniform cylindrical pores with random intersections. The grain model was further developed by Szekely et al. (Szekely et al, 1976), representing the diffuse reaction zone of reacting porous solids and considering a solid made up of individual grains of equal size which could be spheres, long cylinders or flat plates. In this model, the solid surface area decreases nonlinearly with increasing the reaction degree. On the other hand, Bhatia & Perlmutter (Bhatia & Perlmutter, 1980) presented the random pore model which allows for arbitrary pore size distributions. In this model, the reaction surface changes due to two competing processes: (1) the effect of pore growth during gasification; and (2) the destruction of pores due to coalescence of neighboring pores. The model subsumes several earlier treatments as special cases. Other published models are the random capillary model (Gavalas, 1980), the discrete random pore model (Bhatia & Vartak, 1996), and the modified discrete random pore model (Srinivasalu et al., 2000).

In the analysis of gasification experiments with Argentine solid fuels, two different models were applied: (1) the grain model; and (2) the random pore model, and the mathematical formalisms are described briefly below.

Assuming separation of variables, the kinetic expression for the reaction rate is given by:

$$X(t) = \frac{n_{CO}(t)}{n_{CO}(t_f)} \tag{5}$$

being $X(t)$ the degree of reaction; $k(T)$ and $G(C_g)$ include the effects of temperature and gaseous reactant concentration in the reaction rate, respectively; and $f(X)$ accounts for the changes in physical or chemical properties of reacting solid with reaction degree.

The temperature dependence, i.e. the apparent reaction constant $k(T)$, is given by an Arrhenius equation with k_0 being the pre-exponentional factor and E_a the activation energy. The gas concentration dependence, $G(C_g)$, is given by a power law expression being n the reaction order with respect to gaseous reactant concentration, resulting :

$$k(T)G(C_g) = k_0 e^{-E_a/RT} C_g^n \tag{6}$$

Replacing $f(X)$ by the grain model for spherical grains (GM) and the random pore model (RPM) the expressions for the reaction rate and reaction degree vs. time are the following:

- Grain model:

$$\frac{dX}{dt} = k_0 e^{-\frac{E_a}{RT}} (1-X)^{\frac{2}{3}} \tag{7}$$

$$3\left[1-(1-X)^{1/3}\right] = k_{GM} t \tag{8}$$

- Random pore model:

$$\frac{dX}{dt} = k_0 e^{-\frac{E_a}{RT}} (1-X)\sqrt{\left[1-\psi \ln(1-X)\right]} \tag{9}$$

$$(2/\psi)\left[\sqrt{(1-\psi \ln(1-X))}-1\right] = k_{RPM} t \tag{10}$$

In the random pore model, in addition to the apparent reaction constant k_{RPM} there is another parameter, Ψ, which is related with the pore structure of the initial sample, and can be calculated from the experimental results with the following equation:

$$\psi = \frac{2}{2\ln(1-X_{max})+1} \tag{11}$$

where X_{max} is the value of reaction degree where the reaction rate is maximum.

6. MAIN RESULTS AND DISCUSSION

6.1. Experimental design

Laboratory research activities on gasification reactions of Argentine solid fuels in presence of carbon dioxide and steam comprised a comprehensive theoretical and experimental study on the following two chemical reactions:

$$C + CO_2 \leftrightarrow 2CO \quad \Delta H = 159.7 \text{ kJ/mol} \tag{12}$$

$$C + H_2O \leftrightarrow CO + H_2 \quad \Delta H = 118.9 \text{ kJ/mol} \tag{13}$$

After determining the experimental conditions to get the chemical control regime of gasification reactions in the different experimental setups through the variation of the gaseous flow, sample mass and char particle size, which are detailed in Table 2, the effects of the following parameters were investigated:

1. Composition and rank of feed material, i.e. comparative behaviour of subbituminous coal and asphaltites;
2. Reaction temperature, in the range between 800 and 950 ºC;
3. Partial pressure of gasifying agent, between 30 and 80 %v/v;
4. Conditions of char formation, using chars prepared in the DTF at 850 and 950 ºC, respectively, and chars prepared in a fixed bed reactor at 950 ºC.

Table 2. Experimental conditions to get chemical regime in gasification reactions.

Gasifying agent	Río Turbio Coal		Emanuel Asphaltite	
	TG system	GC system	TG system	GC system
CO_2	Gaseous flow above 7.3 L/h. Sample mass below 10 mg.	Gaseous flow above 4.2 L/h. Sample mass below 16 mg.	Gaseous flow above 5 L/h. Sample mass below 2.5 mg.	na
H_2O	na	na	na	Gaseous flow above 2 L/h. Sample mass below 25 mg.

The main results of the theoretical and experimental research program are given in the following sections.

6.2. Effect of solid fuel composition and rank

Argentine solid fuels were pyrolysed in inert atmosphere (argon) using non-isothermal TGA runs and the resulting TGA curves are presented in Figure 5. The mass losses observed in all cases are due to a mixture of vapors and gases which are released during heating, including CO_2, CO, hydrocarbon species, tars, and so on, and they are in agreement with the corresponding values of volatile matter content given in Table 1.

Figure 5 shows that Fortuna 4 and Toribia asphaltites behaved similarly during pyrolysis, exhibiting a fast volatilization rate between 300 and 500 ºC where about 70 wt% of the total volatile matter was released, replicating the results of the proximate analyses where both materials showed very similar values of fixed carbon, ash and volatile matter. The same agreement between proximate analyses and non-isothermal TGA curves was detected for Susanita asphaltite and Río Turbio coal; in this case, two different portions can be distinguished in TGA curves: (1) a fast volatilization rate between 300 and 500 ºC where about 60 wt% of the volatile matter was released, and (2) a slow volatilization rate above 500 ºC where about 25 wt% of the volatile matter was further released. Finally, Emanuel asphaltite showed a singular behavior with a nearly constant volatilization rate between 300 and 900 ºC.

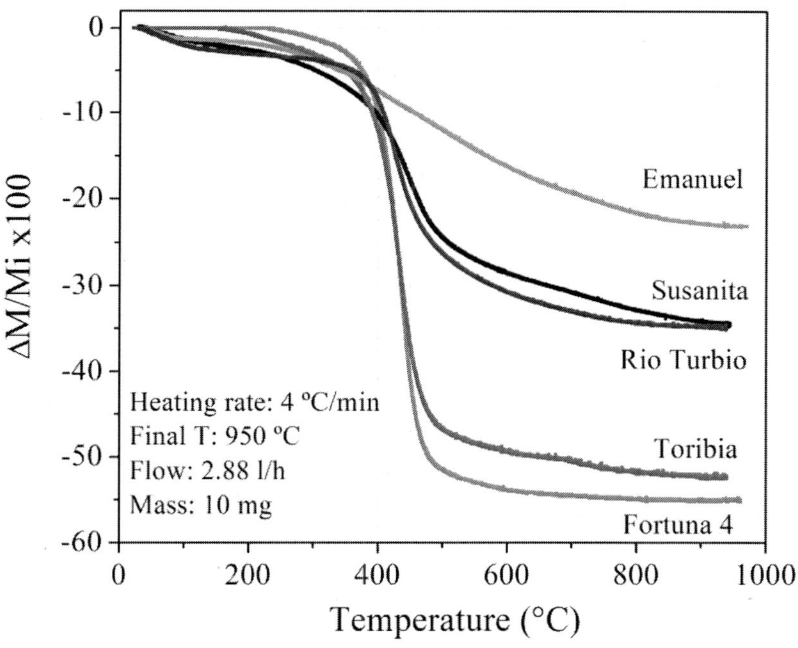

Figure 5. Thermo-gravimetric curves of non-isothermal pyrolysis tests with argon.

The comparative behaviour of Río Turbio coal and asphaltites under CO_2 gasification conditions was studied by performing non-isothermal and isothermal TGA measurements. The non-isothermal TGA curves are presented in Figure 6 and the experimental conditions were: temperature range between room temperature and 950 °C; heating rate of 4 °C/minute; partial pressure of CO_2: 80kPa; and sample mass: 10 mg.

It can be observed that the mass losses measured are due to the release of adsorbed water at low temperature (about 100 ºC) while, at higher temperatures (above 600 ºC), the mass losses corresponded to the gasification reaction of chars with CO_2, producing mainly CO(g). The initial reaction temperatures were: 630, 650, 680, 700, and 730 °C for Susanita, Emanuel, Río Turbio, Toribia, and Fortuna 4, respectively. These temperatures are indicative of the reactivity of chars, meaning that Susanita asphaltite has the highest reactivity and Fortuna 4 asphaltite has the lowest reactivity. Moreover, Toribia and Fortuna 4 asphaltites did not achieve the complete gasification when the temperature reached 950 ºC, showing a reduced reaction rate for the two samples. The other three chars presented a similar mass loss.

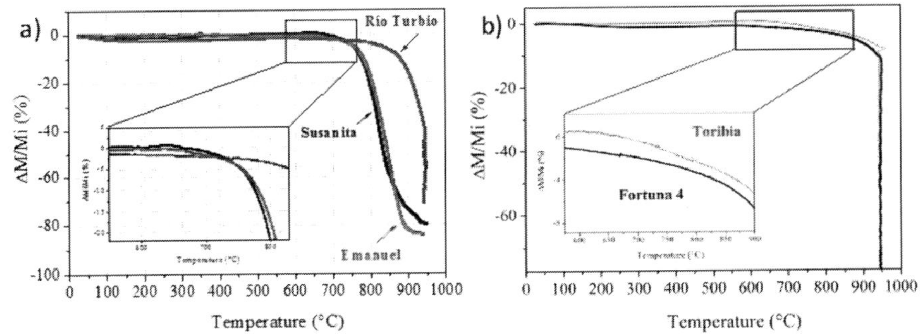

Figure 6. Thermo-gravimetric curves of non-isothermal gasification experiments with CO_2.

In order to study the kinetics of the gasification process, isothermal TGA curves must be obtained. Figure 7 shows the isothermal TGA curves corresponding to the CO_2 gasification of chars obtained in the non-isothermal pyrolysis tests. The experimental conditions were: temperature: 875 ºC; partial pressure of CO_2: 80 kPa; total gaseous flow rate [Ar - CO_2] :10 l/min, and initial sample mass: 25 mg.

The wide range of reaction rates observed for these solid fuels (almost two orders of magnitude between Fortuna 4 and Susanita asphaltites) is indicating that they have different reactivities in presence of CO_2, and the difference may be attributed to the content of fixed carbon and mineral matter. As can be seen in Figure 7, Susanita and Emanuel asphaltites have a similar fixed carbon content, the same happens with Toribia and Fortuna with a lower fixed carbon content, while Río Turbio coal has an intermediate value. Related with the fixed carbon content, the most important effect that can be taken into account is the presence of more amounts of C-C bonds that may signify a greater net energy bond due to the absence or low content of impurities that can produce defects in the solid fuel matrix, increasing the reactivity with the CO_2.

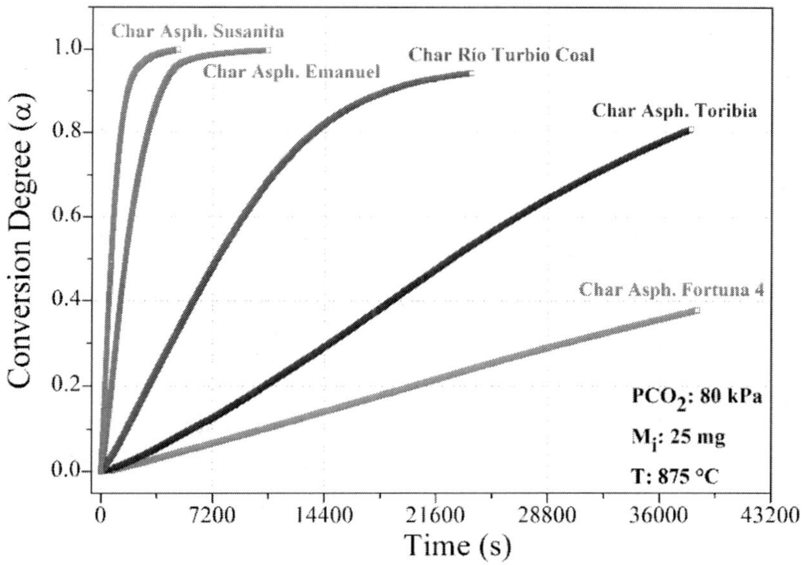

Figure 7. Thermo-gravimetric curves of isothermal gasification experiments with CO_2.

In order to compare the relative influence between the fixed carbon content and the surface area of chars on the gasification rate, the BET areas were determined on chars obtained from Emanuel asphaltite and Río Turbio coal. Even the exposed surface area of Río Turbio char is almost 30 times higher than the Emanuel asphaltite one, the complete gasification reaction occurred after about 6 hours and 1 hour, respectively. This result indicates that the surface area has less influence on the reaction rate compared to the fixed carbon content of the feed material.

Other important characteristics which have a remarkable effect on the gasification rate are the elemental mineral matter and the mineral phases contained in the chars, since it is well-known that mineral matter naturally present in the carbonaceous matrix may act as a catalyst for the gasification reactions. As shown in Table 1, Fortuna 4 and Toribia asphaltites have very small amounts of mineral matter (ash content below 2 wt%), and they presented the lowest reaction rates. Emanuel asphaltite, even having a relatively low mineral matter content and a low BET area (7 wt% of ash and 3.5 m^2/g), has the highest diversity of metals such as V, Mo, Sr, Ni, and Cu among others, and then the catalytic effect of these metals could lead to the high reactivity observed in experiments. The same explanation can be applied to Susanita asphaltite which has a high ash content and also a fairly diversity of metals, showing the highest reactivity in presence of of CO_2. Finally, the reactivity of Río Turbio coal is intermediate between asphaltites of high and low mineral contents. A further catalytic effect that could be observed from the XRD measurements is that those chars containing calcium sulfate show higher reaction rates than those which have calcium forming other compounds.

6.3. Effect of gasification temperature and partial pressure of gasifying agent

From TGA and GC isothermal measurements obtained at several temperatures, partial pressures of reactants, and using CO_2 and steam as gasifying agents, the kinetic parameters of the rate equations of the gasification reactions could be determined.

In the method proposed by Flynn (Basan S. 1986), the activation energy can be determined from equation (5), although $G(C_g)$ and $f(X)$ are unknown functions. Replacing $K(T)$ by an Arrhenius equation and rearranging equation (5):

$$\int_0^{\infty} \frac{dX}{f(X)} = \int_0^{t} G(C_g) A \exp(-E/RT) dt \tag{14}$$

after taking the integral, we have:

$$F(X) = G(C_g) A \exp(-E/RT) t \tag{15}$$

and by taking the logarithm of both sides of equation (15):

$$\ln t = \ln \left[\frac{F(X)}{G(C_g) A} \right] + \frac{E}{RT} \tag{16}$$

The first term in the right hand side of equation (16) is a function of degree of reaction and partial pressure of carbon dioxide. Therefore, if partial pressure of carbon dioxide is keep constant, and the time to attain a certain reaction degree is determined as a function of temperature, equation (16) allows to obtain the activation energy from the slope of the plot ln *t* vs. T^{-1}. Analogous procedure can be applied to obtain the reaction order with respect to gaseous reactant replacing $G(C_g)$ by a power law expression. The reaction order can be obtained from the slope of the plot of ln *t* vs. C_g

By this method (also known as model-free method or iso-convertional method), the activation energies of gasification reactions were obtained and results are shown in Table 3, while the ln *t* vs. T^{-1} plots are shown in Figure 8.

The activation energies calculated are consistent with the fact that the Río Turbio coal showed a lower reaction rate compared to Emanuel asphaltite. Furthermore, the similar values of the activation energies for gasification reactions with CO_2 and steam show that the determining step in the mechanism of these reactions is independent of the gasifying agent used, and it can be associated more with the restructuring of carbon surface than with the gasifying agent accommodation.

Table 3. Activation energies of gasification reactions with CO_2 and steam.

Solid fuel	Activation energy	
	Gasification with CO_2(g)	Gasification with steam
Río Turbio coal	190 ± 10 kJ/mol	na
Emanuel asphaltite	185 ± 10 kJ /mol	186 ± 10 kJ/mol

na: not available

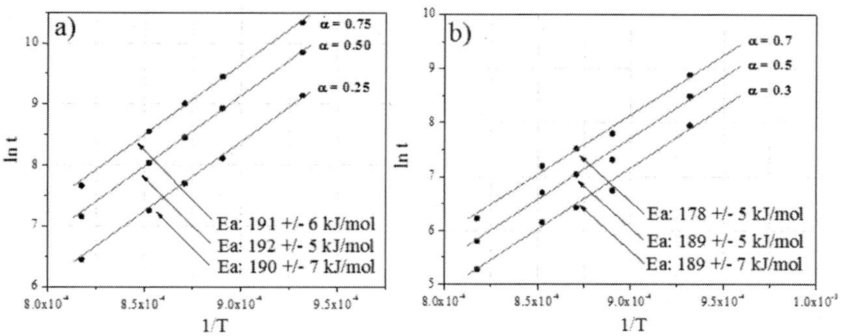

Figure 8. ln t vs. 1/T plot for the calculation of Ea with the Flynn method. (a) Río Turbio coal; (b) Emanuel asphaltite.

The reaction order (n) with respect to the gasifying agent was only determined for CO_2 gasification reactions and results are presented in Table 4 and Figure 9. The reaction order is 1 for the Río Turbio coal and 0.5 for the Emanuel Asphaltite. The first value may indicate that the reaction with CO_2 is produced by occupying only one active site on the surface while, on the opposite, the second value is indicating that a dissociate step on the surface of the particle may be occurring, with the CO_2 molecule being adsorbed and occupying two active sites. The last mechanism requires less activation energy for the breakdown of the C-O chemical bond than the first one, so the gasification reaction is expected to be faster, as it was already shown in the isothermal TGA curves.

Table 4. Reaction order of gasification reactions with CO_2 and steam.

Solid fuel	Reaction order with respect to reactant	
	Gasification with CO_2(g)	Gasification with steam
Río Turbio coal	1	na
Emanuel asphaltite	0.5	1

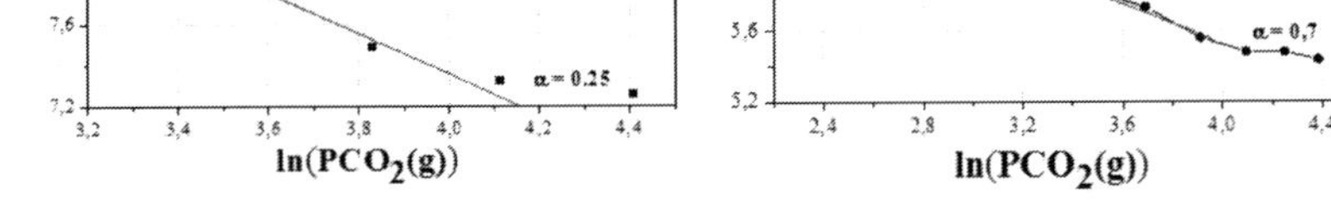

Figure 9. ln t(αi) vs. ln(PCO_2) plot for the calculation of the reaction order with respect to reactant. (a) Río Turbio coal; (b) Emanuel asphaltite.

6.4. Effect of gasifying agent

A comparative analysis of reactivity of Argentine solid fuels in presence of CO_2 and steam was also performed and main results are summarized in Figure 10. Figure 10(a) shows the conversion degree vs. time in TGA curves obtained from Emanuel asphaltite chars gasified at 875 ºC with CO_2 (at a partial pressure of 80 kPa) and steam (at a partial pressure of 20 kPa), respectively. It can be appreciated that the steam reactivity of Emanuel asphaltite char is a little bit higher than CO_2 reactivity even the steam partial pressure used in experiments was lower than the CO_2 partial pressure.

On the other hand, Figure 10(b) shows the conversion degree vs. time in curves obtained from Río Turbio coal chars gasified at 875 ºC in the tubular reactors shown in Figures 3 and 4. In this case, the difference in gasification reactivity between steam and CO_2 is more significant and matches better with the results found in literature (Roberts, D.G. 2000 and Messenbock, R.C. 1999).

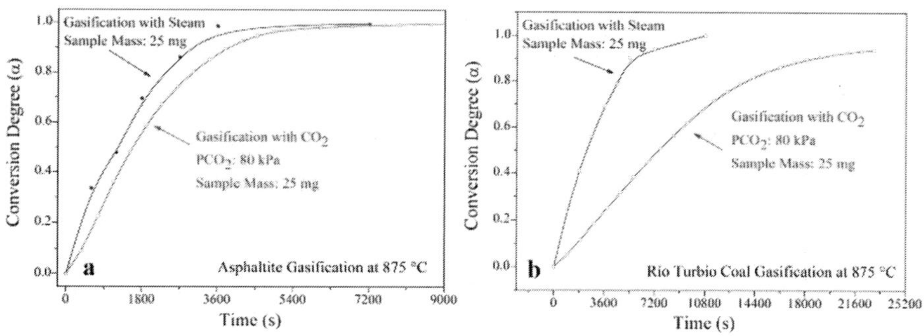

Figure 10. a) TGA curves of Emanuel asphaltite char gasified with steam and CO_2; (b) gasification curves of Río Turbio coal char gasified with steam and CO_2 in tubular reactors.

6.5. Effect of char formation conditions

Chars from Río Turbio coal were prepared in three different conditions: (1) in the DTF at 850 ºC; (2) in the DTF at 950 ºC; and (3) in a Fixed Bed Reactor (FBR) at 950 ºC. Following, 10 mg of each char sample was gasified at 900 ºC in the TGA system with 30% CO_2 partial pressure at the same flow rate, in order to compare their gasification reactivities.

Figure 11 shows the TGA isothermal curves corresponding to the three different chars. It can be observed that both chars pyrolysed at high heating rates in the DTF followed a similar behavior and the gasification reactivity is comparable to reactivities of low-rank coals used in large-scale gasifiers. On the opposite, the TGA curve corresponding to the char prepared at low heating rate in the FBR shows that the gasification reaction progressed much slowly and it was completed after a long time, indicating a very low reactivity in presence of the gasifying agent.

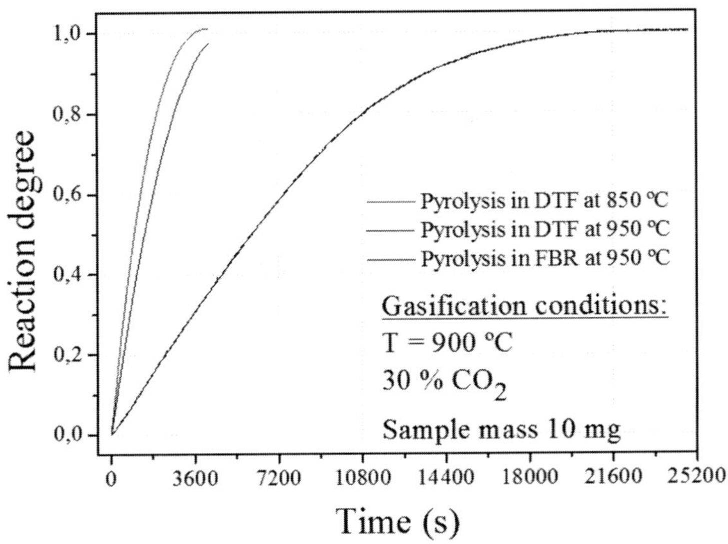

Figure 11. TGA isothermal CO_2 gasification curves of Río Turbio chars prepared in different pyrolysis conditions.

These experimental results are demonstrating that the reactivity of chars to gasifying agents is very dependent on their formation conditions and, then, to get meaningful data about kinetics of gasification reactions, it is very important to produce chars in laboratory at high heating rates and intense gas convection around individual char particles, replicating the real operating conditions of commercial gasification reactors.

Finally, the reaction rate as a function of the conversion degree for those gasification experiments is presented in Figure 12, which also shows the predicted values by the grain model superimposed to the experimental measurements. It can be appreciated that the grain model is expected to well-simulate the gasification behavior of the Río Turbio coal in the temperature range used in experiments. From these fittings, the kinetic parameters of the theoretical models were calculated and they are given in Table 5.

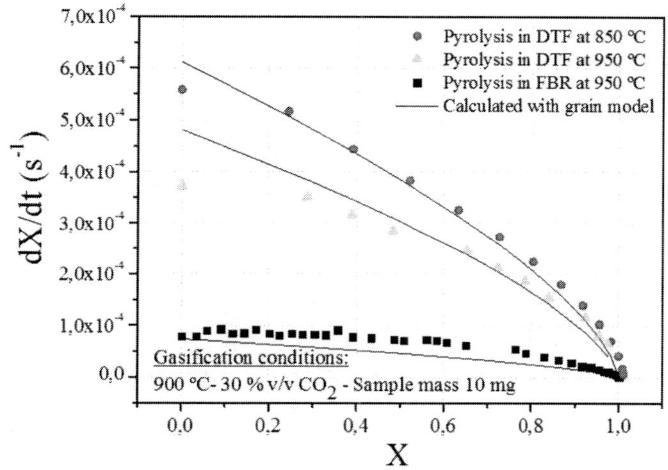

Figure 12. Theoretical and experimental CO_2 gasification rates of Río Turbio chars prepared in different conditions.

Table 5. Kinetic parameters of theoretical models used for simulating the gasification reactions.

Char preparation	Model free method	RPM $\frac{dX}{dt}=k_0e^{-\frac{E_a}{RT}}(1-X)\sqrt{[1-\psi\ln(1-X)]}$			GM $\frac{dX}{dt}=k_0e^{-\frac{E_a}{RT}}(1-X)^{\frac{2}{3}}$	
	Ea kJ/mol	Ea kJ/mol	ko s^{-1}	Ψ	Ea kJ/mol	ko s^{-1}
Pyrolysis in DTF at 850 °C	171 ± 10	165 ± 11	1.15 10^4	2	166 ± 11	1.51 10^4
Pyrolysis in DTF at 950 °C	159 ± 22	158 ± 2	4.07 10^3	2	158 ± 2	5.24 10^3
Pyrolysis in FBR at 950 °C	190 ± 10	na	na	na	195 ± 12	3.6 10^4

It can be observed that the activation energy values of the gasification of chars prepared in the DTF are lower than the value corresponding to the char prepared in the FBR. Meanwhile, the values of the reaction rate constant are similar in all cases, independently of the char preparation method. These values are in agreement with the fact that the gasification reactions of chars prepared in DTFs are faster than the gasification reactions of chars which are pyrolysed in FBRs. Another important aspect is that the activation energy has the most significant effect on the char gasification rate in the present experimental conditions.

7. CONCLUSIONS AND FUTURE WORKS

A comprehensive theoretical and experimental research program is being implemented in Argentina at laboratory scale in the framework of a national strategy for the integral utilization of its domestic coal reserves, addressed to bring together the requirements of a sustainable economic growth with the environmental protection. The research program was designed to simulate in laboratory, as close as possible, the operational conditions of large-scale gasification plants and, then, to provide the necessary information about fundamental mechanisms and kinetics of the gasification reactions for a further scaling up of experimental facilities. For this purpose, specially-designed experimental equipment and test procedures were implemented for gasification experiments using carbon dioxide and steam as gasifying agents.

Experimental program on gasification with carbon dioxide is almost finished and experimental results show that all the Argentine solid fuels studied are amenable to be gasified since their gasification reactivities at high heating rates are comparable with those of low-rank coals used in large-scale gasifiers. Experimental program on steam gasification is just beginning but preliminary experimental results show that the reaction rate is higher than the reaction rate corresponding to the gasification with carbon dioxide.

As it was detected that some mineral phases present in the ashes may have a catalytic effect in gasification reactions, further studies to elucidate this influence are planned for the future, along with the construction of experimental setups for carbon dioxide and steam gasification experiments at higher pressures.

ACKNOWLEDGEMENT

The authors would like to thank to the Agencia Nacional de Promoción Científica y Tecnológica (ANPCyT), Consejo Nacional de Investigaciones Científicas y Técnicas (CONICET), and Universidad Nacional del Comahue (UNComa) for the financial support of this work.

REFERENCES

1. Barrett EP, Soguer LG, Halenda PP.1951The determination of pore volume and area distributions in porous substances. I. Computations from nitrogen isotherms. J. Am. Soc.;7337380
2. S. Basan, O. Güven, 1986Gasification Studies on Argentine Solid FuelsThermochimica Acta106169
3. E. J. Beloff, 1972Gasification Studies on Argentine Solid FuelsFuel51Nº 2, 156159
4. Bhatia S.K. & Perlmutter D.D.1980Gasification Studies on Argentine Solid FuelsAIChE Journal26Nº 3, 379385
5. S. K. Bhatia, B. J. Vartak, 1996Gasification Studies on Argentine Solid FuelsCarbon34Nº 11, 13831391
6. A. E. Bohe, H. E. Nassini, 2011Hydrogen production and applications program in Argentina, Nuclear Hydrogen Prodution Handbook, edited by X. Yan & R. Hino, CRC Press, Taylor & Francis Group, 695723
7. R. N. Carrizo, 2002Mineral solid fuels. Geology and Natural Resources of Santa Cruz, Proceedings of XV Argentine Geological Congress, Santa Cruz, June 2002, 759771in spanish).
8. A. Collot, 2006Gasification Studies on Argentine Solid FuelsInternational Journal of Coal Geology65191212
9. G. De Micco, G. Fouga, A. Bohe, 2010Coal gasification studies applied to hydrogen production, International Journal of Hydrogen Energy, 3560126018
10. G. De Micco, A. Nasjleti, A. Bohe, 2012Gasification Studies on Argentine Solid FuelsFuel95537543
11. G. Domazetis, J. Liesegang, B. D. James, 2005Gasification Studies on Argentine Solid FuelsFuel Processing Technology86463486
12. G. G. Fouga, G. De Micco, A. E. Bohé, 2011Gasification Studies on Argentine Solid FuelsFuel90674680
13. G. R. Gavalas, 1980A random capillary model with application to char gasification at chemically controlled rates, AIChE Journal, 26Nº 4, 577585
14. Higman, and Van der Burgt, M. (2003Gasification, Gulf Professional Publishing, Elsevier, New York.

15. A. Megaritis, R. Messenbock, A. Collot, Y. Zhuo, D. Dugwell, R. Kandiyoti, 1998Internal consistency of coal gasification reactivities determined in bench-scale reactors: Effect of pyrolysis conditions on char reactivity under high-pressure CO2, Fuel, 77Nº 13, 14111420

16. R. C. Messenbock, D. R. Dugwell, R. Kandiyoti, 1999CO2 and steam-gasification in a high-pressure wire-mesh reactor: reactivity of Daw Mill coal and combustion reactivity of its char, Fuel78781793

17. A. J. Minchener, 2005Gasification Studies on Argentine Solid FuelsFuel, 8422222235

18. H. Nassini, G. De Micco, G. Fouga, A. Bohe, 2011Kinetics studies on nuclear-assisted gasification of argentine coals, Proceedings of IAEA´s Technical Meeting/Workshop on Non-Electric Applications of Nuclear Energy, Prague, Czech Republic, 36Ocober, 2011.

19. F. F. Peng, I. C. Lee, R. Y. Yang, 1995Reactivities of in-situ and ex-situ coal chars during gasification in steam at 1000-1400ºC, Fuel Processing Technology, 41233251

20. Pettersen E.E.1957Reaction of porous solids, AIChE Journal, 3Nº 4, 443448

21. D. G. Roberts, D. J. Harris, 2000Gasification Studies on Argentine Solid FuelsEnergy & Fuels, 14483489

22. V. Savelev, A. Golovko, L. Gorbunoa, V. Kamyanov, C. Galvalizi, 2008High-sulphurous Argentinean asphaltites and their thermal liquefaction products, Oil Gas Science Technology, 63Nº 1, 5767

23. G. J. Srinivasalu, S. K. Bhatia, 2000Gasification Studies on Argentine Solid FuelsCarbon384758

24. J. Szekely, J. W. Evans, H. Y. Sohn, 1976Gasification Studies on Argentine Solid FuelsAcademic Press, New York.

25. J. Yu, J. Lucas, T. Wall, 2007Formation of the structure of chars during devolatilization of pulverized coal and its thermo-properties: A review, Progress in Energy and Combustion Science, 33135170

11

CHAPTER

SUPERCRITICAL WATER GASIFICATION OF BIOMASS: A LITERATURE AND TECHNOLOGY OVERVIEW

Onursal Yakaboylu [1,*], John Harinck [1,2], K. G. Smit [2] and Wiebren de Jong [1]

[1] *Faculty of Mechanical, Maritime and Materials Engineering, Process and Energy Department, Delft University of Technology, Leeghwaterstraat 39, NL-2628 CA Delft, The Netherlands*

[2] *Gensos B.V., Wijdenes Spaansweg 57, NL-1764 GK Breezand, The Netherlands*

ABSTRACT

The supercritical water gasification process is an alternative to both conventional gasification as well as anaerobic digestion as it does not require drying and the process takes place at much shorter residence times; a few minutes at most. The drastic changes in the thermo-physical properties of water from the liquid state to the supercritical state make it a promising technology for the efficient conversion of wet biomass into a product gas that after upgrading can be used as substitute natural gas. The earliest research goes back as far as the 1970s and since then, supercritical water has been the subject of many research works in the field of thermochemical conversion of wet biomass. This article reviews the state of the art of the supercritical water gasification technology starting from the thermophysical properties of water and the chemistry of reactions to the process challenges of such a biomass based supercritical water gasification process plant.

Keywords: biomass; supercritical water; gasification; hydrothermal

1. INTRODUCTION

Ever since the industrial revolution, global energy demand and consumption has increased drastically and it is predicted to increase even more in the near future. U.S. Energy Information Administration [1] foresees a 56% increase in the world energy consumption as well as in the natural gas demand in the following 30 years. In OECD Europe, natural gas consumption will increase from 540 to 680 billion cubic meters from 2015 to 2040. In contrast, the natural gas production in OECD Europe will increase only from 254 to 280 billion cubic meters. It is a fact that with such consumption rates, the fossil fuel reserves will eventually deplete.

It is not only the depletion problem that the fossil fuels face. More importantly, fossil fuels are associated with environmental problems. CO_2 emissions already increased from 21.5 to 33.5 billion metric tons from 1990 to 2014, and in 2040 it is expected to reach 45.5 billion metric tons of which 10 billion metric tons originate come from natural gas [1].

Fortunately, the interest in renewable energy is also increasing. Consumption of renewable energy will double and the share of the renewables in the world's energy consumption is expected to increase from 11% in 2010 to 15% in 2040 [1].

Biomass will play an important role among the other renewable energy sources globally as it is already the fourth largest energy resource after coal, oil and natural gas [2]. Furthermore, in particular, non-food ligno-cellulosics are among the most sustainable energy sources which have the potential to decrease fossil fuel consumption. It is possible to obtain gaseous, liquid or solid biofuels from biomass via thermochemical or biochemical conversion routes [3]. Thermochemical conversion consists of pyrolysis, liquefaction, gasification and combustion, whereas biochemical conversion consists of fermentation and digestion [4]. Among them, gasification is one of the most favorable options as the products can serve all types of energy markets: heat, electricity and transportation [3].

However, in case of wet biomass with a high moisture content, it results in a negative impact on the energy efficiency of the gasification process due to the fact that drying costs more energy than the energy content of the product for some biomass types. An alternative method applied for the conversion of wet biomass such as sewage sludge, cattle manure and food industry waste is anaerobic digestion. This process is however characterized by a slow reaction and typical residence times are almost 2–4 weeks. Besides, the fermentation sludge and wastewater from the reactors should further be treated.

The supercritical water gasification process is an alternative to both conventional gasification as well as the anaerobic digestion processes for conversion of wet biomass. This process does not require drying and the process takes place at much shorter residence times; a few minutes at most [3,5]. Supercritical water gasification is therefore considered to be a promising technology for the efficient conversion of wet biomass into a product gas that after upgrading can be used as substitute natural gas. The earliest research goes back as far as the 1970s [6] and since then, supercritical water has been the subject of many research works regarding the thermochemical conversion of wet biomass [7,8,9]. This article reviews the state of the art in supercritical water gasification technology, starting from the thermophysical properties of water and the chemistry of reactions to the process challenges of such a supercritical water gasification of a biomass processing unit.

2. PROPERTIES OF NEAR-CRITICAL AND SUPERCRITICAL WATER

The main reason for the interest in research on supercritical water concerns the favorable physical properties of water and the way they change in the supercritical region which causes water to act as a solvent as well as a catalyst. Furthermore, through hydrolysis reactions, water also acts as a reactant [10].

The critical point for pure water is 374 °C and 22.1 MPa [11]. Above this temperature and pressure, water is in its supercritical phase as shown in Figure 1.

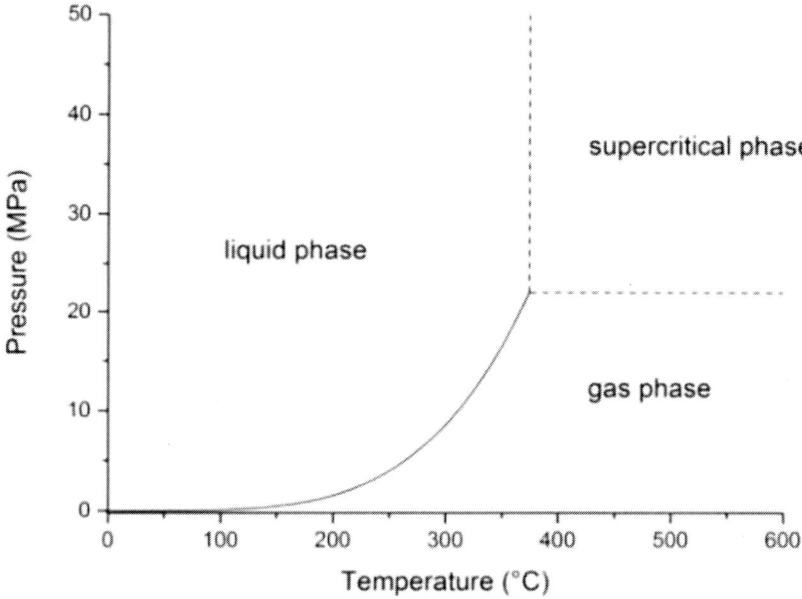

Figure 1. Schematic phase diagram of water; reprinted with permission from [12], copyright 2013 Elsevier.

Gasification of biomass is mainly influenced by the density, viscosity and dielectric constant of water. Above the critical point, physical properties of water drastically change and water behaves as a homogeneous fluid phase. In its supercritical state, water has a gas-like viscosity and liquid-like density, two properties which enhance mass transfer and solvation properties, respectively [3,13]. Figure 2 shows the dielectric constant and Figure 3 shows the density of water at various pressures and temperatures.

Liquid water at standard conditions (25 °C and 0.1 MPa) is an excellent polar solvent due to its high dielectric constant. It has a high solubility for many compounds and electrolytes; however, it is poorly miscible with hydrocarbons and gases. When water enters its supercritical phase, the dielectric constant drastically decreases. Water thus starts to behave like an organic, non-polar solvent which results in poor solubility for inorganics, and complete miscibility with gases and many hydrocarbons. Due to its miscibility, phase boundaries do not exist anymore. This absence leads to fast and complete homogeneous reactions of water with organic compounds [10,14]. Figure 4 shows the solubility of some salts and Figure 5 shows the solubility of benzene in supercritical water.

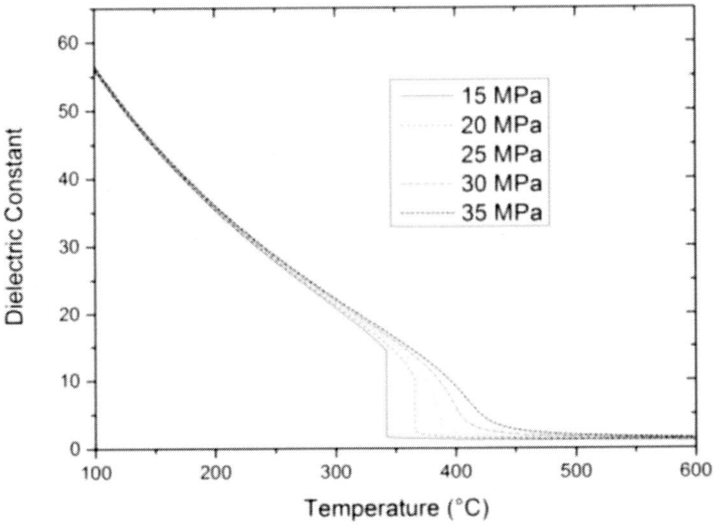

Figure 2. Dielectric constant of water at various temperatures and pressures; reprinted with permission from [12], copyright 2013 Elsevier.

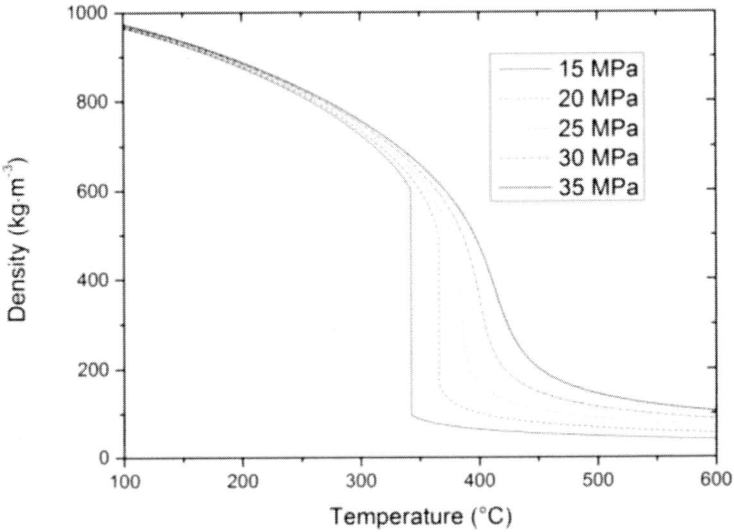

Figure 3. Density of water at various temperatures and pressures; reprinted with permission from [12], copyright 2013 Elsevier.

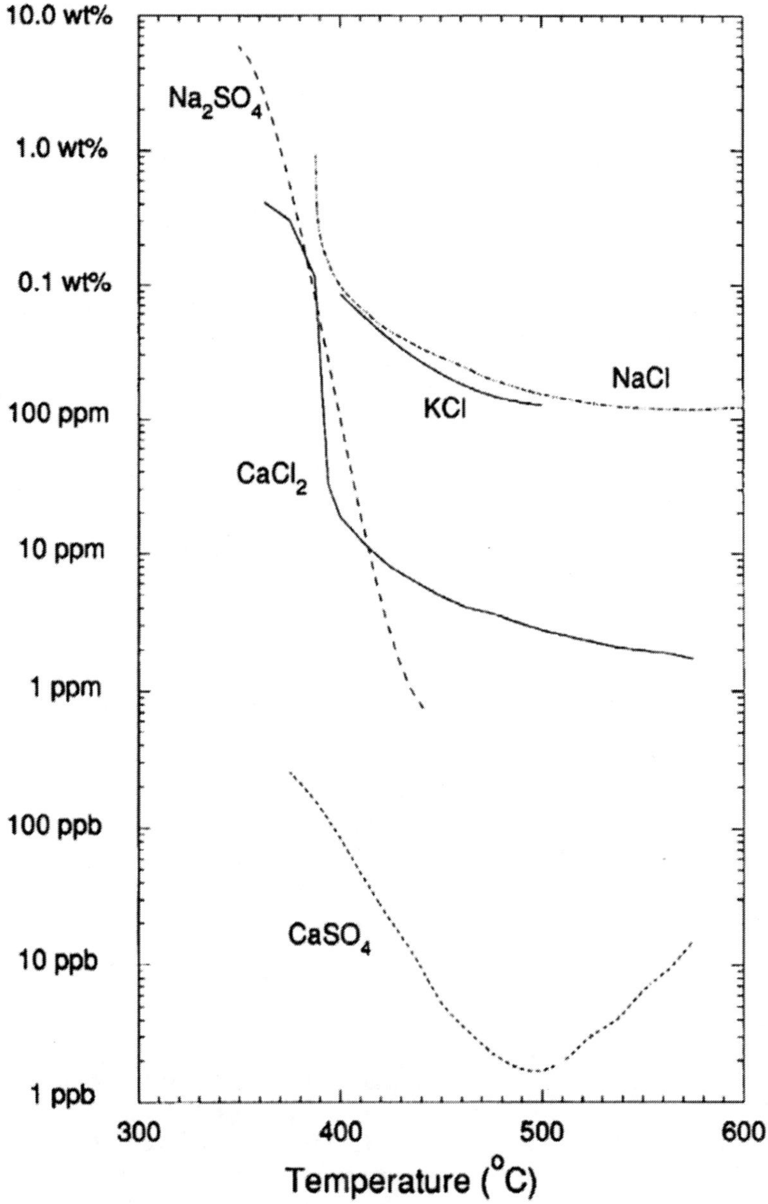

Figure 4. The solubility of limits of various salts at 25 MPa. Reprinted with permission from [9], copyright 2008 Royal Society of Chemistry, original data is from [15].

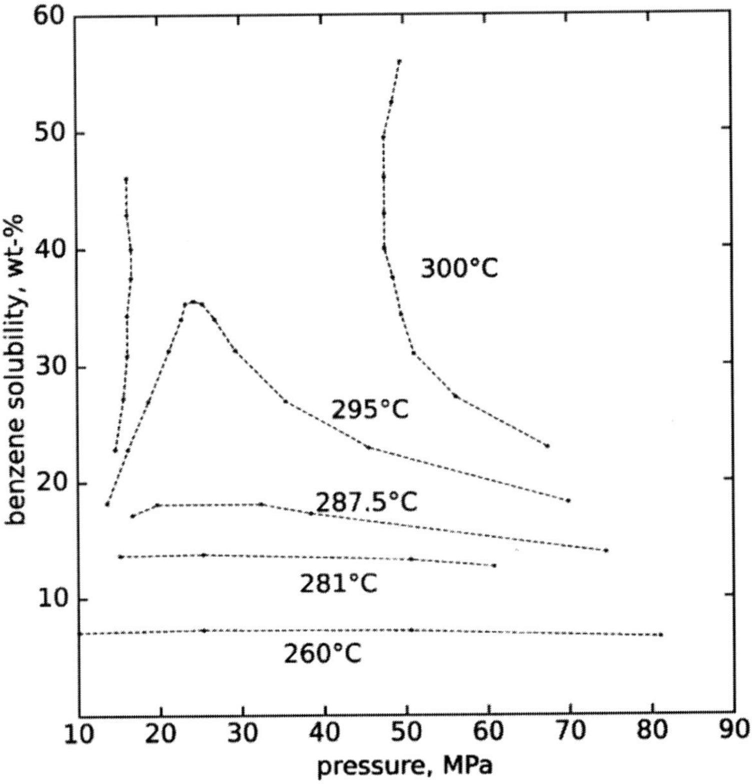

Figure 5. Benzene solubility in high-pressure water. Reprinted with permission from [9], copyright 2008 Royal Society of Chemistry, original data is from [16]. Please note that at 300 °C and above, the phases become completely miscible between 17 and 47 MPa.

Figure 6 shows the ionic product of water at different conditions. The ionic product of water increases with an increase in the pressure, however, the temperature shows a more complicated effect. At 25 MPa, the ionic product of water increases with temperature and reaches its highest value of 10^{-11} at a temperature of around 250 °C. Starting from 250 °C, it decreases slightly till the critical temperature, then reaches a value of 10^{-13} and it subsequently decreases drastically with the increase in temperature reaching a value of 10^{-23} at 600 °C [17]. When the ionic product of water is relatively high, water acts as an acid or base catalyst due to the high concentration of H_3O^+ and OH^- ions. At these conditions (liquid water, high pressure supercritical water and probably the dense gas phase), the main reaction pathways are ionic. However, when the ionic product of water is low (steam and less dense supercritical water), the main reactions are radical. Around the critical point of water, both ionic and radical reactions take place and compete with each other. It is concluded that when the ionic product is higher than 10^{-14}, aqueous phase ionic reactions preferably take place and when the ionic product is much lower than 10^{-14}, free radical gas phase reactions become dominant. Below the critical temperature, the rate of ionic reactions generally increases with an increase in temperature until the critical temperature is reached. Near the critical point, the reaction rate decreases drastically and shows a strong and characteristic non-Arrhenius behavior. At the critical point, the reaction rates can decrease or increase drastically depending on the chemistry [10,14,18].

Another interesting phenomenon that is observed concerning the thermophysical property of supercritical water is its change of the isobaric heat capacity, denoted with C_p. Figure 7 shows the isobaric heat capacity of water at various temperatures and pressures. Please note that at the critical point, C_p of water tends to approach infinity and becomes not measurable (the highest value of C_p given by [19] is at 373.91 °C with a value of 2.4355 × 10^5 J·mol^{-1}·K^{-1}). When the water shifts to its supercritical phase from liquid phase, its latent heat tends to zero. Instead, there is a high C_p value which mimics the high energy demand for the phase change. However, the increase in the pressure results in a decrease of the C_p value.

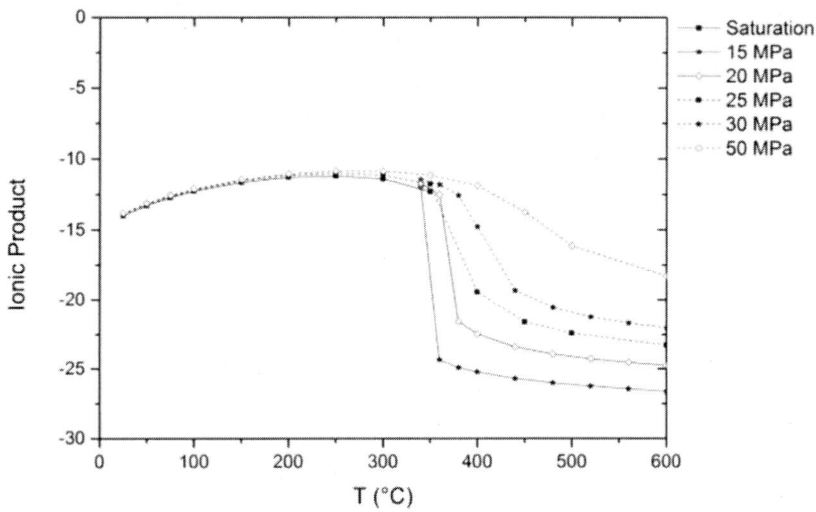

Figure 6. Ionic product of water at various temperatures and pressures; data taken from [17].

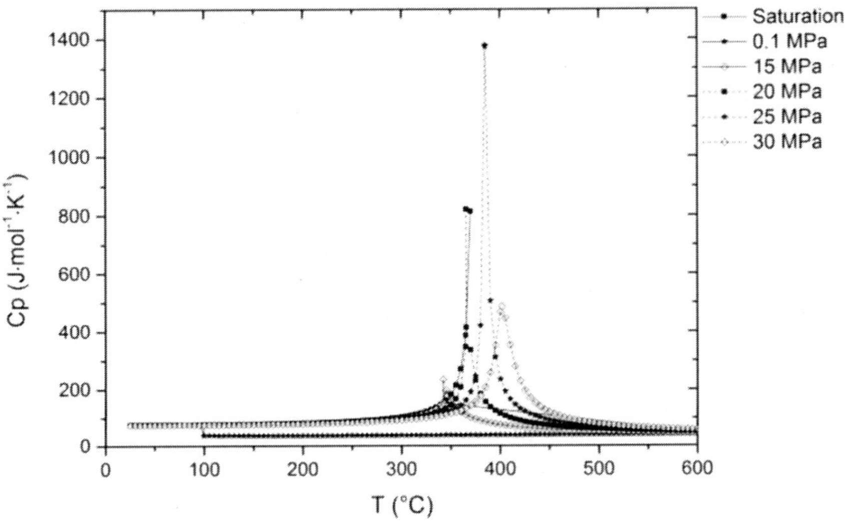

Figure 7. Isobaric heat capacity of water at various temperatures and pressures; data taken from [19].

The changes in the thermophysical properties of water, especially the disappearance of the phase boundaries and non-polar like solvent behavior, gives the opportunity for salt separation and tar-free gasification of biomass due to the salt precipitation, and dissolution and conversion of tar-precursors in a hydrothermal medium. The term hydrothermal refers to an aqueous system at temperatures and pressures near or above the critical point of water [13].

3. HYDROTHERMAL CONVERSION OF ORGANIC FEEDSTOCKS

Hydrothermal conversion of biomass can be classified as carbonization, oxidation, liquefaction and gasification depending on the process parameters. Each of these routes results in different products varying from biochar to hydrogen rich gas [20,21,22,23,24]. Hereafter, they are briefly summarized.

3.1. Carbonization

Hydrothermal carbonization (HTC) is the (pre)treatment of lignocellulosic biomass in hot (180–280 °C) compressed water at residence times varying from minutes to hours. The solid product is hydrophobic and reported to be similar to lignite by means of ultimate analysis. Besides, it can be easily pelletized. By-products include aqueous sugars, acids, carbon dioxide and water [21].

3.2. Oxidation

Supercritical water oxidation (SCWO) has been the main research topic in the 1990s regarding the hydrothermal conversion of biomass. It is an effective technology for the treatment of organic compounds resulting in CO_2 and H_2O as the main products as well as N_2 when nitrogen containing compounds are treated. The process takes place at very short residence times (less than 50 s at around 650 °C). The main application of SCWO is the destruction of wastewaters and sludges [25,26].

3.3. Liquefaction

Hydrothermal liquefaction (HTL) is a technology which enables the conversion of biomass into clean liquid fuels (so called "biocrude" or "biooil"). The process takes place with the presence of water or water-containing solvent/co-solvent. The most efficient conversion can be obtained in the presence of a catalyst (mostly alkaline based) at a temperature of 200–400 °C and at a pressure 5–25 MPa. The process's oily product can be defined as a viscous crude oil replacement. However, it has been reported to show important differences as compared to conventional crude such as having a significantly higher oxygen content which (may) require an upgrading process [9,24].

3.4. Gasification

Gasification of organic feedstocks in a hydrothermal medium can be performed at different temperatures and pressures, or in the presence of a catalyst. It can be a catalytic gasification process at subcritical conditions (225–265 °C and 2.9–5.6 MPa) [27], a low temperature catalytic gasification process at supercritical conditions (near critical temperatures to ~500 °C) and a high temperature non-catalytic gasification process (at temperatures higher than 500 °C) [9]. It has significant advantages over the other biomass conversion routes by means of heat utilization when the moisture content of the feedstocks exceeds 30 wt%. Figure 8 shows the comparison of supercritical water gasification (SCWG) with the other biomass conversion routes on heat utilization efficiency basis. Another comparison between conventional gasification and SCWG can be made based on the works of Gassner et al. [28] and van der Meijden et al. [29]. Here, both of the authors investigated the energy efficiencies of wood gasification processes. The results of Gassner et al. [28] showed that the overall energy efficiency for a wood gasification process targeting the production of synthetic natural gas (SNG) in SCWG is 70%. The overall energy efficiency is given in Equation (1) as:

$$\varepsilon = \frac{\Delta h^{0}_{SNG}\dot{m}^{-}_{SNG} + \dot{E}^{-} + \dot{Q}^{-}}{\Delta h^{0}_{biomass}\dot{m}^{+}_{biomass,daf} + \dot{E}^{+}} \tag{1}$$

where $\Delta h0SNG$ refers to the lower heating value (LHV), $m^{\cdot}$ refers to the mass flow, $E^{\cdot}$ refers to the mechanical or electrical power, $Q^{\cdot}$ refers to the heat flow and daf refers to the dry ash free basis. The superscripts – and + refers to the flows leaving the system and flows entering the system, respectively. The same net efficiency for a SNG production process from wood were found to be 54.3% for an entrained flow gasifier operating at 3 MPa, 58.1% for a circulating fluidized bed operating at 1 MPa and 66.8% for an allothermal gasifier operating at 0.1 MPa on LHV basis [29].

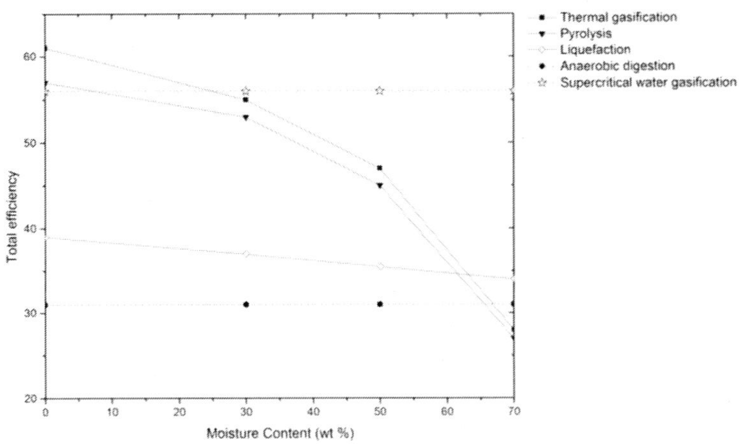

Figure 8. Total efficiency of heat utilization processes versus biomass moisture content; data taken from [30]. The total efficiency is defined as the HHV based energy content of the product divided by the energy content of all energy inputs to the process.

Supercritical water gasification of biomass can be performed under different process conditions with different feedstocks. The feedstock can be varied from simple model biomass compounds to complex real biomass feedstocks with or without the presence of catalyst at batch or continuous conditions. The experimental and modelling approaches of supercritical water gasification of biomass are described in detail in the next sections.

4. SUPERCRITICAL WATER GASIFICATION OF BIOMASS: EXPERIMENTAL APPROACHES

4.1. Understanding the Chemistry

Understanding the gasification reaction pathways of a real biomass is quite challenging due to the complex nature of a real biomass feedstock. However, performing experiments with representative model biomass compounds gives the opportunity to understand how the main ingredients of a real biomass could behave under the same process conditions. Cellulose, hemicellulose and lignin content of the biomass is generally represented by their monomers such as glucose, xylose and guaiacol, whereas the protein content of the biomass is represented by some major amino acids. Effects of inorganic content of the biomass are tested by performing model biomass experiments in the presence of some salt solutions.

4.1.1. Biopolymers

Cellulose conversion in supercritical water has been the subject of studies presented by Sasaki et al. [31,32] and Resende et al. [33]. The results show that cellulose undergoes a rapid hydrolysis and decomposes to its monomer glucose at very short residence times as short as 3 s at 400 °C. At higher temperatures, intermediate species formation starts and from these intermediates further gaseous compounds are formed. Real biomass derived hemicellulose conversion has not been investigated so far, however lignin conversion has been investigated by Yong and Matsumura [34,35] and Resende et al. [36]. The results show that lignin is mainly decomposed to guaiacol which further decomposes to other intermediates. These intermediates then form permanent gas products at higher temperatures. Char can be formed directly from lignin or through these intermediates.

4.1.2. Monomers

Glucose has been the main research focus among the other model biomass derived sugar compounds. The earliest research has been conducted by Amin et al. [6] in 1975. Lee et al. [37], Hao et al. [38], Williams and Onwudili [39], Goodwin and Rorrer [40], and Güngören Madenoğlu et al. [41] all performed parametric studies on the gasification of glucose in supercritical water. As expected, the increase in the temperature and residence time resulted in increased glucose conversion to gaseous products enhancing hydrogen production at higher temperatures.

Kabyemela et al. [42] and Matsumura et al. [43] examined the glucose decomposition pathways in supercritical water. They have proposed reaction pathways and determined the reaction rate constants for various temperatures. Kabyemela et al. [42] proposed that glucose undergoes an isomerization reaction to form fructose and then forms intermediates such as acids and 5-Hydroxymethylfurfural (5-HMF) at a temperature interval of 300–400 °C in which the decomposition reactions are proposed to be first order reactions. The intermediate products are proposed to be further converted to gas products at higher temperatures or longer residence times by different authors such as Goodwin and Rorrer [40] and Chuntanapum and Matsumura [44]. The latter research group proposed a general reaction mechanism for glucose to gas conversion. The reaction pathway is given in Figure 9. Similar to these researchers, Kruse and Gawlik [45] proposed that decomposition of glucose or fructose to furfurals is preferred under ionic conditions (characterized by a temperature lower than 374 °C), whereas a reaction pathway to acids or aldehydes is preferred at free radical conditions (characterized by temperatures higher than 374 °C).

Unlike glucose, there are only limited research works on the xylose reactions in supercritical water. Aida et al. [46] examined the xylose decomposition reactions at temperatures between 350 and 400 °C and Goodwin and Rorrer [47] examined its conversion at temperatures in the range of 450–650 °C. Similar to glucose reactions, xylose undergoes isomerization and decomposition reactions. The smaller organic compounds then further decompose to produce gas compounds at temperatures higher than 400 °C. Figure 10 shows the reaction mechanism of xylose in supercritical water within a temperature interval of 450 and 650 °C proposed by Goodwin and Rorrer [47].

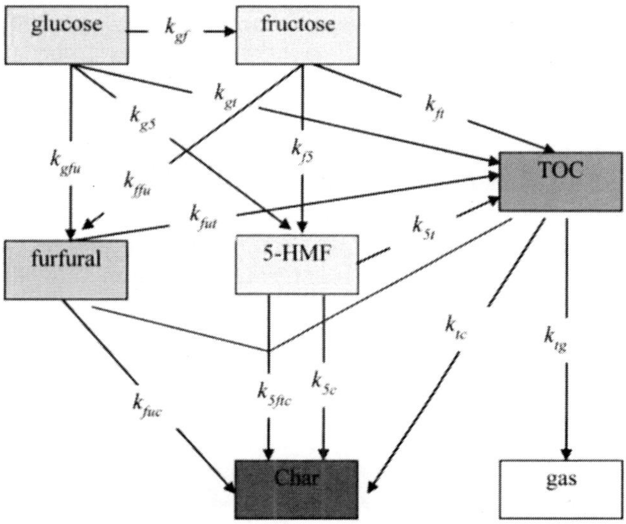

Figure 9. Proposed reaction pathway for glucose to gas conversion at temperatures between 300 and 400 °C. Reprinted with permission from [44], copyright 2010 American Chemical Society. Please note that TOC refers to total organic compounds in liquid phase such as acids.

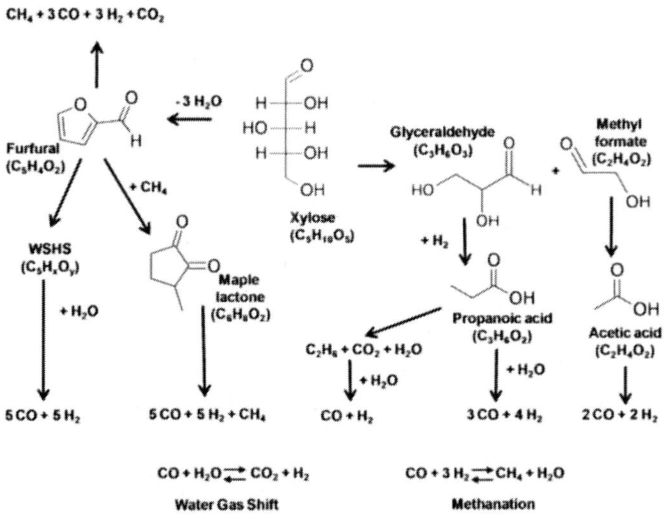

Figure 10. The reaction mechanism of xylose in supercritical water at a temperature interval of 450 and 650 °C proposed by Goodwin and Rorrer [47]. Reprinted with permission from [47], copyright 2010 Elsevier. WSHS refers to water soluble humic substances.

As a lignin model component, guaiacol conversion in supercritical water has been examined by Wahyudiono et al. [48,49] and Dileo et al. [50]. In addition, Yong and Matsumura [34,35] investigated the lignin decomposition pathway through guaiacol. The results indicate that guaiacol first decomposes to o-cresol, catechol, phenol and other liquid products. These intermediates further decompose to smaller compounds which then subsequently decompose to form gases. Figure 11 shows the lignin and guaiacol reaction mechanism at supercritical water at temperatures between 390 and 450 °C. The reaction rates are assumed to be first order.

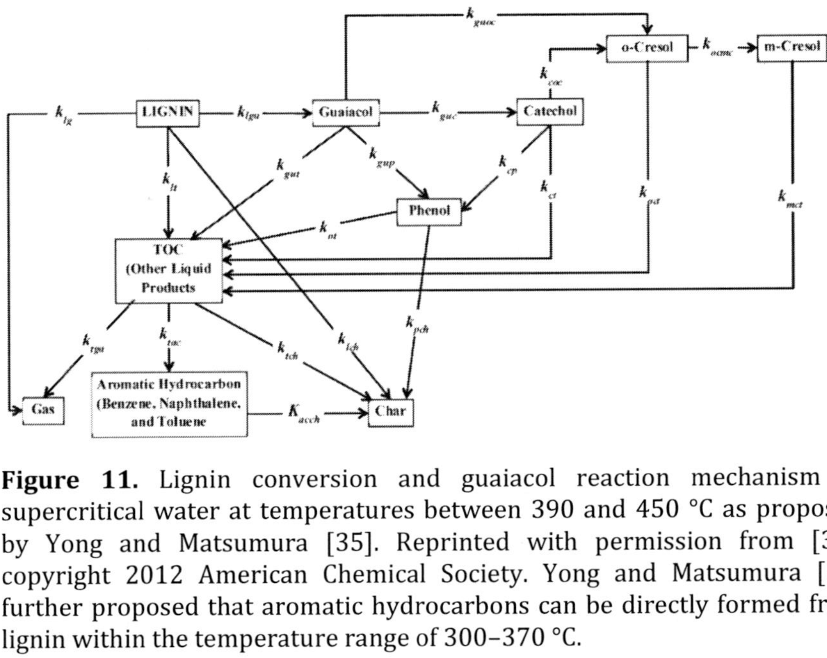

Figure 11. Lignin conversion and guaiacol reaction mechanism in supercritical water at temperatures between 390 and 450 °C as proposed by Yong and Matsumura [35]. Reprinted with permission from [35], copyright 2012 American Chemical Society. Yong and Matsumura [34] further proposed that aromatic hydrocarbons can be directly formed from lignin within the temperature range of 300–370 °C.

4.1.3. Amino Acids

Klingler et al. [51] examined the decomposition of two amino acids in sub- and supercritical water: alanine and glycine. The investigated temperature range was between 250 and 450 °C. The results indicate that both of the amino acids hydrolyze and release NH_3 to form acid intermediates as well as amine compounds. The acid compounds further decompose to produce smaller intermediates and gases. Figure 12 shows the reaction mechanism of amino acids in supercritical water proposed by the group. However, Dileo et al. [52] also observed solid, black product in addition to liquids and gases at 500 °C.

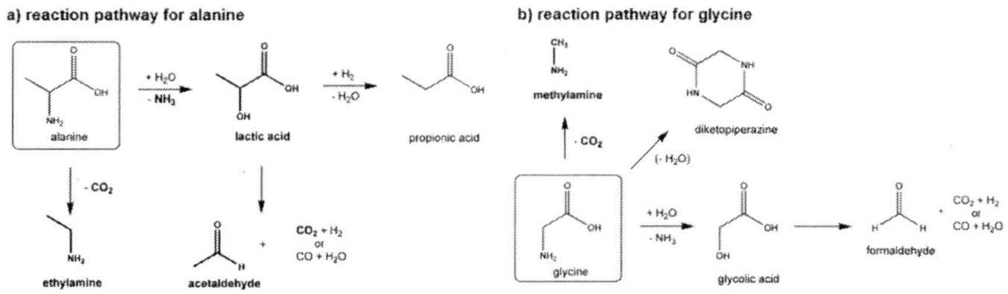

Figure 12. The reaction mechanism of (**a**) amino acids alanine and (**b**) glycine in supercritical water at 250–450 °C proposed by Klingler et al. [51]. Reprinted.

4.1.4. Intermediates

The conversion of intermediates in supercritical water has also been investigated by many researchers. Zhang et al. [53] and Shin et al. [54] examined the decomposition of formic acid in supercritical water. The results point to a mechanism in which formic acid undergoes dehydration and decarboxylation reactions to form carbon monoxide and carbon dioxide. Aida et al. [55] and Mok et al. [56] studied the decomposition of lactic acid in supercritical water. The results are similar to formic acid reactions: lactic acid also undergoes dehydration and decarboxylation reactions and forms smaller intermediates such as acetaldehyde and acrylic acid. These intermediates further decompose to form gases. Chuntanapum and Matsumura [44,57,58] investigated the behavior of 5-HMF and its importance in char formation mechanism in supercritical water gasification reactions. Their results indicate that 5-HMF can undergo polymerization reactions and form char only at subcritical conditions. The pathway for 5-HMF is depicted in Figure 9.

4.1.5. Mixtures

In order to understand how real biomass could behave in gasification reactions, mixtures of two or more model biomass compounds have been investigated. Yoshida and Matsumura [59] investigated the gasification performance of mixtures of cellulose, xylan and lignin at 400 °C. They have proposed a correlation to predict the gasification and hydrogen production efficiency based on the amount of cellulose, xylan and lignin. The results indicate that the presence of lignin decreases the gas production as well as the production of H_2 from the intermediates of cellulose and xylan. Yoshida et al. [60] further investigated the gasification performance of sawdust and rice straw. They have compared the gasification efficiency and hydrogen production with their correlation. Gasification and hydrogen production yields were lower than those calculated values which led the authors to suggest that even in real biomass, interactions between each component occurred. Similar results have been observed by other authors: Weiss-Hortala et al. [61] observed a significant decrease in gas yields even when a small portion of phenol (a lignin decomposition compound) was blended with glucose. Goodwin and Rorrer [62] could not attain a 100% carbon gasification efficiency for xylose and phenol mixtures even at 750 °C. However, Castello and Fiori [63] reported that at 400 °C, phenol did not appear to behave as an inert compound, but contributed to gas production to a limited extent and reacted in liquid phase to form other compounds, possibly tar and/or solids.

4.1.6. Effect of Salts and the Role as Homogeneous Catalysts

The effect of inorganic constituents in biomass was investigated by gasifying model biomass compounds in the presence of salts. Sinag et al. [64] investigated the influence of K_2CO_3 on the gasification of glucose. Similarly, Kruse and Faquir [65] and Kruse et al. [66] tested the influence of other potassium containing alkali compounds such as KOH and $KHCO_3$. The authors found that the presence of alkali salts increases the total gas yield as well as hydrogen yield and the amount of phenols. Sinag et al. [64] proposed a reaction pathway for hydrogen formation in the presence of K_2CO_3 as:

$$K_2CO_3 + H_2O \rightarrow KHCO_3 + KOH \quad (2)$$

$$KOH + CO \rightarrow HCOOK \quad (3)$$

$$HCOOK + H_2O \rightarrow KHCO_3 + H_2 \quad (4)$$

$$2KHCO_3 \rightarrow H_2O + K_2CO_3 + CO_2 \quad (5)$$

$$H_2O + CO \leftrightarrow HCOOH \leftrightarrow H_2 + CO_2 \quad (6)$$

The positive effects of other (earth) alkali salts such as (NaOH, $Ca(OH)_2$, Na_2CO_3, $NaHCO_3$) on the gasification yield have also been reported by other authors [67,68,69,70,71] for different kinds of biomass feedstocks. Yildiz Bircan et al. [72] investigated the addition of $Ca(OH)_2$ on the gasification performance of an amino acid, l-cysteine. The results showed that the addition of $Ca(OH)_2$ enhanced the H_2 formation, but most importantly, it was observed that $Ca(OH)_2$ was consumed throughout the reactions and reacted with CO_2 to form $CaCO_3$.

4.1.7. Effect of Heterogeneous Catalysts

Catalysts are used in order to achieve higher conversion at lower temperatures (<500 °C), especially when the desired product is methane [73,74]. The presence of a catalyst changes the reaction pathway of the biomass, resulting in the reduction of tar and coke production and an increase in the gas yields. Figure 13 shows the influence of catalyst on the reaction pathway of wood gasification under supercritical conditions. Ruthenium, nickel and activated carbon based heterogeneous catalysts have been widely used in the literature [75,76,77,78,79] for both model and real biomass experiments enabling full carbon gasification efficiency to be obtained even at temperatures around 400 °C.

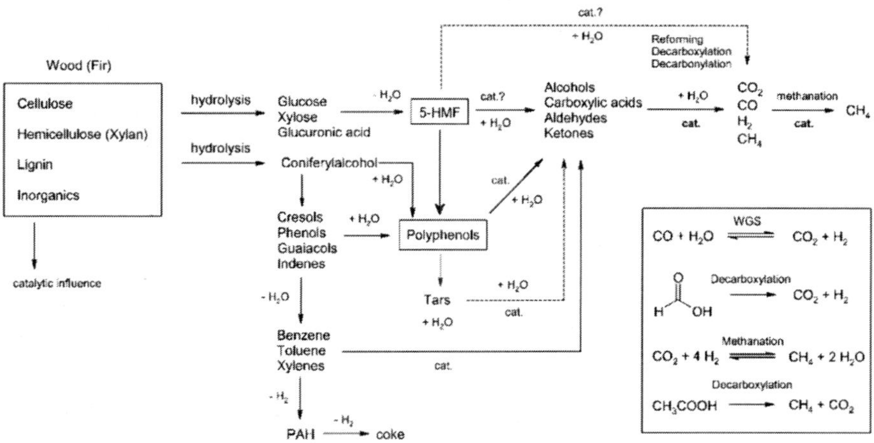

Figure 13. The influence of catalyst on the reaction pathway of wood gasification under supercritical conditions proposed by Waldner and Vogel [78]. Reprinted with permission from [78], copyright 2005 American Chemical Society. The term "cat." designates reaction pathways influenced by the presence of a catalyst, whereas the term "cat.?" denotes pathways that are only assumed to be promoted by a catalyst.

Yanik et al. [80] have investigated natural minerals like red mud (a by-product from alumina production plant containing Fe_2O_3 (37.7%), Al_2O_3 (17.3%), SiO_2 (17.1%), TiO_2 (4.8%), Na_2O (7.1%) and CaO (4.5%) by wt%) and trona ($NaHCO_3 \cdot Na_2CO_3 \cdot 2H_2O$). Both of these minerals have been found to significantly increase the H_2 yield and slightly promote the total gas yield for different kinds of carbohydrate based feedstocks.

It has been reported that the reactor wall material may also behave like a heterogeneous catalyst. Boukis et al. [81] studied the methanol reforming in a nickel-based alloy Inconel 625. They concluded that the conversion rate as well as the product gas composition is influenced by the presence of heavy metals on the inner surface of the reactor and the catalytic effect remained high even after more than a thousand operation hours. Similarly, Gadhe and Gupta [82] observed a catalytic effect of the wall material from Inconel 600, which is an alloy of Ni, Cr and Fe, for methanol reforming. Lee et al. [37] performed glucose gasification in a reactor made of Hastelloy C-276 and noted a catalytic effect and enhancement in the hydrogen yields. Similar results were observed by Yu et al. [83]; these authors tested Inconel, new Hastelloy and corroded Hastelloy for glucose hydrothermal gasification. Their results indicated that corroded Hastelloy significantly increased the hydrogen and total gas yields. Castello et al. [84] have investigated the effect of wall material for stainless steel and Inconel 625 for both glucose and sawdust gasification in micro-autoclaves without the presence of catalyst at temperatures of 350 °C and 400 °C. They have observed that stainless steel favored hydrogen production, whereas Inconel 625 enhanced CO consuming methanation and light hydrocarbon formation under supercritical conditions. Under subcritical conditions, the two reactors gave similar results.

4.1.8. Behavior of Heteroatoms

Yildiz Bircan et al. [72] investigated the behavior of heteroatoms such as S, N and P in supercritical water gasification using l-cysteine containing hetero-atoms S and N, and o-phospho-dl-serine containing P as feedstocks. They have performed experiments in an oven which was kept at 400 °C and at 26–27 MPa. After a residence time of 50 min, the oven was cooled to 30 °C and the components were analyzed. The results showed that sulphur partitioned to SO_2 and H_2S in the gas phase, and SO_3^{2-}, SO_4^{2-} and S^{2-} ions in the liquid phase. When $Ca(OH)_2$ was used as an additive, the gas phase sulphur compounds concentrations decreased by about one tenth and partitioned to the liquid phase sulphur. However, for both of the cases more than 90% of the sulphur partitioned to the solid phase. For nitrogen compounds, NO and NO_2 were detected in trace levels and NH_3 was not detectable. A total of 95.7% percent of nitrogen was converted to NH_4^+ in the liquid phase. This proportion decreased to 82% when $Ca(OH)_2$ was used as an additive. When it comes to phosphorus compounds, 93.3% of the phosphorus partitioned to ionic compounds in the liquid phase. When $Ca(OH)_2$ was used as an additive, more than 90% of the phosphorus precipitated as a solid phase.

4.1.9. Gas Phase Reactions

It has been reported by many authors (such as [47,85,86]) that water gas shift and methanation reactions are assumed to be the main gas phase reactions which may result in an equilibrium state. Equations (7) and (8) show the water gas shift reaction and methanation reactions, respectively.

$$CO + H_2O \leftrightarrow CO_2 + H_2 \tag{7}$$

$$CO + 3H_2 \leftrightarrow CH_4 + H_2O \tag{8}$$

In the water gas shift reaction, carbon monoxide and water form formic acid which further decomposes to carbon dioxide and hydrogen through the Equations (9) and (10) [87,88].

$$CO + H_2O \leftrightarrow HCOOH \tag{9}$$

$$HCOOH \leftrightarrow CO_2 + H_2 \tag{10}$$

Savage's group [85,86] performed sensitivity analyses on the reaction rates for non-catalytic supercritical water gasification of biomass processes. It can be concluded from the results that the main reactions which determine the hydrogen and methane formation are not the water gas shift and methanation reactions in the gas phase, but the reactions which directly result in the gas formation from the feedstock and/or intermediates. It has also been reported in the literature [87,88,89,90] that the rates of non-catalytic water gas shift and methanation reactions in supercritical water are quite slow, resulting in only 5% conversion even after several minutes.

4.2. Real Biomass Experiments

Antal et al. [75] state that the first real biomass experiments in supercritical water have been performed by Modell et al. [91] in 1985 using maple wood sawdust as a feedstock. The results indicated that the sawdust quickly decomposed to tars and gases without the formation of char. Since the first experiments in 1985, real biomass has also been the subject of supercritical water gasification.

4.2.1. Carbohydrates

In their work, Antal et al. [75] examined gasification performance of different biomass feedstocks such as corn- and potato-starch gels, wood sawdust suspended in a cornstarch gel, and potato wastes in three different tubular reactor which had activated carbon as a catalyst. They have performed experiments at temperatures higher than 650 °C and at a pressure of 28 MPa. Their results were in a good agreement with the equilibrium calculations resulting in an extraordinary gas yield higher than 2 L/g (liters of gas under 25 °C and 0.1 MPa per gram of organic matter in the feed) with a high yield of hydrogen (57%). Irrespective of the reactor geometry and method of heating, they have observed plugging problems after 1–2 h of operation time with feedstocks that contained approximately 15 wt% of organic material.

D'Jesús et al. [92] gasified corn starch, clover grass and corn silage in supercritical water. They investigated the influence of pressure, temperature, residence time and alkali addition on gasification performance. They reported that the change in the pressure did not alter the gasification yield; however, the increase in the temperature significantly increased the gasification yield. Longer residence time resulted in higher gasification yields till the system attained a maximum. Potassium addition (in the form of $KHCO_3$) was found to increase the carbon gasification efficiency for corn starch. Although, no significant difference was observed for clover grass and corn silage as they naturally contained potassium.

4.2.2. Algae

Stucki et al. [93] gasified a microalgae species, Spirulina platensis, in the presence of different catalysts. One hundred percent of the carbon content of the microalgae was gasified using a Ru/C catalysts at a temperature of around 400 °C and at a pressure around 32 MPa. Their results were in very good agreement with the equilibrium predictions. They proposed that 60%–70% of the heating value of the algae can be recovered in the form of methane.

Chakinala et al. [94] gasified Chlorella Vulgaris in supercritical water. They have observed more than 80% carbon gasification efficiency at 700 °C with a 7.3 wt% dry mass concentration at a pressure of 24 MPa with a residence time of 2 min. As expected, gasification efficiency decreased with a decrease in temperature and residence time. The increase in the dry mass concentration had a negative impact on the carbon gasification efficiency. They have also tested the effect of different catalysts on the carbon gasification efficiency. The results indicated that the presence of a catalyst increased the carbon gasification efficiency, reaching even a 100% with a Ru/TiO_2 catalyst at 600 °C with a residence time of 2 min.

Guan et al. [95] gasified another microalgae sample, Nannochloropsis sp., in supercritical water at 450–550 °C at around 25 MPa. They have concluded that higher gas yields were obtained by higher temperatures, longer reaction times (>20 min), higher water densities, and lower algae loadings. Their findings also show that when the biomass loading reduced from 15 wt% to 1 wt%, H_2 yield more than tripled.

Onwudili et al. [71] gasified three different types of microalgae samples (Chlorella vulgaris, Spirulina platensis and Saccharina latissima) at 500 °C and 36 MPa in an Inconel batch reactor for 30 min. They also tested the influence of NaOH and/or Ni-Al_2O_3 on the gasification performance. Their results indicated that the presence of NaOH more than doubled the hydrogen yield and decreased the tar yield up to 71%. Presence of the nickel catalyst had also decreased the tar formation. A similar study has been performed by Bagnoud-Velásquez et al. [96] on the effluent recycling and the catalytic gasification of a microalgae sample, Phaeodactylum tricornutum. In a continuous reactor, for a 6.5 wt% algae concentration, they were able to obtain 31% carbon gasification efficiency at 420 °C and 32.3 MPa in the presence of Ru/C catalysts. They observed rapid deactivation of the catalyst due to sulphur poisoning and coke formation.

4.2.3. Sludge

Xu and Antal [97] gasified digested sewage sludge, corn starch and poplar wood sawdust in the presence of coconut shell activated carbon catalyst. They have mixed sewage sludge (up to 7.69 wt%) with corn starch gel (up to 7.69 wt%) in order to form a viscous gel. They have performed experiments at 650 °C and at 28 MPa. Even with a lower sewage sludge loading (2.1 dry wt% digested sewage sludge with dry 5.1 wt% corn starch), the reactor had plugging problems after 1–2 h of operation time due to high ash content of sewage sludge.

Chen et al. [98] used a fluidized bed to overcome reactor plugging problems in sewage sludge gasification. They have also tested the effect of operating parameters as temperature, concentration of the feedstock, alkali catalysts and catalyst loading on gaseous products and carbon distribution. The authors have performed experiments at a pressure of 25 MPa and at a temperature range of 480–540 °C with a biomass concentration varying from 4 to 12 wt%. As expected, the increase in temperature and decrease in feed concentration increased the carbon gasification efficiency. Addition of alkali catalyst was found to enhance the hydrogen production.

Zhai et al. [99] gasified digested sewage sludge in a batch reactor at temperatures up to 425 °C at a pressure varying from 25 to 35 MPa, with a residence time of 10–15 min and a dry matter content ranging from 5 to 25 wt%. In the best case, they were able to obtain 3.5% carbon gasification efficiency where the presence of K_2CO_3 increased this value up to 26%.

4.2.4. Manure

Waldner [13] gasified real biomass samples such as wood and swine manure in a batch reactor in the presence of different catalysts. Using a Raney Ni 2800 catalyst, 100% carbon gasification efficiency was obtained at a temperature around 400 °C and at a pressure of 31 MPa for wood. The gas composition was in agreement with the equilibrium predictions. Similar conditions resulted in more than 75% carbon gasification efficiency for swine manure.

Nakamura et al. [77] investigated the gasification of chicken manure in the presence of suspended fine activated carbon catalysts in an experimental pilot plant. At 600 °C and 25 MPa with a residence time of 1.7 min, they were able to completely gasify 2 wt% and 10 wt% chicken manure in the presence of 0.4 wt% and 5 wt% activated carbon, respectively. However, when there was no catalyst present, 80% carbon gasification efficiency was observed with a 2 wt% feed concentration. Their setup included a liquefaction process with a residence time of 26.7 min of which the temperature was 180 °C and pressure was 1.2 MPa.

Yong and Matsumura [100] investigated the gasification behavior of poultry manure and eucalyptus wood in a flow reactor and the effect of activated carbon on the gasification performance. Their results indicate that addition of 0.1 wt% wood to 0.5 wt% manure increased the carbon gasification efficiency (CGE) at temperatures between 550 and 650 °C and at a pressure of 25 MPa. However, the effect of wood decreased at 0.2 wt% and became almost insignificant at 0.3 wt%. They stated that cellulose and hemicellulose containing wood biomass are more easily decomposed in SCW than poultry manure. The increase in the loading of wood in the mixture resulted in the deceleration of the gas producing pathways in the overall reactions. It was also found that the usage of activated carbon in the feedstock mixture improved the gasification efficiency. An overview of the real biomass experiments is presented in Table 1.

5. SUPERCRITICAL WATER GASIFICATION OF BIOMASS: MODELING APPROACHES

Modeling enables simulating real cases in order to predict products and equipment performances and overcome possible challenges for different process conditions.

Table 1. An overview of the real biomass experiments.

Feedstock	wt%	Temperature (°C)	Pressure (MPa)	Reactor type	Residence time	CGE (%)	Mole fraction of gas products (%)				Reference
							CH_4	CO_2	H_2	CO	
Corn Starch	10.4	~700	28	Tubular with a carbon catalyst	2.18 h	91	22	38	37	2	[75]
Microalgae (*Chlorella Vulgaris*)	7.3	600	24	Quartz Capillary	2 min	53	25	26	7	22	[94]
Sewage Sludge	10	540	25	Fluidized Bed	N/A	32	13	43	26	13.5	[98]
Swine Manure	13.2	405	30.1	Batch reactor with nickel catalyst	36 min	75.8	46	43	10.7	0.1	[13]
Chicken Manure	10	600	25	Flow reactor with carbon catalyst	1.7 min	90	21.2	45.4	28.7	0	[77]

5.1. Kinetic Modeling

Studies aimed at elucidating the reaction kinetics under hydrothermal gasification conditions have started with simple sugar compounds and have been extended later to real biomass feedstocks. Matsumura's group [43,44,57,58] and Kabyemela et al. [42] studied the glucose, fructose and glucose derived intermediate decompositions in sub- and supercritical water. Aida et al. [46] and Goodwin and Rorrer [47] investigated the reactions of D-xylose in sub- and supercritical water. Wahyudiono et al. [49] studied the thermal decomposition kinetics of guaiacol as a lignin derived material. Sasaki et al. [31,101] further studied the kinetics of cellulose and cellobiose conversion in sub- and supercritical water. Regarding modelling, Resende and Savage [86] developed a kinetic model for non-catalytic supercritical water gasification of cellulose and lignin at high temperatures, and Guan et al. [85] proposed a reaction mechanism for the supercritical water gasification of a real biomass feedstock; microalgae. However, these kinetic models and parameters have limited applicability as:

- in most of these works, all of the reactions are assumed to be first order reactions;
- the temperature effect on the rate constants has not been sufficiently investigated;
- most of these works only incorporate the decomposition rates and the formation of intermediates, but do not show the whole gasification routes;

- the models which include entire gasification routes incorporate the lumped kinetic model, which neglects the composition of intermediates.

Table 2. An overview of the kinetic parameters determined for selected conversion pathways of biomass constituent compounds during supercritical water gasification. Please note that TOC refers to water-soluble organic species such as 1,2,4-benzenetriol, 1,4-benzenediol, 5-methyl-2-furaldehyde, levulinic acid, and formic acid.

Compound	Type of Reaction	Activation Energy ($kJ \cdot mol^{-1}$)	Pre-Exp. Factor (s^{-1})	Temperature Range (°C)	Pressure (Mpa)	Reference
Cellulose	Conversion	145.9	$1 \times 10^{11.9}$	320–370	25	[31]
Cellulose	Conversion	547.9	$1 \times 10^{44.6}$	370–400	25	[31]
Cellobiose	Conversion	96.4	1.28×10^{8}	300–350	25	[110]
Cellobiose	Conversion	96.4	1.48×10^{8}	350–400	30	[110]
Cellobiose	Hydrolysis	108.6	1.10×10^{9}	300–350	25	[110]
Cellobiose	Hydrolysis	108.6	1.15×10^{9}	350–400	30	[110]
Cellobiose	Pyrolysis to glucosyl-eryhtrose and glycoaldehyde	30.4	17.68	300–350	25	[110]
Cellobiose	Pyrolysis to glucosyl-eryhtrose and glycoaldehyde	30.4	57.13	350–400	30	[110]
Cellobiose	Pyrolysis to glucosyl-glycoaldehyde and eryhtrose	69.3	82.730.74	300–350	25	[110]
Cellobiose	Pyrolysis to glucosyl-glycoaldehyde and eryhtrose	69.3	83.456.10	350–400	30	[110]
Glucose	Decomposition	121	1.33×10^{10}	175–400	25	[43]
Glucose	Decomposition	96	1.23×10^{8}	300–350	25	[111]
Glucose	Isomerization to fructose	112.75	2.99×10^{9}	300–400	25–30	[42]
Fructose	Decomposition to acids	130.94	7.48×10^{10}	300–400	25–30	[42]
Glucose	Conversion to 5-HMF	114.40	1.49×10^{8}	300–400	25	[44]
Glucose	Conversion to non-furfural organics	137.40	2.05×10^{11}	300–400	25	[44]
TOC	Conversion to gas products	27.92	0.78	300–400	25	[44]
TOC	Conversion to char	17.29	0.04	300–400	25	[44]
Xylose	Conversion to furfural	120.1	1.20×10^{12}	450–650	25	[47]
Furfural	Conversion to WSHS	55.6	5.70×10^{3}	450–650	25	[47]
WSHS	Conversion to gas	138.9	1.60×10^{8}	450–650	25	[47]
Guaiacol	Conversion	40.0	1.87×10^{3}	380–400	30	[48]

Another interesting approach to kinetic modeling is the detailed kinetic modeling. Here, the whole reaction mechanisms consist of hundreds of elementary reactions. Most of these attempts have been made for SCWO of simple compounds such as methanol, carbon monoxide, methane and hydrogen [18,102,103,104,105,106]. Detailed kinetic modeling has also been applied for supercritical water gasification systems. Ederer et al. [107] have investigated the pyrolysis of tert-butylbenzene in supercritical water, Bühler et al. [108] investigated the glycerol decomposition in supercritical water and Castello and Fiori [109] used the detailed kinetic models for the supercritical water oxidation of methanol to model the gasification of methanol in supercritical water. Even though the results are promising, the application of detailed kinetic modeling for real biomass is not viable yet due to the complex nature of real biomass feedstocks. Table 2 shows an overview of the kinetic parameters for the selected conversion pathways of biomass constituent compounds during supercritical water gasification.

5.2. Computational Fluid Dynamics Modeling

Computational fluid dynamics (CFD) modeling helps to simulate and design a reactor [112]. Yoshida and Matsumura [112] were the first researchers who performed CFD analysis for a SCWG reactor. The authors used Fluent 6.3® software to perform simulations for 4.9 wt% glucose conversion in the presence of hydrogen peroxide as oxidation agent in supercritical water at 400 °C and 25.4 MPa. However, the authors ignored all the reactions and calculated only the flow of water and inert particles as virtual char products.

Goodwin and Rorrer [113] performed CFD simulations for xylose gasification in a Hastelloy micro-channel reactor at 650 °C and 25 MPa. The authors used Fluent 6.3.26® for the simulations and the fluid was modeled by the Navier-Stokes equations for laminar flow. They performed simulations using a six-reaction based kinetic model which they developed in a previous paper [47]. The authors used Peng-Robinson EoS for the pure gas species for the calculation of the fugacities; however, the mixing interactions in the reacting fluid were taken into account. The authors found that CFD simulations accurately predicted the gas yields. Besides, it was found that the endothermic xylose gasification reactions had little influence on the reaction temperature, thus not affecting the gas yield or H_2 selectivity.

Wei et al. [114] performed CFD simulations with a Eulerian model which incorporates the mechanical kinetic theory for solid particles to predict the solid distribution and the residence time distribution of the feeding materials for a fluidized bed reactor. Based on the simulation results, they proposed a feeding pipe with an angle of 45° to enable a uniform solid distribution and long residence time among the reactor.

A different approach using CFD has been performed by Withag et al. [115,116]. Here, the authors performed CFD simulations to calculate heat transfer characteristics of supercritical water to a flow reactor for both 1D and 2D cases. Further, Withag [117] also performed simulations for a reacting flow of methanol in supercritical water for a 2D case. Based on the simulations, it was concluded that methanol is mainly converted near the wall due to the high fluid temperatures resulting in high local reaction rates.

5.3. Thermodynamic Equilibrium Modeling

In addition to the kinetic models, supercritical water gasification of biomass compounds has also been modelled following a thermodynamic equilibrium modelling approach. Antal et al. [75] were the first researchers who used the thermodynamic equilibrium approach. They compared the experimental results with the thermodynamic equilibrium predictions using STANJAN and HYSIM of which the first one uses ideal gas assumption as an equation of state (EoS) and the latter one Peng–Robinson EoS. Tang and Kitagawa [118] performed thermodynamic analysis for various model and real biomass compounds using Peng-Robinson EoS describing the non-ideal properties of the compounds. The authors used a Gibbs free energy minimization method (also known as the non-stoichiometric method) in order to predict the equilibrium amounts of gases. More authors followed the same approach [119,120,121,122,123,124]. All of these works have been developed for the supercritical region and take only the gas phase compounds into account. Castello and Fiori [125] extended the gas phase products and introduced graphitic carbon as an additional compound, and investigated the application of thermodynamic constraints such as solid carbon formation and process heat duty.

An alternative to Gibbs free energy minimization was given in the works of Letellier et al. [126] and Marias et al. [127]. Here, the authors used the discrete reaction equilibria method (also known as stoichiometric approach) to predict the equilibrium composition of the gas phase compounds. Furthermore, they also modeled the flash column for the separation of product gases from water and other organic compounds.

The aforementioned developed models do not cover the subcritical region of the process and the behavior of inorganics, and the partitioning behavior of elements. However, Yanagida et al. [79] investigated the equilibrium behavior of inorganic elements in poultry manure during supercritical water gasification using a software; HSC Chemistry 6.12. Seven inorganic elements, namely N, Ca, K, P, S, Cl and Si, as well as C, H, O content of manure and possible compounds that could be formed by these elements, have been taken into account for the thermodynamic equilibrium calculations. The simulation was run first at 32 MPa and 600 °C and, after subtracting the amount of gases formed, the simulation then was run at room temperature conditions to observe the phase and compound behavior of the elements. The results have been compared with experimental works and found to be in agreement. A similar approach has been conducted by Yakaboylu et al. [12]. Here, the authors used the software packages of FactSage 5.4.1 and SimuSage 1.12 for the prediction of the equilibrium partitioning of 12 elements on the basis of compounds and phases for a mixture of pig and cow manure sample. The simulations have been performed for a wide range of temperatures (100–580 °C), pressures (25–30 MPa) and dry mass concentrations (5–20 wt%). The same research group [128] has also developed a multi-phase thermodynamic model for the prediction of equilibrium state compounds for supercritical water gasification of biomass systems. The authors validated the model by comparing their predictions with experimental results and model predictions already published. They also performed a case study for a microalgae sample and investigated the behavior of gaseous compounds and elements at different pressures and temperatures as well as dry matter concentrations. Table 3 shows an overview of the thermodynamic equilibrium modeling papers. There are other research works in the literature which use commercial software packages for the prediction of product compounds following a thermodynamic equilibrium calculation approach. However, most of these works concentrate on process modeling which is discussed in detail in the following section.

It is a fact that an arbitrary multi-component multi-phase system at a given temperature, pressure and chemical composition will always tend to minimize its total Gibbs free energy till it reaches the global minimum state. However, the predicted compounds at that minimum state are the "unconstrained" equilibrium state compounds. Most of the real conversion systems do not reach that state due to the natural constraints (such as slow kinetics or reaction barrier conditions) that could keep the system away from it [129]. In that sense, for most of the cases, predictions using a thermodynamic equilibrium approach will not show the real case results but will give an insight in the process limits as well as the formation behavior of the compounds.

Table 3. An overview of the thermodynamic equilibrium modeling of supercritical water gasification of biomass papers.

Investigated compounds	Type of approach	Type of EoS or software used for the calculations	Considered phases	Significant contribution	Reference
Methanol, glucose, cellulose, starch, sawdust	Gibbs free energy minimization (GFEM)	Peng-Robinson EoS	Only gas phase	The first paper with a thermodynamic model	[118]
Glucose	GFEM	Duan EoS	Only gas phase	The introduction of additional constraints	[124]
Wood sawdust	GFEM	Duan EoS	Gas phase and solid carbon	The introduction of solid carbon	[122]
Poultry manure	GFEM	HSC chemistry 6.12	Multi-Phase	The first paper which took the inorganic elements into account	[79]
Methanol, ethanol, glycerol, glucose and cellulose	GFEM	Ideal gas	Gas phase and solid carbon	No need for initial guesses for the mole amount of product species	[123]
Methanol, glucose, sewage sludge	Reaction equilibria	Peng-Robinson EoS	Gas phase and solid carbon	The first paper which used the reaction equilibria method	[126]
Glycerol and microalgae	GFEM	Peng-Robinson EoS	Gas phase and solid carbon	Char formation conditions have been examined	[125]
Glucose and cellulose	GFEM	Virial EoS and ideal gas	Only gas phase	Introduction of the entropy maximization	[120]
Pig-Cow manure mix	GFEM	FactSage 5.4.1 and SimuSage 1.12	Multi-Phase	The first paper which investigated the partitioning behavior of elements	[12]
Microalgae	GFEM	Peng-Robinson EoS for gases and revised HKF EoS for the aqueous compounds	Multi-Phase	The first paper which proposed a multi-phase mathematical model for both subcritical and supercritical regions	[128]

5.4. Process Modeling

Even though thermodynamic equilibrium models predict the (major) product compounds formed in reactors, one needs to model the performance of the other equipment in such a biomass based supercritical water gasification processing plant in order to gain insight into the whole process performance. Feng et al. [130] were the first researchers who performed a thermodynamic analysis of a supercritical water biomass gasification plant. They have analyzed the phase behavior and phase equilibria in the reactor and separators; in addition, they designed an optimized heat exchange network and performed a system exergy analysis. The authors used the statistical association fluids theory (SAFT) EoS for the thermodynamic calculations. Their results indicated that a cellulose gasification plant with reactor operating conditions of 600 °C and 35 MPa has an exergy efficiency of 40.6%. The exergy efficiency in this work is defined as the ratio of the exergy flows of the inlet (cellulose as being the feedstock, providing heat and electricity for the process) and outlet (hydrogen-rich) products.

Luterbacher et al. [131] performed process simulations and life cycle assessment (LCA) for wood and manure as a feedstock in supercritical water gasification plant targeting to produce SNG. Using AspenPlus™ 2004.1 software for the process simulations, their results showed that 62% of the manure's lower heating value (LHV) is converted to SNG and 71% of wood's LHV is converted to SNG. Similarly, Gassner et al. [28,132] offered optimal process design for polygeneration of SNG, power and heat from various biomass sources varying from a typical lignocellulosic material (modeled as $CH_{1.35}O_{0.63}$) to manure and sewage sludge. The authors made an extensive analysis from both thermo-economic and optimization points of view concerning the processes. The results indicate that the SNG conversion efficiency of wood process was found to reach up to 70%. The same value was found to reach up to 75% for microalgae, 70% for manure and 60% for undigested sewage sludge where the SNG conversion efficiency is given in Equation (11) as:

$$\varepsilon = \frac{\Delta h^0_{SNG} \dot{m}^-_{SNG}}{\Delta h^0_{biomass} \dot{m}^+_{biomass,daf}} \quad (11)$$

where *Δh0SNG* refers to the LHV, *m*˙ refers to the mass flow and daf refers to the dry ash free basis. The superscripts – and + refer to the flows leaving the system and flows entering the system, respectively.

Gutiérrez Ortiz et al. [133] performed process simulations using AspenPlus™ for gasification of glycerol in supercritical water. They investigated the effect of process conditions on the hydrogen yield. Using predictive Soave-Redlich-Kwong EoS for Gibbs free energy minimization in the reactor to predict the equilibrium composition, the authors concluded that the highest yield for hydrogen could be obtained at 900 °C with 1 mol

% glycerol in the feed. The authors reported that the pressure did not have an influence on the results.

A similar research has been conducted by Fiori et al. [134]. The authors performed process simulations using AspenPlus™ aiming at hydrogen production and energetic self-sustainability of different biomass feedstocks of glycerol, phenol, microalgae Spirulina, sewage sludge and grape marc. Their results indicated that biomass feed share should be at a minimum of 15–25 wt% in order to establish an energetic self-sustainable process (11.4 wt% for phenol and 22.9 wt% for Spirulina for a reactor operating at 700 °C and 30 MPa). The results also indicate that the thermal energy required for the microalgae process is around 0.6 MJ·(kg-total-feed)$^{-1}$ for a reactor operating at 500 °C and 30 MPa when the dry biomass concentration in the feed is 10 wt%. This value is around 1.5 MJ·(kg-total-feed)$^{-1}$ for a reactor operating at 700 °C with the same process conditions.

Another process simulation concerning a system model based on methanol gasification has been performed by Withag et al. [135]. Here, the authors also used AspenPlus™ and tested the validity of different EoS concluding that the property methods based on the Peng–Robinson and Soave Redlich-Kwong EoS all give a prediction of the H_2 mole fraction within a bandwidth of 3.5% compared to experimental results. They have also performed simulations for different cases to predict the thermal efficiencies for methanol, glucose and cellulose conversion. Similar to the findings of Fiori et al. [134], the authors concluded that the dry mass concentration of the feed should be at least 14.5 wt% for methanol and ~25 wt% for cellulose and glucose to have a thermal efficiency of 60% for a reactor operating at 600 °C and 30 MPa.

6. PROCESS CHALLENGES AND REACTOR TECHNOLOGY ASPECTS FOR INDUSTRIAL APPLICATIONS

Most of the experiments have been conducted in batch reactors which are not suitable for industrial applications. In contrast, industry rather demands continuous processes. However, continuous experimental approaches have some challenges which need to be overcome in order to scale up the process and be applicable in industry.

Pumpability of the biomass slurry is a problem in continuous processes. The dry matter content of the biomass slurry should not exceed the pumpability limits due to possible clogging problems. On the other hand, a high dry matter content results in higher gas production which increases the energy efficiency and the profitability of the plant. Yet, the pumpability of the biomass slurry depends on the nature of the biomass [136]. Antal et al. [75] used a cement pump to be able to pump a suspension of 4 wt% starch gel mixed with wood sawdust and other particulate biomass. On the other hand, sewage sludge containing 40% dry matter can be successfully pumped [137]. A pretreatment of reducing the particle size of the biomass [136] and obtaining a suspended feed [75] can be used to overcome the possible pumpability problems.

Corrosion is another problem for industrial scale applications. The process takes place in harsh conditions which lead to corrosion of the reactor construction material. In addition, gases and minerals which are formed during the gasification lead to corrosion as well [136]. Marrone and Hong [138] propose some corrosion control approaches which are (i) vortex/circulating flow reactor to prevent corrosive species from reaching a solid surface; (ii) use of high corrosion resistant materials such as nickel based alloys or stainless steel or (iii) reducing the temperature to 400 °C instead of 600 °C to improve the energy balance which would enable using other types of corrosion control methods such as using liners or coatings that decrease the heat transfer.

Working at low temperatures might have a positive effect on solving corrosion and energy efficiency problems, however, carbon gasification efficiency decreases due to the slower reaction rates. One way to overcome such problems is to use catalysts to enhance the reaction rates. Waldner and Vogel [78] have already obtained a full carbon gasification for wood samples and sufficiently high carbon efficiency (>80%) for swine manure experiments at low temperatures (~400 °C) using catalysts. Similar results have been observed by other researchers [93] as well. However, in addition to increasing the process costs—catalysts are often reported to be prone to deactivation problems [73,139,140,141] caused by (i) the poisoning of the active surface of the catalyst by sulfur; (ii) fouling of the catalyst surface by precipitated salts and other minerals and (iii) coke formation and deposition on the catalyst surface [142]. Coke formation may also cause plugging problems in flow reactors [143].

Zöhrer et al. [144] state that mixing the unheated, relatively concentrated organic solution with the preheated water avoids unwanted reactions and coke formation caused by the slow heating of the organic feed. However, this option could be feasible only for model biomass compounds as the concentrated biomass slurry would not be suitable for pumping as discussed earlier. For real biomass processes, rapid heating of the feed can overcome such coke and tar formation behavior as the positive effect of rapid heating on less tar production and higher gas yields are reported by other researchers [145,146] as well. Another possible solution to coke and tar formation was given by Kruse and Faquir [65]. The authors proposed a reaction process in which a continuously stirred tank reactor (CSTR) is followed by a tubular reactor in order to benefit from the "active hydrogen" formation in the CSTR due to its backmixing. "Active hydrogen", which is referred as the in statu nascendi hydrogen formed through the intermediate stages of the water gas shift reaction or by hydrogen, is reported to suppress tar and coke formation reactions. Furthermore, the gas yields increases when the CSTR is combined with a tubular reactor.

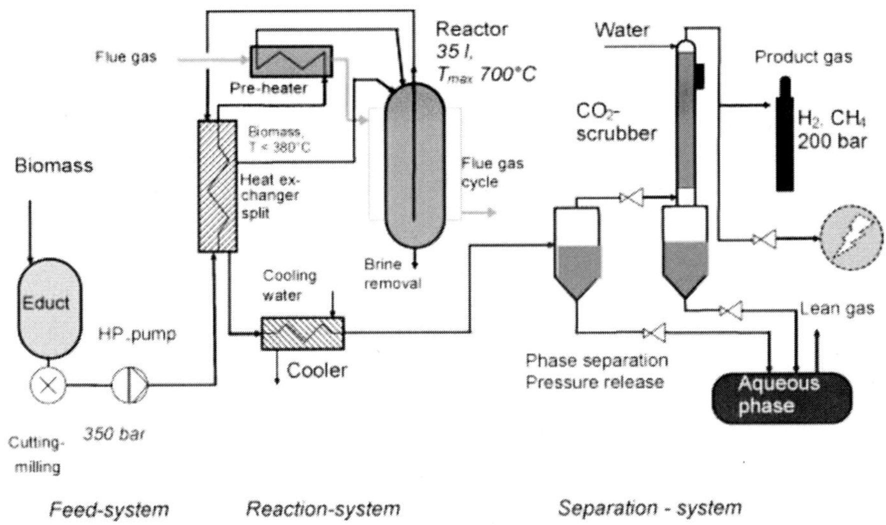

Figure 14. The process flow diagram of VERENA pilot plant. Reprinted with permission from [136], copyright 2009 Elsevier.

Obtaining sufficient process thermal efficiency is another challenge for supercritical water gasification of biomass systems. Thermal efficiency should be high enough to make the process viable for industrial applications. The efficiency becomes more important when the desired reactor temperature increases [136]. Proposed process demonstration units [73] as well as the VERENA pilot plant in Karlsruhe Institute of Technology [147,148] benefit from the high temperature of the product stream leaving from the reactor in order to preheat the feed using a heat exchanger. However, as reported earlier, if the heating rate of the heat exchanger is low, coke and tar formation will take place. To overcome such problems, VERENA pilot plant mixes the concentrated biomass, which was heated up to the critical temperature, with high temperature pure water [147]. The process flow diagram of VERENA pilot plant is given in Figure 14. Another approach was given by researchers in State Key Laboratory of Multiphase Flow in Power Engineering (SKLMF) of China [149,150,151,152]. Here, the researchers proposed solar energy based heating for the reactor and successfully applied this energy source for supercritical water gasification of biomass compounds. The Dutch company Gensos B.V. has patented a process that incorporates extensive heat recovery which enables a high process thermal efficiency [153].

The most important challenge in supercritical water gasification of biomass processes remains the salt precipitation due to the rapid decrease in the solubility of salts in the supercritical region (see Figure 5). The precipitated salts may lead to plugging problems if the tube diameters are small, crystals are large or sticky, heterogeneous catalysts are used, and the velocity of the flow is low [136]. The first possible solutions to salt precipitation problem were offered in the 1980s [154] and 1990s [155,156] as the salt precipitation was also a major problem for supercritical water oxidation processes. Hong et al. [154] patented a reverse flow reactor (known as MODAR reactor) in which the feedstock is fed from the top of the reactor and the cold water from the bottom. The upper part of the reactor is kept at supercritical temperatures, whereas the bottom part is kept at subcritical temperatures. Supercritical fluid leaves the reactor from the top and a brine effluent leaves from the bottom. Daman [155,156] patented a transpiring wall reactor in which solute-free water flows through a porous reactor liner in order to keep salt deposition as well as corrosion problems away from the reactor wall [157]. Xu et al. [158] combined these two approaches for a sewage sludge supercritical water oxidation process which is expected to prevent reactor plugging and corrosion problems. However, Marrone and Hong [138] stated that the transpiring wall reactor has limited usefulness for gasification systems due to adverse effect of dilution on the system's energy balance. On the other hand, Schubert et al. [159] used a salt separation vessel which the design is similar to MODAR's reverse flow reactor. The feed inlet and effluent to the reactor are from the top, whereas the brine effluent is from the bottom. The top of the reactor is at supercritical and the bottom part is at subcritical temperatures. In contrast to MODAR reactor which acts as both salt separator and reactor, salt separation vessel proposed by Schubert et al. [159] is a kind of pretreatment unit before the reactor. Zöhrer et al. [141] used this salt separator to perform catalytic gasification and liquefaction of fermentation residue samples. They observed almost no salt separation at 430 °C. At 450 °C, the salt separation efficiency was higher. However, at 470 °C massive tar formation which led to blockage in the rig was observed. Similar to MODAR reactor, Boukis et al. [147,148] used a reactor in the VERENA pilot plant which acts as both reactor and salt separator. The lower part of the reactor has a cooler [160] that enables the brines and solids leaving the reactor via a third output from the lower part of the reactor.

Another possible solution for plugging problems in the reactor was proposed by Matsumura and Minowa [5]. The authors speculated on the potential benefits of applying a fluidized bed reactor to the supercritical water gasification process. They investigated the fundamental design parameters of such a fluidized bed. Potic et al. [161] performed fluidization experiments for supercritical water in micro cylindrical quartz reactor with an internal diameter of 1 mm. Lu et al. [162] were the first researchers who performed supercritical water gasification experiments in a fluidized bed reactor at SKLMF. The reactor material was constructed from 316 mm stainless steel. The bed diameter was 30 mm, freeboard diameter was 40 mm, and the total length was 915 mm. The authors stated that no reactor plugging was observed. Other researchers in SKLMF [98,163] continued to perform gasification experiments in the same fluidized bed reactor. Real biomass experiments (2 wt% sewage sludge + 2 wt% sodium carboxymethyl cellulose) resulted in more than 40% carbon gasification efficiency at a temperature of 540 °C and at a pressure of 25 MPa. Gensos B.V. has patented a process and apparatus that incorporates a fluidized bed reactor [164,165].

7. CONCLUSIONS

Supercritical water gasification is a promising technology for the efficient conversion of wet biomass into a product gas that after upgrading can be used as substitute natural gas or hydrogen rich gas. The gasification behavior and the chemistry of the decomposition reactions of the model biomass compounds have been well examined throughout the decades; this work made a survey of mechanisms proposed and associated reaction kinetics. Thermodynamic modeling has been performed by several groups with quite some success and an overview of these attempts was given in this paper. Attempts have also been made for real biomass feedstocks varying from manure to sewage sludge. The results show the great potential of using supercritical water for gasification purposes. However, the process still faces some challenges which need to be overcome in order to enter the market. The main challenges seem to be thermal efficiency, plugging and corrosion problems. Fortunately, new reactor technologies and process concepts are being developed for the aforementioned problems.

ACKNOWLEDGMENTS

This research project is carried out within the Agentschap NL funded project "Superkritische vergassing van natte reststromen" (contract EOSLT10051). The authors would like to thank to Güçhan Yapar for his help in the preparation of Table 2.

AUTHORS CONTRIBUTIONS

Please prepare a short, one paragraph statement giving the individual contribution of each co-author to the reported research and writing of the paper.

CONFLICTS OF INTEREST

The authors declare no conflict of interest.

REFERENCES

1. U.S. Energy Information Administration International Energy Outlook 2013; Report Number: DOE/EIA-0484(2013). U.S. Department of Energy: Washington, DC, USA, 2013.
2. Process: The potential for biomass in the energy mix. Filtr. Sep. 2006, 43, 28–30.
3. Basu, P.; Mettanant, V. Biomass gasification in supercritical water—a review. Int. J. Chem. React. Eng. 2009, 7, 1–61.
4. European Commission Biomass—Green Energy for Europe; Report Number: EUR21350; European Commission: Brussels, Belgium, 2005.
5. Matsumura, Y.; Minowa, T. Fundamental design of a continuous biomass gasification process using a supercritical water fluidized bed. Int. J. Hydrog. Energy 2004, 29, 701–707.
6. Amin, S.; Reid, R.; Modell, M. Reforming and decomposition of glucose in an aqueous phase. In Proceedings of the Intersociety Conference on Environmental Systems, San Francisco, CA, USA, 21–24 July 1975.
7. Barner, H.E.; Huang, C.Y.; Johnson, T.; Jacobs, G.; Martch, M.A.; Killilea, W.R. Supercritical water oxidation: An emerging technology. J. Hazard. Mater. 1992, 31, 1–17.

8. Brunner, G. Near critical and supercritical water. Part I. Hydrolytic and hydrothermal processes. J. Supercrit. Fluids 2009, 47, 373–381.
9. Peterson, A.A.; Vogel, F.; Lachance, R.P.; Fröling, M.; Antal, M.J.J.; Tester, J.W. Thermochemical biofuel production in hydrothermal media: A review of sub- and supercritical water technologies. Energy Environ. Sci. 2008, 1, 32–65.
10. Kruse, A.; Dinjus, E. Hot compressed water as reaction medium and reactant: Properties and synthesis reactions. J. Supercrit. Fluids 2007, 39, 362–380.
11. Wagner, W.; Pruß, A. The IAPWS formulation 1995 for the thermodynamic properties of ordinary water substance for general and scientific use. J. Phys. Chem. Ref. Data 2002, 31, 387–535.
12. Yakaboylu, O.; Harinck, J.; Gerton Smit, K.G.; de Jong, W. Supercritical water gasification of manure: A thermodynamic equilibrium modeling approach. Biomass Bioenergy 2013, 59, 253–263.
13. Waldner, M.H. Catalytic Hydrothermal Gasification of Biomass for the Production of Synthetic Natural Gas. Ph.D. Thesis, Eidgenössische Technische Hochschule/Paul Scherrer Institute, Zürich, Switzerland, 2007.
14. Masaru Watanabe, T.S. Chemical reactions of C(1) compounds in near-critical and supercritical water. Chem. Rev. 2004, 104, 5803–5821.
15. Armellini, F.J. Phase Equilibria and Precipitation Phenomena of Sodium Chloride and Sodium Sulfate in Sub- and Supercritical Water. Ph.D. Thesis, Massachusetts Institute of Technology, Cambridge, MA, USA, 1993.
16. Connolly, J.F. Solubility of hydrocarbons in water near the critical solution temperatures. J. Chem. Eng. Data 1966, 11, 13–16.
17. Marshall, W.L.; Franck, E.U. Ion product of water substance, 0–1000 °C, 1–10,000 bars New International Formulation and its background. J. Phys. Chem. Ref. Data 1981, 10.
18. Webley, P.A.; Tester, J.W. Fundamental kinetics of methane oxidation in supercritical water. Energy Fuels 1991, 5, 411–419.
19. Thermophysical Properties of Fluid Systems. Available online: http://webbook.nist.gov/chemistry/fluid/ (accessed on 6 June 2014).
20. Castello, D. Supercritical Water Gasification of Biomass. Ph.D. Thesis, University of Trento, Trento, Italy, 2013.
21. Coronella, C.J.; Lynam, J.G.; Reza, M.T.; Uddin, M.H. Hydrothermal carbonization of lignocellulosic biomass. In Application of Hydrothermal Reactions to Biomass Conversion; Jin, F., Ed.; Springer: Berlin, Germany, 2014; pp. 275–311.

22. Onwudili, J.A. Hydrothermal gasification of biomass for hydrogen production. In Application of Hydrothermal Reactions to Biomass Conversion; Jin, F., Ed.; Springer: Berlin, Germany, 2014; pp. 219–246.
23. Toor, S.S.; Rosendahl, L.A.; Hoffmann, J.; Pedersen, T.H.; Nielsen, R.P.; Søgaard, E.G. Hydrothermal liquefaction of biomass. In Application of Hydrothermal Reactions to Biomass Conversion; Jin, F., Ed.; Springer: Berlin, Germany, 2014; pp. 189–217.
24. Xu, C.C.; Shao, Y.; Yuan, Z.; Cheng, S.; Feng, S.; Nazari, L.; Tymchyshyn, M. Hydrothermal Liquefaction of Biomass in Hot-Compressed Water, Alcohols, and Alcohol-Water Co-solvents for Biocrude Production. In Application of Hydrothermal Reactions to Biomass Conversion; Jin, F., Ed.; Springer: Berlin, Germany, 2014; pp. 171–187.
25. Bermejo, M.D.; Cocero, M.J. Supercritical water oxidation: A technical review. AIChE J. 2006, 52, 3933–3951.
26. Hodes, M.; Marrone, P.A.; Hong, G.T.; Smith, K.A.; Tester, J.W. Salt precipitation and scale control in supercritical water oxidation—Part A: Fundamentals and research. J. Supercrit. Fluids 2004, 29, 265–288.
27. Cortright, R.D.; Davda, R.R.; Dumesic, J.A. Hydrogen from catalytic reforming of biomass-derived hydrocarbons in liquid water. Nature 2002, 418, 964–967.
28. Gassner, M.; Vogel, F.; Heyen, G.; Maréchal, F. Optimal process design for the polygeneration of SNG, power and heat by hydrothermal gasification of waste biomass: Process optimisation for selected substrates. Energy Environ. Sci. 2011, 4, 1742–1758.
29. Van der Meijden, C.M.; Veringa, H.J.; Rabou, L.P.L.M. The production of synthetic natural gas (SNG): A comparison of three wood gasification systems for energy balance and overall efficiency. Biomass Bioenergy 2010, 34, 302–311.
30. Yoshida, Y.; Dowaki, K.; Matsumura, Y.; Matsuhashi, R.; Li, D.; Ishitani, H.; Komiyama, H. Comprehensive comparison of efficiency and CO2 emissions between biomass energy conversion technologies—Position of supercritical water gasification in biomass technologies. Biomass Bioenergy 2003, 25, 257–272.
31. Sasaki, M.; Adschiri, T.; Arai, K. Kinetics of cellulose conversion at 25 MPa in sub- and supercritical water. AIChE J. 2004, 50, 192–202.
32. Sasaki, M.; Kabyemela, B.; Malaluan, R.; Hirose, S.; Takeda, N.; Adschiri, T.; Arai, K. Cellulose hydrolysis in subcritical and supercritical water. J. Supercrit. Fluids 1998, 13, 261–268.
33. Resende, F.L.P.; Neff, M.E.; Savage, P.E. Noncatalytic gasification of cellulose in supercritical water. Energy Fuels 2007, 21, 3637–3643.

34. Yong, T.L.-K.; Matsumura, Y. Kinetic analysis of lignin hydrothermal conversion in sub- and supercritical Water. Ind. Eng. Chem. Res. 2013, 52, 5626–5639.
35. Yong, T.L.-K.; Matsumura, Y. Reaction kinetics of the lignin conversion in supercritical water. Ind. Eng. Chem. Res. 2012, 51, 11975–11988.
36. Resende, F.L.P.; Fraley, S.A.; Berger, M.J.; Savage, P.E. Noncatalytic gasification of lignin in supercritical water. Energy Fuels 2008, 22, 1328–1334.
37. Lee, I.-G.; Kim, M.-S.; Ihm, S.-K. Gasification of glucose in supercritical water. Ind. Eng. Chem. Res. 2002, 41, 1182–1188.
38. Hao, X.H.; Guo, L.J.; Mao, X.; Zhang, X.M.; Chen, X.J. Hydrogen production from glucose used as a model compound of biomass gasified in supercritical water. Int. J. Hydrog. Energy 2003, 28, 55–64.
39. Williams, P.T.; Onwudili, J. Composition of products from the supercritical water gasification of glucose: A model biomass compound. Ind. Eng. Chem. Res. 2005, 44, 8739–8749.
40. Goodwin, A.K.; Rorrer, G.L. Conversion of glucose to hydrogen-rich gas by supercritical water in a microchannel reactor. Ind. Eng. Chem. Res. 2008, 47, 4106–4114.
41. Güngören Madenoğlu, T.; Sağlam, M.; Yüksel, M.; Ballice, L. Simultaneous effect of temperature and pressure on catalytic hydrothermal gasification of glucose. J. Supercrit. Fluids 2013, 73, 151–160.
42. Kabyemela, B.M.; Adschiri, T.; Malaluan, R.M.; Arai, K. Glucose and fructose decomposition in subcritical and supercritical water: Detailed reaction pathway, mechanisms, and kinetics. Ind. Eng. Chem. Res. 1999, 38, 2888–2895.
43. Matsumura, Y.; Yanachi, S.; Yoshida, T. Glucose decomposition kinetics in water at 25 MPa in the temperature range of 448–673 K. Ind. Eng. Chem. Res. 2006, 45, 1875–1879.
44. Chuntanapum, A.; Matsumura, Y. Char formation mechanism in supercritical water gasification process: A study of model compounds. Ind. Eng. Chem. Res. 2010, 49, 4055–4062.
45. Kruse, A.; Gawlik, A. Biomass conversion in water at 330–410 °C and 30–50 MPa. Identification of key compounds for indicating different chemical reaction pathways. Ind. Eng. Chem. Res. 2003, 42, 267–279.
46. Aida, T.M.; Shiraishi, N.; Kubo, M.; Watanabe, M.; Smith, R.L. Reaction kinetics of D-xylose in sub- and supercritical water. J. Supercrit. Fluids 2010, 55, 208–216.

47. Goodwin, A.K.; Rorrer, G.L. Reaction rates for supercritical water gasification of xylose in a micro-tubular reactor. Chem. Eng. J. 2010, 163, 10–21.
48. Wahyudiono, W; Kanetake, T.; Sasaki, M.; Goto, M. Decomposition of a lignin model compound under hydrothermal conditions. Chem. Eng. Technol. 2007, 30, 1113–1122.
49. Wahyudiono, W; Sasaki, M.; Goto, M. Thermal decomposition of guaiacol in sub- and supercritical water and its kinetic analysis. J. Mater. Cycles Waste Manag. 2011, 13, 68–79.
50. Dileo, G.J.; Neff, M.E.; Savage, P.E. Gasification of guaiacol and phenol in supercritical water. Energy Fuels 2007, 21, 2340–2345.
51. Klingler, D.; Berg, J.; Vogel, H. Hydrothermal reactions of alanine and glycine in sub- and supercritical water. J. Supercrit. Fluids 2007, 43, 112–119.
52. Dileo, G.J.; Neff, M.E.; Kim, S.; Savage, P.E. Supercritical water gasification of phenol and glycine as models for plant and protein biomass. Energy Fuels 2008, 22, 871–877.
53. Zhang, Y.; Zhang, J.; Zhao, L.; Sheng, C. Decomposition of formic acid in supercritical water. Energy Fuels 2010, 24, 95–99.
54. Shin, K.S.; Cho, H.Y.; Nam, Y.W.; Lee, D.S. Hydrothermal decomposition of formic acid in sub- and supercritical water. Environ. Eng. Res. 1998, 3, 61–66.
55. Aida, T.M.; Ikarashi, A.; Saito, Y.; Watanabe, M.; Smith, R.L.; Arai, K. Dehydration of lactic acid to acrylic acid in high temperature water at high pressures. J. Supercrit. Fluids 2009, 50, 257–264.
56. Mok, W.S.; Antal, M.J.; Jones, M. Formation of acrylic acid from lactic acid in supercritical water. J. Org. Chem. 1989, 54, 4596–4602.
57. Chuntanapum, A.; Matsumura, Y. Formation of tarry material from 5-HMF in subcritical and supercritical water. Ind. Eng. Chem. Res. 2009, 48, 9837–9846.
58. Chuntanapum, A.; Yong, T.L.-K.; Miyake, S.; Matsumura, Y. Behavior of 5-HMF in subcritical and supercritical water. Ind. Eng. Chem. Res. 2008, 47, 2956–2962.
59. Yoshida, T.; Matsumura, Y. Gasification of cellulose, xylan, and lignin mixtures in supercritical water. Ind. Eng. Chem. Res. 2001, 40, 5469–5474.
60. Yoshida, T.; Oshima, Y.; Matsumura, Y. Gasification of biomass model compounds and real biomass in supercritical water. Biomass Bioenergy 2004, 26, 71–78.
61. Weiss-Hortala, E.; Kruse, A.; Ceccarelli, C.; Barna, R. Influence of phenol on glucose degradation during supercritical water gasification. J. Supercrit. Fluids 2010, 53, 42–47.

62. Goodwin, A.K.; Rorrer, G.L. Conversion of xylose and xylose-phenol mixtures to hydrogen-rich gas by supercritical water in an isothermal microtube flow reactor. Energy Fuels 2009, 23, 3818–3825.
63. Castello, D.; Kruse, A.; Fiori, L. Supercritical water gasification of glucose/phenol mixtures as model compounds for ligno-cellulosic biomass. Chem. Eng. Trans. 2014, 37, 193–198.
64. Sınağ, A.; Kruse, A.; Schwarzkopf, V. Key compounds of the hydropyrolysis of glucose in supercritical water in the presence of K2CO3. Ind. Eng. Chem. Res. 2003, 42, 3516–3521.
65. Kruse, A.; Faquir, M. Hydrothermal biomass gasification—effects of salts, backmixing and their interaction. Chem. Eng. Technol. 2007, 30, 749–754.
66. Kruse, A.; Meier, D.; Rimbrecht, P.; Schacht, M. Gasification of pyrocatechol in supercritical water in the presence of potassium hydroxide. Ind. Eng. Chem. Res. 2000, 39, 4842–4848.
67. Guo, Y.; Wang, S.; Wang, Y.; Zhang, J.; Xu, D.; Gong, Y. Gasification of two and three-components mixture in supercritical water: Influence of NaOH and initial reactants of acetic acid and phenol. Int. J. Hydrog. Energy 2012, 37, 2278–2286.
68. Minowa, T.; Zhen, F.; Ogi, T. Cellulose decomposition in hot-compressed water with alkali or nickel catalyst. J. Supercrit. Fluids 1998, 13, 253–259.
69. Muangrat, R.; Onwudili, J.A.; Williams, P.T. Influence of alkali catalysts on the production of hydrogen-rich gas from the hydrothermal gasification of food processing waste. Appl. Catal. B Environ. 2010, 100, 440–449.
70. Muangrat, R.; Onwudili, J.A.; Williams, P.T. Alkali-promoted hydrothermal gasification of biomass food processing waste: A parametric study. Int. J. Hydrog. Energy 2010, 35, 7405–7415.
71. Onwudili, J.A.; Lea-Langton, A.R.; Ross, A.B.; Williams, P.T. Catalytic hydrothermal gasification of algae for hydrogen production: Composition of reaction products and potential for nutrient recycling. Bioresour. Technol. 2013, 127, 72–80.
72. Yildiz Bircan, S.; Kamoshita, H.; Kanamori, R.; Ishida, Y.; Matsumoto, K.; Hasegawa, Y.; Kitagawa, K. Behavior of heteroatom compounds in hydrothermal gasification of biowaste for hydrogen production. Appl. Energy 2011, 88, 4874–4878.
73. Matsumura, Y.; Minowa, T.; Potic, B.; Kersten, S.; Prins, W.; van Swaaij, W.; Vandebeld, B.; Elliott, D.; Neuenschwander, G.; Kruse, A.; et al. Biomass gasification in near- and super-critical water: Status and prospects. Biomass Bioenergy 2005, 29, 269–292.

74. Zöhrer, H. Hydrothermal gasification of fermentation residues for SNG-production. Ph.D. Thesis, Swiss Federal Institute of Technology in Zurich, Zürich, Switzerland, 2013.
75. Antal, M.J.; Allen, S.G.; Schulman, D.; Xu, X.; Divilio, R.J. Biomass gasification in supercritical water. Ind. Eng. Chem. Res. 2000, 39, 4040–4053.
76. Byrd, A.J.; Pant, K.K.; Gupta, R.B. Hydrogen production from glycerol by reforming in supercritical water over Ru/Al2O3 catalyst. Fuel 2008, 87, 2956–2960.
77. Nakamura, A.; Kiyonaga, E.; Yamamura, Y.; Shimizu, Y.; Minowa, T.; Noda, Y.; Matsumura, Y. Gasification of catalyst-suspended chicken manure in supercritical water. J. Chem. Eng. Jpn. 2008, 41, 433–440.
78. Waldner, M.H.; Vogel, F. Renewable production of methane from woody biomass by catalytic hydrothermal gasification. Ind. Eng. Chem. Res. 2005, 44, 4543–4551.
79. Yanagida, T.; Minowa, T.; Nakamura, A.; Matsumura, Y.; Noda, Y. Behavior of inorganic elements in poultry manure during supercritical water gasification. J. Jpn. Inst. Energy 2008, 87, 731–736.
80. Yanik, J.; Ebale, S.; Kruse, A.; Saglam, M.; Yüksel, M. Biomass gasification in supercritical water: II. Effect of catalyst. Int. J. Hydrog. Energy 2008, 33, 4520–4526.
81. Boukis, N.; Diem, V.; Habicht, W.; Dinjus, E. Methanol reforming in supercritical water. Ind. Eng. Chem. Res. 2003, 42, 728–735.
82. Gadhe, J.B.; Gupta, R.B. Hydrogen production by methanol reforming in supercritical water: Suppression of methane formation. Ind. Eng. Chem. Res. 2005, 44, 4577–4585.
83. Yu, D.; Aihara, M.; Antal, M.J. Hydrogen production by steam reforming glucose in supercritical water. Energy Fuels 1993, 7, 574–577.
84. Castello, D.; Kruse, A.; Fiori, L. Biomass gasification in supercritical and subcritical water: The effect of the reactor material. Chem. Eng. J. 2013, 228, 535–544.
85. Guan, Q.; Wei, C.; Savage, P.E. Kinetic model for supercritical water gasification of algae. Phys. Chem. Chem. Phys. 2012, 14, 3140–3147.
86. Resende, F.L.P.; Savage, P.E. Kinetic model for noncatalytic supercritical water gasification of cellulose and lignin. AIChE J. 2010, 56, 2412–2420.
87. Araki, K.; Fujiwara, H.; Sugimoto, K.; Oshima, Y.; Koda, S. Kinetics of water-gas shift reaction in supercritical water. J. Chem. Eng. Jpn. 2004, 37, 443–448.

88. Rice, S.F.; Steeper, R.R.; Aiken, J.D. Water density effects on homogeneous water-gas shift reaction kinetics. J. Phys. Chem. A 1998, 102, 2673–2678.
89. Sato, T.; Kurosawa, S.; Smith, R.L., Jr.; Adschiri, T.; Arai, K. Water gas shift reaction kinetics under noncatalytic conditions in supercritical water. J. Supercrit. Fluids 2004, 29, 113–119.
90. Kruse, A.; Dinjus, E. Hydrogen from methane and supercritical water. Angew. Chem. Int. Ed. 2003, 42, 909–911.
91. Modell, M. Gasification and liquefaction of forest products in supercritical water. In Fundamentals of Thermochemical Biomass Conversion; Overend, R., Milne, T., Mudge, L., Eds.; Elsevier: Amsterdam, The Netherlands, 1985; pp. 95–119.
92. D'Jesús, P.; Boukis, N.; Kraushaar-Czarnetzki, B.; Dinjus, E. Gasification of corn and clover grass in supercritical water. Fuel 2006, 85, 1032–1038.
93. Stucki, S.; Vogel, F.; Ludwig, C.; Haiduc, A.G.; Brandenberger, M. Catalytic gasification of algae in supercritical water for biofuel production and carbon capture. Energy Environ. Sci. 2009, 2, 535–541.
94. Chakinala, A.G.; Brilman, D.W.F.; van Swaaij, W.P.M.; Kersten, S.R.A. Catalytic and non-catalytic supercritical water gasification of microalgae and glycerol. Ind. Eng. Chem. Res. 2010, 49, 1113–1122.
95. Guan, Q.; Savage, P.E.; Wei, C. Gasification of alga Nannochloropsis sp. in supercritical water. J. Supercrit. Fluids 2012, 61, 139–145.
96. Bagnoud-Velásquez, M.; Brandenberger, M.; Vogel, F.; Ludwig, C. Continuous catalytic hydrothermal gasification of algal biomass and case study on toxicity of aluminum as a step toward effluents recycling. Catal. Today 2014, 223, 35–43.
97. Xu, X.; Antal, M.J. Gasification of sewage sludge and other biomass for hydrogen production in supercritical water. Environ. Prog. 1998, 17, 215–220.
98. Chen, Y.; Guo, L.; Cao, W.; Jin, H.; Guo, S.; Zhang, X. Hydrogen production by sewage sludge gasification in supercritical water with a fluidized bed reactor. Int. J. Hydrog. Energy 2013, 38, 12991–12999.
99. Zhai, Y.; Wang, C.; Chen, H.; Li, C.; Zeng, G.; Pang, D.; Lu, P. Digested sewage sludge gasification in supercritical water. Waste Manag. Res. 2013, 31, 393–400.
100. Yong, T.L.-K.; Matsumura, Y. Catalytic gasification of poultry manure and eucalyptus wood mixture in supercritical water. Ind. Eng. Chem. Res. 2012, 51, 5685–5690.
101. Sasaki, M.; Furukawa, M.; Minami, K.; Adschiri, T.; Arai, K. Kinetics and mechanism of cellobiose hydrolysis and retro-aldol

condensation in subcritical and supercritical water. Ind. Eng. Chem. Res. 2002, 41, 6642–6649.

102. Brock, E.E.; Savage, P.E. Detailed chemical kinetics model for supercritical water oxidation of C1 compounds and H2. AIChE J. 1995, 41, 1874–1888.

103. Brock, E.E.; Savage, P.E.; Barker, J.R. A reduced mechanism for methanol oxidation in supercritical water. Chem. Eng. Sci. 1998, 53, 857–867.

104. Brock, E.E.; Oshima, Y.; Savage, P.E.; Barker, J.R. Kinetics and mechanism of methanol oxidation in supercritical water. J. Phys. Chem. 1996, 100, 15834–15842.

105. Helling, R.K.; Tester, J.W. Oxidation kinetics of carbon monoxide in supercritical water. Energy Fuels 1987, 1, 417–423.

106. Tester, J.W.; Webley, P.A.; Holgate, H.R. Revised global kinetic measurements of methanol oxidation in supercritical water. Ind. Eng. Chem. Res. 1993, 32, 236–239.

107. Ederer, H.J.; Kruse, A.; Mas, C.; Ebert, K.H. Modelling of the pyrolysis of tert-butylbenzene in supercritical water. J. Supercrit. Fluids 1999, 15, 191–204.

108. Bühler, W.; Dinjus, E.; Ederer, H.J.; Kruse, A.; Mas, C. Ionic reactions and pyrolysis of glycerol as competing reaction pathways in near- and supercritical water. J. Supercrit. Fluids 2002, 22, 37–53.

109. Castello, D.; Fiori, L. Kinetics modeling and main reaction schemes for the supercritical water gasification of methanol. J. Supercrit. Fluids 2012, 69, 64–74.

110. Kabyemela, B.M.; Takigawa, M.; Adschiri, T.; Malaluan, R.M.; Arai, K. Mechanism and Kinetics of Cellobiose Decomposition in Sub- and Supercritical Water. Ind. Eng. Chem. Res. 1998, 37, 357–361.

111. Kabyemela, B.M.; Adschiri, T.; Malaluan, R.M.; Arai, K. Kinetics of glucose epimerization and decomposition in subcritical and supercritical water. Ind. Eng. Chem. Res. 1997, 36, 1552–1558.

112. Yoshida, T.; Matsumura, Y. Reactor development for supercritical water gasification of 4.9 wt% glucose solution at 673 K by using computational fluid dynamics. Ind. Eng. Chem. Res. 2009, 48, 8381–8386.

113. Goodwin, A.K.; Rorrer, G.L. Modeling of supercritical water gasification of xylose to hydrogen-rich gas in a hastelloy microchannel reactor. Ind. Eng. Chem. Res. 2011, 50, 7172–7182.

114. Wei, L.; Lu, Y.; Wei, J. Hydrogen production by supercritical water gasification of biomass: Particle and residence time distribution in fluidized bed reactor. Int. J. Hydrog. Energy 2013, 38, 13117–13124.

115. Withag, J.A.M.; Sallevelt, J.L.H.P.; Brilman, D.W.F.; Bramer, E.A.; Brem, G. Heat transfer characteristics of supercritical water in a tube: Application for 2D and an experimental validation. J. Supercrit. Fluids 2012, 70, 156–170.
116. Sallevelt, J.L.H.P.; Withag, J.A.M.; Bramer, E.A.; Brilman, D.W.F.; Brem, G. One-dimensional model for heat transfer to a supercritical water flow in a tube. J. Supercrit. Fluids 2012, 68, 1–12.
117. Withag, J.A.M. On the Gasification of Wet Biomass in Supercritical Water. Ph.D. Thesis, University of Twente, Enschede, The Netherlands, 2013.
118. Tang, H.; Kitagawa, K. Supercritical water gasification of biomass: Thermodynamic analysis with direct Gibbs free energy minimization. Chem. Eng. J. 2005, 106, 261–267.
119. Freitas, A.C.; Guirardello, R. Thermodynamic analysis of supercritical water gasification of microalgae biomass for hydrogen and syngas production. Chem. Eng. Trans. 2013, 32, 553–558.
120. Freitas, A.C.D.; Guirardello, R. Supercritical water gasification of glucose and cellulose for hydrogen and syngas production. Chem. Eng. Trans. 2012, 27, 361–366.
121. Freitas, A.C.D.; Guirardello, R. Comparison of several glycerol reforming methods for hydrogen and syngas production using Gibbs energy minimization. Int. J. Hydrog. Energy 2014, 39, 17969–17984.
122. Lu, Y.; Guo, L.; Zhang, X.; Yan, Q. Thermodynamic modeling and analysis of biomass gasification for hydrogen production in supercritical water. Chem. Eng. J. 2007, 131, 233–244.
123. Voll, F.A.P.; Rossi, C.C.R.S.; Silva, C.; Guirardello, R.; Souza, R.O.M.A.; Cabral, V.F.; Cardozo-Filho, L. Thermodynamic analysis of supercritical water gasification of methanol, ethanol, glycerol, glucose and cellulose. Int. J. Hydrog. Energy 2009, 34, 9737–9744.
124. Yan, Q.; Guo, L.; Lu, Y. Thermodynamic analysis of hydrogen production from biomass gasification in supercritical water. Energy Convers. Manag. 2006, 47, 1515–1528.
125. Castello, D.; Fiori, L. Supercritical water gasification of biomass: Thermodynamic constraints. Bioresour. Technol. 2011, 102, 7574–7582.
126. Letellier, S.; Marias, F.; Cezac, P.; Serin, J.P. Gasification of aqueous biomass in supercritical water: A thermodynamic equilibrium analysis. J. Supercrit. Fluids 2010, 51, 353–361.
127. Marias, F.; Letellier, S.; Cezac, P.; Serin, J.P. Energetic analysis of gasification of aqueous biomass in supercritical water. Biomass Bioenergy 2011, 35, 59–73.

128. Yakaboylu, O.; Harinck, J.; Smit, K.G.; de Jong, W. Supercritical water gasification of biomass: A thermodynamic model for the prediction of product compounds at equilibrium state. Energy Fuels 2014, 28, 2506–2522.
129. Kozeschnik, E. A numerical model for evaluation of unconstrained and compositionally constrained thermodynamic equilibria. Calphad 2000, 24, 245–252.
130. Feng, W.; van der Kooi, H.J.; de Swaan Arons, J. Biomass conversions in subcritical and supercritical water: Driving force, phase equilibria, and thermodynamic analysis. Chem. Eng. Process. Process Intensif. 2004, 43, 1459–1467.
131. Luterbacher, J.S.; Fröling, M.; Vogel, F.; Maréchal, F.; Tester, J.W. Hydrothermal gasification of waste biomass: Process design and life cycle asessment. Environ. Sci. Technol. 2009, 43, 1578–1583.
132. Gassner, M.; Vogel, F.; Heyen, G.; Maréchal, F. Optimal process design for the polygeneration of SNG, power and heat by hydrothermal gasification of waste biomass: Thermo-economic process modelling and integration. Energy Environ. Sci. 2011, 4, 1726–1741.
133. Gutiérrez Ortiz, F.J.; Ollero, P.; Serrera, A.; Sanz, A. Thermodynamic study of the supercritical water reforming of glycerol. Int. J. Hydrog. Energy 2011, 36, 8994–9013.
134. Fiori, L.; Valbusa, M.; Castello, D. Supercritical water gasification of biomass for H2 production: Process design. Bioresour. Technol. 2012, 121, 139–147.
135. Withag, J.A.M.; Smeets, J.R.; Bramer, E.A.; Brem, G. System model for gasification of biomass model compounds in supercritical water—A thermodynamic analysis. J. Supercrit. Fluids 2012, 61, 157–166.
136. Kruse, A. Hydrothermal biomass gasification. J. Supercrit. Fluids 2009, 47, 391–399.
137. Tolman, R. Process for Converting Sewage Sludge and Municipal Solid Wastes to Clean Fuels; Report Number: P600-01-12; California Energy Commission: Sacramento, CA, USA, 2001.
138. Marrone, P.A.; Hong, G.T. Corrosion control methods in supercritical water oxidation and gasification processes. J. Supercrit. Fluids 2009, 51, 83–103.
139. Elliott, D.C. Catalytic hydrothermal gasification of biomass. Biofuels Bioprod. Biorefin. 2008, 2, 254–265.
140. Guo, Y.; Wang, S.Z.; Xu, D.H.; Gong, Y.M.; Ma, H.H.; Tang, X.Y. Review of catalytic supercritical water gasification for hydrogen production from biomass. Renew. Sustain. Energy Rev. 2010, 14, 334–343.

141. Zöhrer, H.; de Boni, E.; Vogel, F. Hydrothermal processing of fermentation residues in a continuous multistage rig—Operational challenges for liquefaction, salt separation, and catalytic gasification. Biomass Bioenergy 2014, 65, 51–63.
142. Zöhrer, H.; Vogel, F. Hydrothermal catalytic gasification of fermentation residues from a biogas plant. Biomass Bioenergy 2013, 53, 138–148.
143. Karayıldırım, T.; Sınağ, A.; Kruse, A. Char and Coke Formation as Unwanted Side Reaction of the Hydrothermal Biomass Gasification. Chem. Eng. Technol. 2008, 31, 1561–1568.
144. Zöhrer, H.; Mayr, F.; Vogel, F. Stability and performance of ruthenium catalysts based on refractory oxide supports in supercritical water conditions. Energy Fuels 2013, 27, 4739–4747.
145. Matsumura, Y.; Harada, M.; Nagata, K.; Kikuchi, Y. Effect of heating rate of biomass feedstock on carbon gasification efficiency in supercritical water gasification. Chem. Eng. Commun. 2006, 193, 649–659.
146. Sınağ, A.; Kruse, A.; Rathert, J. Influence of the heating rate and the type of catalyst on the formation of key intermediates and on the generation of gases during hydropyrolysis of glucose in supercritical water in a batch reactor. Ind. Eng. Chem. Res. 2004, 43, 502–508.
147. Boukis, N.; Galla, U.; Müller, H.; Dinjus, E. Biomass gasification in supercritical water. Experimental progress achieved with the VERENA pilot plant. In Proceedings of the 15th European Conference & Exhibition, 7–11 May 2007; Volume 7, pp. 1013–1016.
148. Boukis, N.; Galla, U.; D'Jesus, P.; Müller, H.; Dinjus, E. Gasification of wet biomass in supercritical water. Results of pilot plant experiments. In Proceedings of the 14th European Biomass Conference, Paris, France, 17–21 October 2005; pp. 964–967.
149. Chen, J.; Lu, Y.; Guo, L.; Zhang, X.; Xiao, P. Hydrogen production by biomass gasification in supercritical water using concentrated solar energy: System development and proof of concept. Int. J. Hydrog. Energy 2010, 35, 7134–7141.
150. Liao, B.; Guo, L.; Lu, Y.; Zhang, X. Solar receiver/reactor for hydrogen production with biomass gasification in supercritical water. Int. J. Hydrog. Energy 2013, 38, 13038–13044.
151. Lu, Y.; Zhao, L.; Guo, L. Technical and economic evaluation of solar hydrogen production by supercritical water gasification of biomass in China. Int. J. Hydrog. Energy 2011, 36, 14349–14359.
152. Xiao, P.; Guo, L.; Zhang, X.; Zhu, C.; Ma, S. Continuous hydrogen production by biomass gasification in supercritical water heated by

molten salt flow: System development and reactor assessment. Int. J. Hydrog. Energy 2013, 38, 12927–12937.
153. Harinck, J.; Smit, K.G. A Process and a Reaction Apparatus for the Gasification of Wet Biomass. WO Application WO/2013/030026, 7 March 2013.
154. Hong, G.T.; Killilea, W.R.; Thomason, T.B. Method for Solids Separation in a Wet Oxidation Type Process. U.S. Patent 4,822,497, 18 April 1989.
155. Daman, E.L. Process and Apparatus for Supercritical Water Oxidation. U.S. Patent 5,723,045, 3 March 1998.
156. Daman, E.L. Process and Apparatus for Supercritical Water Oxidation. U.S. Patent 5,571,423, 5 November 1996.
157. Marrone, P.A.; Hodes, M.; Smith, K.A.; Tester, J.W. Salt precipitation and scale control in supercritical water oxidation—Part B: Commercial/full-scale applications. J. Supercrit. Fluids 2004, 29, 289–312.
158. Xu, D.H.; Wang, S.Z.; Gong, Y.M.; Guo, Y.; Tang, X.Y.; Ma, H.H. A novel concept reactor design for preventing salt deposition in supercritical water. Chem. Eng. Res. Des. 2010, 88, 1515–1522.
159. Schubert, M.; Regler, J.W.; Vogel, F. Continuous salt precipitation and separation from supercritical water. Part 1: Type 1 salts. J. Supercrit. Fluids 2010, 52, 99–112.
160. Karlsruhe, F. Device for treating Flowable Materials in Super-Critical Water, e.g., for Treating Effluent, Comprises Cylindrical Reactor with Pressure Lines for Introducing Educt and Removing Product. German Patent DE20220307, 30 April 2003.
161. Potic, B.; Kersten, S.R.A.; Ye, M.; van der Hoef, M.A.; Kuipers, J.A.M.; van Swaaij, W.P.M. Fluidization with hot compressed water in micro-reactors. Chem. Eng. Sci. 2005, 60, 5982–5990.
162. Lu, Y.; Jin, H.; Guo, L.; Zhang, X.; Cao, C.; Guo, X. Hydrogen production by biomass gasification in supercritical water with a fluidized bed reactor. Int. J. Hydrog. Energy 2008, 33, 6066–6075.
163. Cao, C.; Guo, L.; Jin, H.; Guo, S.; Lu, Y.; Zhang, X. The influence of alkali precipitation on supercritical water gasification of glucose and the alkali recovery in fluidized-bed reactor. Int. J. Hydrog. Energy 2013, 38, 13293–13299.
164. Harinck, J.; Smit, K.G. A Reaction Apparatus and a Process for the Gasification of Wet Biomass. WO Application WO/2013/030027, 7 March 2013.
165. Harinck, J.; Smit, K.G. A Process for the Gasification of Wet Biomass. WO Application WO/2013/030028, 7 March 2013.

INDEX